CHEMICAL GROUTING

CIVIL ENGINEERING

A Series of Textbooks and Reference Books

1. Bridge and Pier Protective Systems and Devices, *Kenneth N. Derucher and Conrad P. Heins, Jr.*
2. Structural Analysis and Design, *Conrad P. Heins, Jr., and Kenneth N. Derucher*
3. Bridge Maintenance Inspection and Evaluation, *Kenneth R. White, John Minor, Kenneth N. Derucher, and Conrad P. Heins, Jr.*
4. Salinity and Irrigation in Water Resources, *Dan Yaron*
5. Urban Storm Drainage Management, *John R. Sheaffer and Kenneth R. Wright with William C. Taggart and Ruth M. Wright*
6. Chemical Grouting, *Reuben H. Karol*
7. Sealants in Construction, *Jerome M. Klosowski*
8. Chemical Grouting, Second Edition, Revised and Expanded, *Reuben H. Karol*

Other Volumes in Preparation

CHEMICAL GROUTING

Second Edition, Revised and Expanded

REUBEN H. KAROL

Consulting Engineer
Princeton, New Jersey

MARCEL DEKKER, INC. New York and Basel

Library of Congress Cataloging-in-Publication Data

Karol, R. H. (Reuben H.), [date]
Chemical grouting / Reuben H. Karol. --2nd ed., rev. and expanded.
p. cm. --(Civil engineering ; 8)
Includes bibliographical references.
ISBN 0-8247-7888-X (alk. paper)
1. Grouting. 2. Chemicals. I. Title II. Series
TA1.C4523 vol. 8.
[TA755]
624 s--dc20 89-23422
[624.1'5] CIP

This book is printed on acid-free paper.

MARCEL DEKKER, INC.
270 Madison Avenue, New York, New York 10016

Current printing (last digit):
10 9 8 7 6 5 4 3 2 1

PRINTED IN THE UNITED STATES OF AMERICA

Preface to the Second Edition

In the time since publication of the First Edition, many changes have occurred involving the materials and practice of chemical grouting. First and foremost has been the growing acceptance of chemical grouting as a preventive measure, as well as a remedial measure. Together with this acceptance has come a significant increase in the number of small contractors working with chemical grouts. This positive growth has been fueled by a better understanding of the properties and behavior of grouts, and by a spread of general knowledge of how to exploit those properties. These are the factors that have prompted the writing of a second edition of this book. While the original material is still applicable, this edition expands the various topics presented earlier, and adds new topics that have either gained in importance or have been developed since the publication of the first edition.

In Chapter 1, additional information has been given on competitive methods: compressed air, freezing, and slurry walls. The subsection on history has been updated and expanded. In Chapter 2, data has been added to better define soils for grouting purposes, and in Chapter 3, correlation between theory and grouting acceptance is discussed.

Chapter 4 has been greatly expanded to include advances in knowledge of existing grouts, as well as properties of new materials. More emphasis is given to the silicates, still the major grout in the United States, and details of the available acrylates (the growing acrylamide replacement) are given. Many charts and tables have been added.

In Chapter 5, the discussion of flow of grout through soils has been expanded. In Chapters 6 and 7, sections have been added concerning instrumentation and its relationships to the use of short gel times in the field.

Chapters 8 through 12, which deal primarily with various field applications, have been expanded by the addition of new case histories to each, and references to other articles.

In Chapter 13, excerpts have been included of recent specifications for chemical grouts. Chapter 14 has been updated.

Much new material has been added to the appendixes. Microfine cements are covered in Appendix A. These materials are not chemical grouts, but they rival the chemical grouts in penetrability and strength. Setting times, however, are very long. This opens the future to growing use of mixtures of chemical grouts and microfine cements to optimize the better properties of each.

Other appendixes list a computer program for determination of optimum grout hole spacing, a test procedure for determining design strength of grouted soils, and a glossary of terms.

Reuben H. Karol

Preface to the First Edition

As this book is being written, we are on the threshold of a new era in chemical grouting. The federal government and many universities have begun to take serious interest in the engineering applications of chemical grouts. The research now underway and that planned for the future will go far in improving the reliability and efficiency of a field grouting project. At the same time, concerns over environmental pollution and personnel health hazards threaten to eliminate some of the most versatile chemical grouts and have spurred a search for new and safer materials.

In the United States, large-scale use of cement grouts began at the turn of the century, when federal agencies began treating dam foundation sites. The practices and specifications developed for those purposes quickly became the unofficial grouting standards for virtually all grouting projects in the United States. In later years this was to prove a deterrent to the developing chemical grouts because of the tendency to force cement grouting practices on materials with properties and capabilities vastly different from cement.

Applicators with previous cement experience (and who didn't have some?) insisted on handling chemical grouts as if they were low-viscosity, expensive cement grouts.

In particular, pumping with long gel times to a pressure refusal, common practice with cement, is very wasteful and inefficient with chemicals. Also, payment by the volume of grout placed, common practice with cement, tends to stifle engineering design of a grouting operation. The biggest technical offender was the batch pumping system, since it precluded taking advantage of the special properties of chemical grouts, particularly short, controllable gel times. The most imposing mental obstacle for cement grouters was the acceptance of the fact that chemical grouts could be pumped into a formation for periods much longer than the setting time.

Over the past two decades, chemical grouting technology gained acceptance as a bona fide construction tool. Current practice makes use of sophisticated multipump grout plants and grout pipes, with accurate controls and monitors that permit full exploitation of the unique properties of available grouting materials. Further, the engineering profession also has accepted the fact that a technology exists and that there are reasonable and reliable methods of applying engineering principles to the design of a grouting operation. As we enter the 1980s, chemical grouting is taking its place alongside other accepted water control and strengthening techniques such as well pointing and underpinning. This book deals primarily with the materials and techniques from the 1950s on, with emphasis on current practices when they have superseded earlier developments.

Reuben H. Karol
Rutgers University
New Brunswick, New Jersey

Acknowledgments

Many people have knowingly or unknowingly contributed to this book. They include virtually all the field personnel with whom I have worked on jobs as well as those with whom I have served over the past two decades on the ASCE and ASTM Grouting Committee and boards of consultants for various projects. I am grateful for the help and encouragement that many of these people have given me in compiling the information that comprises this book. In particular, my thanks to Herb Parsons and John Gnaedinger, who taught me the first things I ever learned about chemical grouting; to Ed Graf, whose two-decade role of devil's advocate has helped us both; and to Joseph Welsh, Wallace Baker, and Bruce Lamberton, whose willingness to share their expertise has helped the whole profession. I must also note my appreciation to individuals and companies who so kindly furnished job and research reports, including Wayne Clough, Jim Mitchell, Ray Krizek, Woodward-Clyde Consultants, Hayward Baker Company, American Dewatering Corp., and Florida Power Corp.

Finally, my gratitude also goes to my associates in the Division of Continuing Education at Rutgers University and my family, whose encouragement and understanding helped me through this year.

Contents

1

Introduction

1.1 INTRODUCTION

Under the action of gravity, surface water and groundwater always tend to flow from higher to lower elevations. Surface water will flow over solid and through permeable formations, and its volume and velocity are a function of the available supply and the fluid head. Groundwater can move only through a pervious material (fractured or fissured rock or soils with interconnected open voids), so its flow characteristic is also a function of formation permeability. Groundwater elevation varies as the supply source varies and can be raised or lowered locally by increasing or decreasing the local supply (by nature as through precipitation or artificially as through pumping a well or irrigating). In general, over a large surface area, groundwater surface is a subdued replica of ground surface.

Many construction projects require the lowering of the natural land surface to provide for foundations, basements, and other low-level facilities. Other projects such as tunnels and shafts require underground construction of long, open tubes. Whenever such excavations go below groundwater surface, they disrupt the existing flow patterns by creating a zone of low pressure potential. Groundwater begins to flow radially toward and into the excavation. The situation is further aggravated by the fact that construction procedures generally enlarge existing fissures and voids and create new ones in the vicinity of the excavation.

Contractors anticipate infiltration when the excavation is planned to go below groundwater level and generally make provisions for diverting the flow of water before it reaches the excavation or removing it before or after it enters. Water problems during construction, which carry a cost penalty, occur when the provisions to handle groundwater prove ineffective or inadequate. The scope of water problems can range from nuisance value to actual retardation of the construction schedule to complete shutdown.

Water problems may also occur after the completion of construction. Seepage that may have been tolerable during construction may become intolerable during facility operations. Post construction seepage may increase to intolerable levels due to termination of construction seepage control procedures. Unanticipated water problems may occur because the structural elements cause long-term modification of surface drainage patterns or subsurface seepage patterns. Unusual amounts of precipitation may raise normal ground water levels. Occasionally, shrinkage cracks in, and settlement of foundation elements may result in postconstruction seepage problems.

The presence of unanticipated groundwater (either static or flowing) may lower the design value of bearing capacity. If higher values were used, based on dry conditions, water must be kept permanently from the foundation area. The presence of water in basement areas may prevent use of such areas.

The contractor has at his or her disposal many field procedures to prevent seepage or to control it after it reaches intolerable amounts. Some of these procedures are briefly discussed in Sec. 1.2.

1.2 METHODS OF GROUNDWATER CONTROL

Drainage

An interconnected system of open ditches surrounding an excavated area may be dug to remove limited volumes of inflowing water from a construction site. Drainage ditches and channels are generally feasible only in soils with cohesion. (In granular materials the ditch slopes become too wide to be practical, and excavation in rock is often too costly.) If local topography permits, the entire drainage system is operated by gravity and discharged at an elevation and location that prohibits discharge reentry to the drained area. Such procedures are seldom 100% effective but are often adequate for surface water. They are not usually effective for groundwater infiltration unless the flow is very small.

Pumping from Sumps

When the volume of water is very large, when gravity discharge of the system is not feasible, and when groundwater is seeping into the excavation, sump pumps may be used at one or more locations to remove the water from the construction site. When it becomes

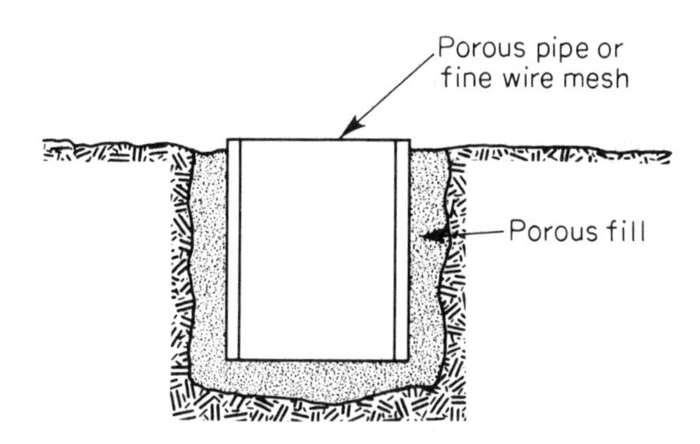

FIGURE 1.1 Typical sump construction.

necessary to use pumps, they must generally be operated around the clock, with adequate standby equipment. Costs become very much higher than for simple gravity systems. Typical sump construction is shown in Fig. 1.1.

Caulking

Surface coatings sprayed on porous granular deposits are generally ineffective for seepage control, although they may be useful in keeping dust to a minimum and in reducing wind and surface water erosion. In fractured and fissured rock, caulking from the exposed surface can be effective in sealing the cracks against water flow. Lead wool, hemp, or other suitable materials are wedged into the crack using a hammer and chisels. As in any method where open flow channels are closed, caulking of leaks may divert the flow to fissures that were previously dry.

Caulking can also be done in cracks and construction joints in concrete. Shutting off local seepage in this fashion can be done as a temporary expedient to facilitate water control by other methods.

Drainage by gravity, sump pumps, and caulking are far less costly than the more sophisticated procedures that follow in the text, and these methods should be used wherever they can be effective.

Compressed Air

The construction of tunnels and shafts for the recovery of natural resources dates back to biblical times—King Solomon's mines, for example. In those days, if water seeped into an excavation, it had to be removed by direct physical labor; there was no effective way of stopping the inflow.

More recently, the development of sophisticated seepage control systems and the universal availability of power have facilitated the removal of water inflow from excavations, so that they may go deeper than was previously possible. Together with improvements in water removal methods came alternative procedures and concern for the safety of workmen using those methods.

As indicated in Sec. 1.1, excavation below the water table creates a zone of low pressure potential, resulting in inflow. To reduce or stop the inflow, it is only necessary to increase the pressure in the zone of low potential. This may be done by sealing the open end of the tunnel or shaft and applying a fluid pressure. The fluid used may be liquid or gas. Liquids are used in applications not requiring handwork, such as slurry trenches and drill holes. When people must be located within the excavation, gas is used, almost always compressed air.

In theory, the applied air pressure need only equal the pore water pressure in order to halt inflow. In practice, it isn't practical to balance pressures that closely, so the air pressure always exceeds the actual or anticipated pore water pressure by some finite amount. In fine-grained soils the loss of air is small and creates no problems. In very coarse formations or open rock masses the loss may be so great as to make the use of air impractical.

Considerable cost is involved in setting up the locks, compressors, and piping necessary for a compressed air system. This makes the use of this method uneconomic on small jobs. Also, while there is no theoretical depth limit to a compressed air system, there is a practical limit to the pressures in which people can work. These pressures are set by law in many places and in practice generally limit the use of compressed air to shallow depths.

When workmen go from standard atmospheric pressure to a pressurized zone, additional air will dissolve in the blood to reach new equilibrium conditions. Upon leaving the pressurized zone, excess dissolved air will leave the blood. If this process occurs too rapidly, air bubbles will form in the blood vessels, restricting flow and resulting in possible serious injury or death. This condition is known as the "bends," and the amount of time required to go from maximum pressure to atmospheric (decompression) with safety is well documented. For example, at 40 psi (50 psi is normally the maximum allowable value), a decompression time of 143 minutes is needed. Allowing the same time for compression, only about 3 hours of an eight hour working day are actually spent in productive work. Thus, labor on a compressed air job is very costly, since hourly rates also rise as the pressure rises.

Freezing

Controlled ground freezing for mining and construction applications has been done for more than a century, with the first application credited in 1862 in a Welsh coal mine. Only in recent years, however, have significant improvements in technique occurred, in concert with advanced refrigeration technology.

Ground freezing is effective in soils through which ground water is moving at less than about 3 inches per hour. This process involves the use of a refrigeration system to convert in situ pore water to ice. If the frozen formation was saturated, it becomes impermeable after ice forms. If the formation was less than saturated, the initial permeability is reduced after ice forms. In either case, the ice will bond adjacent particles together, adding shear strength to the formation. For particles of sand size and smaller, the added strength may be a significant design factor.

Ice is the only factor that makes frozen ground different from unfrozen ground. It is often the key component of the ice–ground system. Frozen ground behaves as a viscoplastic material with strength properties primarily dependent on the ice content, duration of loading, and temperature.

Freezing of subsurface formations is done by placing a series of pipes into the formation. Each pipe is closed at the bottom and has a smaller open-ended tube or pipe inside. Cold fluid, a liquid such as brine or liquid nitrogen, is circulated to remove heat from the formation and discharge it above the ground surface. (Brine is normally sent through a surface refrigeration plant, to be cooled and re-used. Liquid nitrogen is usually wasted as a gas.) The pore water around each refrigerant pipe begins to freeze and, with continued cooling, the ice layer grows radially to meet the ice spreading from adjacent pipes. Thus, a continuous frozen structure is formed.

Freezing may be accomplished in any formation regardless of structure, grain size, or permeability. Further, there are no depth limitations to the method, except for the possible inability at great depths to place pipe with sufficient accuracy. However, the initial cost of a freezing plant and brine circulation system is high, and the freezing process is slow (weeks or months), precluding economical use on small projects. Use of liquid nitrogen as a coolant overcomes these limitations. Freezing (using nitrogen) is now universally considered as a possible solution for small field jobs involving seepage.

The problems associated with the freezing method occur primarily during the thawing period. They are mainly related to the volume decrease when ice changes to water. Uneven thawing around a tunnel or shaft can cause structural misalignment and buckling, with possible damage to the shaft or tunnel lining. In addition, concrete poured against a frozen formation may be of poor quality. Particularly in deep shafts where groundwater pressure is high, significant water inflow through porous concrete may occur after thawing. It may be necessary to use other water control methods to make a final seal.

Well Pointing

When water is withdrawn locally from a point below the ground water surface, a concavity is formed in that surface above the withdrawal point. If sufficient volume of water is withdrawn, the concavity will reach down to the point of withdrawal. Figure 1.2 shows the initial and final elevation of groundwater in the vicinity of a small well that is being pumped at a constant rate. The radius R is related to the pumping and recharging rates and represents an equilibrium condition between these two factors. If the pumping rate is increased, the distance R will increase by an amount m; the water level in the well, or drawdown, will also lower, and an additional volume of soil represented by the crosshatched area will become dry. Since the drawdown is linear and the soil volume is a cubic function, increasing the pumping rate from one well is an inefficient procedure for drying a large area. The rate of groundwater slope change is greatest near the well, so closely spaced wells are most efficient. Actual spacing is based on economic considerations in which individual well efficiency is balanced against well cost. (These

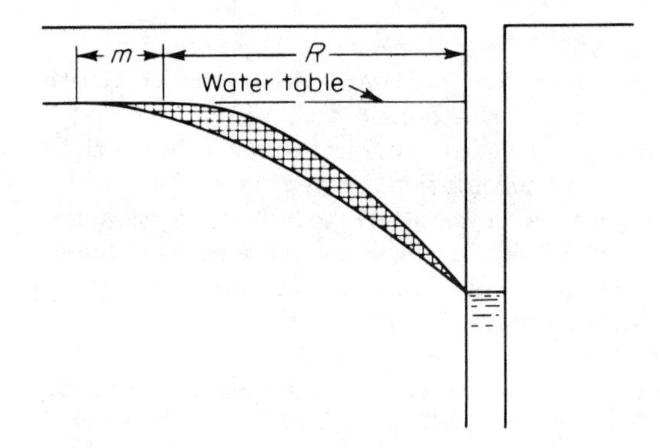

FIGURE 1.2 Depression in groundwater surface due to pumping from well.

same factors, in reverse, dictate the spacing of grout pipes. This is discussed in detail in later sections.)

When pumping equipment is at the surface, the depth of wells that can be pumped is limited to the lift supplied by air pressure. In practice, because of equipment efficiencies, gravity lifts seldom go above 15 ft. Only the suction pipe need be placed down the well, and therefore the well can be very small in diameter. The lower end of the suction pipe may carry a long screen for greater efficiency, since this permits the use of a graded filter. Such a combination of modifications is called well pointing.

The practice of well pointing most probably grew from the attempts to empty sumps by pumping. The earliest recorded use in construction was for the Kilsby railroad tunnel in England in 1838. Pumping from large vertical shafts was done to lower the local groundwater surface elevation. Extensive use of and development of dewatering systems began about 50 years ago, leading to significant refinements in equipment and techniques. Installation of well-point systems is now done primarily by specialists in the field.

In construction practice, well pointing systems are widely used for dewatering large areas to shallow depths. Dewatering to depths below that accomplished by a single gravity lift is possible by placing a second or third well-point stage in the dewatered zone of the first stage.

Well points are placed by jetting whenever soil conditions make this practical. In augered or drilled holes, the annular space between the well point and the hole is filled with sand. This permits drainage of all the pervious strata through which the well point passes. Groups of well points are connected together to horizontal pipes called headers, and these in turn lead to pumps. In some soils, sealing the upper end of the hole casing increases the pumping yield. It is also possible to increase flow into the well point by applying vacuum directly to the hole casing. Typical spacing of well points is shown by the design charts in Figs. 1.3 and 1.4, which give trial spacings for various field conditions. (For greater detail on well-point systems, see Ref. 3.)

In fine silts and fine varved deposits well points may not be able to function efficiently. An alternative is electro-osmosis, which may be considered a special form of well pointing. Direct current, applied through a pair of electrodes driven into the soil, places a negative charge on capillary water adsorbed on soil particles. Adjacent water particles, thus positively charged, migrate toward the cathode carrying the bulk of the pore water. Well points are used for the cathode and can be pumped as water accumulates. In some cases, osmotic pressure is great enough to discharge water from the top of the well point without the use of external pumps.

Required electrical potentials may be as high as 100V, with as much as 30 A required per cathode. Power costs are therefore

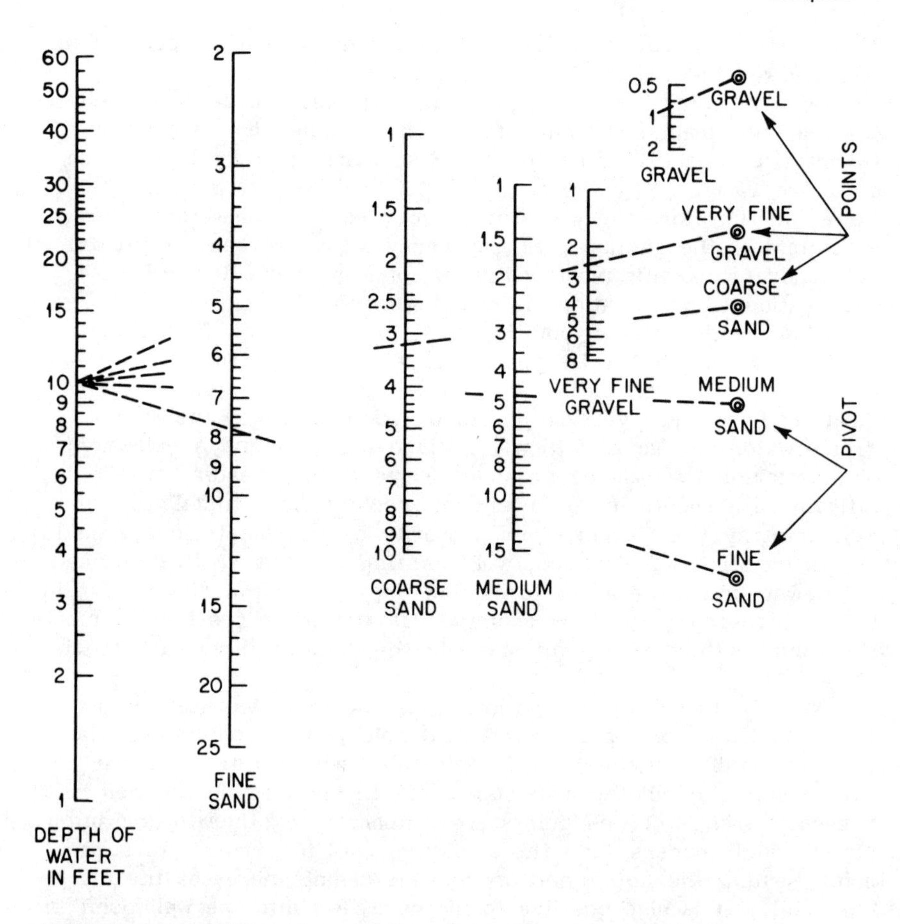

FIGURE 1.3 Well point spacing for permeable sand and gravel-clean uniform soil. (From Ref. 2.)

large. In practice electro-osmotic flow is slow, often less than 100 gpd (gallons per day). However, the method can be effective in soils too fine to be dewatered by other well-pointing systems.

The relationship between the various types of well pointing and the range of grain sizes in which each is applicable is shown in Fig. 1.5. For comparison, bars for other methods are also shown.

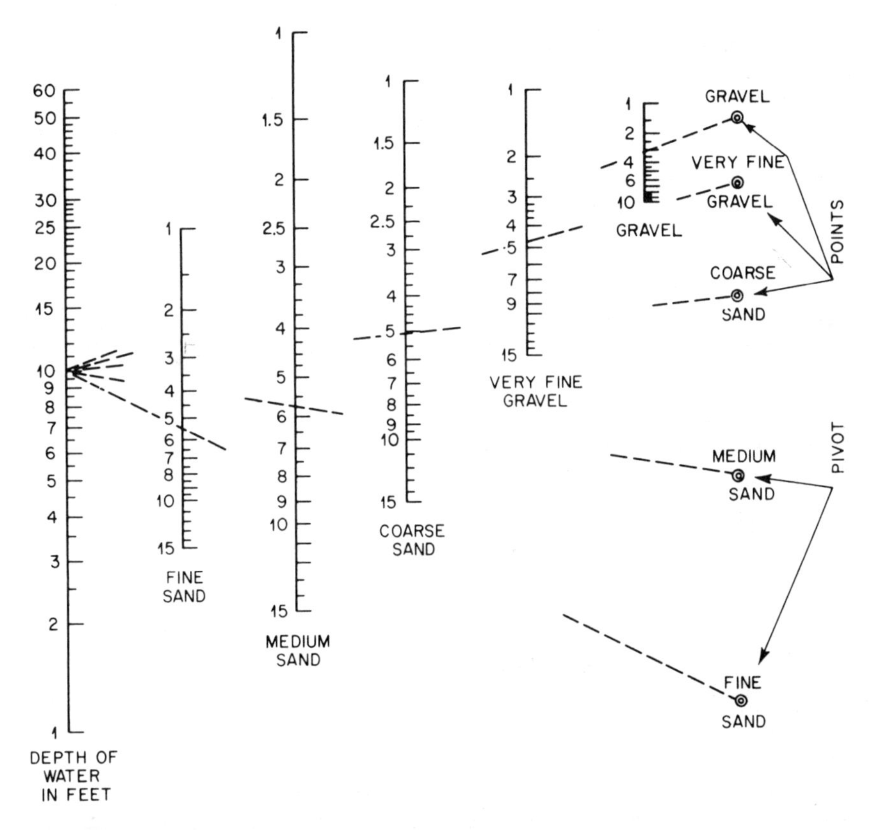

Figure 1.4 Well point spacing for permeable sand and gravel-variable soil. (From Ref. 2.)

Slurry Trenches

The last decade has seen a major growth in construction use of a new technique based on older principles. The slurry trench is an outgrowth of the use of mud to keep drill holes open.

A suspension of clay in water, when placed in a drill hole to at least the elevation of the water table, will prevent groundwater from flowing into the drill hole. Since the suspension has greater density than groundwater, there is in fact a tendency for the suspension to penetrate the side walls of the hole. Water flows into the ground, leaving a mud cake on the walls of the hole. The mud sheathing helps the hole from caving.

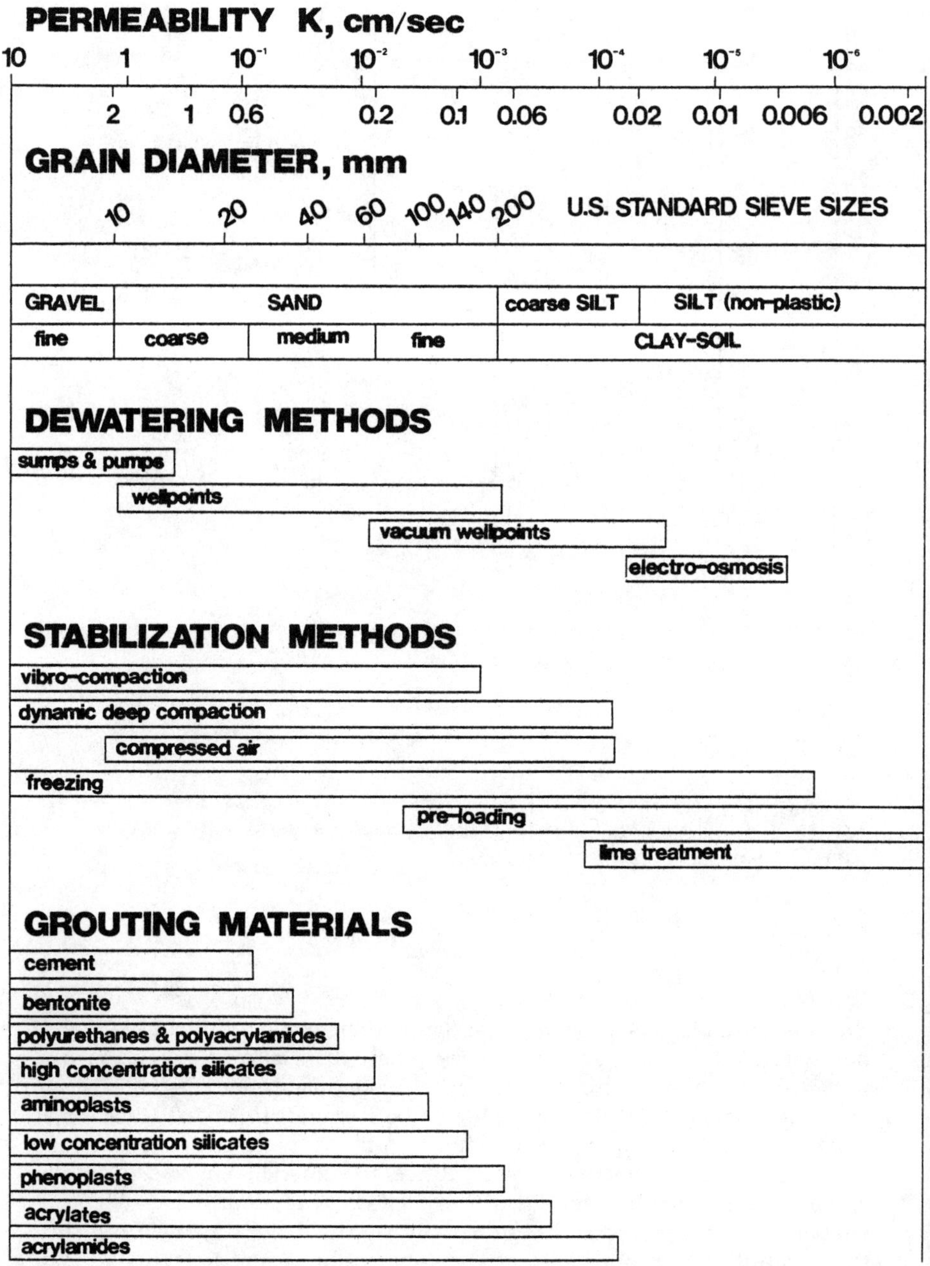

PERMEABILITY K, cm/sec

10 1 10^{-1} 10^{-2} 10^{-3} 10^{-4} 10^{-5} 10^{-6}

2 1 0.6 0.2 0.1 0.06 0.02 0.01 0.006 0.002

GRAIN DIAMETER, mm

10 20 40 60 100 140 200 U.S. STANDARD SIEVE SIZES

GRAVEL | SAND | coarse SILT | SILT (non-plastic)

fine | coarse | medium | fine | CLAY-SOIL

DEWATERING METHODS

sumps & pumps

wellpoints

vacuum wellpoints

electro-osmosis

STABILIZATION METHODS

vibro-compaction

dynamic deep compaction

compressed air

freezing

pre-loading

lime treatment

GROUTING MATERIALS

cement

bentonite

polyurethanes & polyacrylamides

high concentration silicates

aminoplasts

low concentration silicates

phenoplasts

acrylates

acrylamides

FIGURE 1.5 Dewatering methods related to grain size.

The same principle can apply to much larger excavations such as deep, narrow trenches dug by a backhoe or clamshell. Such excavations may remain open in granular materials as long as they are kept full of slurry (generally bentonite and water) at the proper density. The lateral pressure of the slurry against the excavation wall must exceed the combined lateral and fluid pressure of the formation. As excavation depth increases, the pressure differential may decrease and eventually become negative. For long excavations this would cause collapse of the soil. For short excavations, such as normally used in construction practice, arching action contributes to stability. Although failures have occurred in the field, empirical data derived from such failures have led to more reliable construction practice.

The slurry is displaced in small lateral increments by an impervious fill of soil or concrete. When clay, or a mixture of clay with silt or sand is used as a filler, the resulting construction is referred to as a slurry trench. When soil-cement or concrete is used, the structure is called a slurry wall. Such walls have load carrying capacity as well as serving as water barriers.

In very coarse materials, such as gravels, slurry trench construction may not be practical because of excess loss of slurry into the formation. When depths of trenches exceed 100 to 150 ft, special equipment and techniques are needed, which may make the method economically unfeasible. Within those limits, however, slurry trenching has begun to displace grouting for the purpose of forming cutoff walls.

1.3 DEFINITIONS

The 1964 edition of *Webster's Dictionary* defined mortar as a "plastic building material (as a mixture of cement, lime or gypsum plaster with sand and water) that hardens and is used in masonry or plastering." The same edition defined grout as a "thin mortar."

The 1974 edition extends the definition of grout to a "thin mortar used for filling spaces (as the joints in masonry), *also*: any of various other materials (as a mixture of cement and water or chemicals that solidify) used for a similar purpose."

The grouter, however, defines what he/she does as the practice of filling the fissures, pores, and voids in natural or synthetic materials in order to alter the physical properties of the treated mass. A grout may then be simply defined as a material used for grouting. The Grouting Committee, Geotechnical Engineering Division of the American Society of Civil Engineers, in its "Glossary of Terms Related to Grouting," defines grout as follows: "in soil and rock grouting, a material injected into a soil or rock formation to change the physical characteristics of the formation." Chemical grout is defined as "any grouting material characterized by being a pure solution; no particles in suspension" [1]. (A selection of terms from this glossary, specifically related to chemical grouting, appears in Appendix B.)

The key phrase in the definition is "in order to alter the physical properties." This is the purpose of grouting, and to qualify for definition as a grout, materials used for grouting must have that capability.

The definition is actually very broad. The formation changes desired are always related to strength and/or permeability. Virtually any solid has the capability of plugging formation voids under some conditions. Materials such as bran, oat hulls, straw, and sawdust have been used as grouts (primarily by drilling crews trying to plug a zone in a hole and recover drill water circulation). More common materials include sand, clay, and cement.

All the specific materials mentioned so far are solids that do not dissolve in water. When used as grouting materials, they are mixed with water to form a suspension. The water acts as the moving vehicle which carries the solid particles into the formation until the solids drop out of suspension. All these materials fall into the category of suspended-solids or particulate grouts, often referred to as suspension grouts.

The other broad category of grouts comprises those composed of solids which are soluble in water and are handled as solutions, and other materials that may naturally be liquids. These materials, which in themselves contain no suspended solid particles, are called chemical grouts. (In practice, suspended solids are often added to chemical grouts to modify the solution properties, but these materials are considered additives, and the operation is still considered to be chemical grouting.) Although chemical grouts are often referred to in terms of the solids content, this is generally understood to mean the percent solids in the solution.

The major functional difference between particulate grouts and chemical grouts is that penetrability of the former is a function of particle size, while for the latter it is a function of solution viscosity.

1.4 HISTORY OF GROUTING

Chemical grouting is a relatively recent technology, its modern era beginning in the early 1950s. Only in the past decade have the materials and techniques gained universal acceptance in the construction industry. Even so, there are many practicing construction engineers who retain doubts about the selection and use of chemical grouts. As recently as 1984, a federal government publication [3] contained the following statements: (1) "There is considerable literature on the subject of chemical grouting, but it is diverse, unorganized and often outdated." (2) "In selecting a chemical grout, it is difficult, when reviewing the literature to find anything which states which grout is probably best for a given application or how to go about making such a decision."

In contrast to these somewhat negative statements, the same publication four pages later lists a number of government publications that contain excellent details of grouting materials and procedures. This list is reproduced in Appendix E.

The invention of pressure grouting is generally credited to the French engineer Charles Berigny. In 1802, using techniques he named the "injection process," Berigny repaired deteriorated masonry walls in the port city Dieppe by pumping a suspension of clay and lime into them.

The earliest use of portland cement as a grout is variously credited to Marc Brunel in 1838 (used on the first Thames tunnel in England), to W. R. Kinipple (introduction of the injection process in England in 1856), and to Thomas Hawksley (first use of cement grout to consolidate rock in 1876).

Between 1880 and 1905, cement grout was used to control water inflow from fissured, water-bearing rock strata encountered during shaft sinking for coal in France and Belgium. During this interval great progress was made in equipment design and application techniques. These advances were soon applied to tunnels and dam foundations.

Large-scale use of cement grouting in the United States dates back to the turn of the century, when federal agencies began treating dam foundation sites. Cement and clay–cement grouts were used in huge volumes to consolidate foundation strata and to create cutoff walls (grout curtains). During this period, specifications and practices were developed in detail for this specialized type of grouting. These quickly became the unofficial grouting standards for the United States.

The first chemical grout is credited to a European, Jeziorsky, who was granted a patent in 1886 based on injecting concentrated sodium silicate into one hole and a coagulant into another (nearby) hole. In 1909, Lemaire and Dumont patented a single-shot process consisting of a mixture of dilute silicate and acid solutions. Shortly thereafter, A. Francois used a mixture of sodium silicate and aluminum sulfate solutions brought together at the injection hole.

Francois found that the use of silicate grouts facilitated the subsequent pumping of cement grout. He concluded that the silicate was acting as a lubricant. The use of sodium silicate as a "lubricant" persisted on a small scale until several decades ago. Actually, it is more probable that either (1) the grouting pressure fractured the formation making for larger voids to be filled by cement or (2) the silicate grout gelled in the smaller voids, preventing these voids from filtering the water from the grout.

A Dutch engineer, H. J. Joosten, is credited with the earliest demonstration of the reliability of the chemical grouting process in 1925. Joosten used concentrated sodium silicate injected into one hole and a strong calcium chloride solution injected under high pressure into an adjacent hole. This process, known by the name of the man who originally demonstrated its value, is still in use today, although on a very limited scale, both with and without modification. In fact, from the first use in the late 1800s until the early 1950s, sodium silicate was synonymous with chemical grouting, and all chemical grouts used during that interval were sodium silicate based.

Other silicate formulations developed soon after Joostens original work. Between 1930 and 1940, field work using sodium bicarbonate, sodium aluminate, hydrochloric acid, and copper sulphate as reagents was successfully performed.

A new era in chemical grouting started in the United States at about mid century. Since its introduction, research aimed at reducing the Joosten process to a reliable single-shot injection system had been ongoing. The breakthrough came as a result of advances in polymer chemistry and culminated in the early 1950s with the marketing of AM-9 (trademark, American Cyanamid Company), a mixture of organic monomers that were polymerized in situ after any selected time interval. The rapid development of new markets for chemical grouts was given great impetus by Cyanamid's marketing decision, which included the establishment of a research center (initially called Soils Engineering Research Center and later Engineering Chemicals Research Center, located in Princeton, New Jersey. From 1956 to 1967, this center published over 1000 pages of technical reports related to chemical grouts and grouting) to develop grouting techniques and technology.

At about the same time, chrome–lignin grouts (lignosulfonate solutions catalyzed with chromate salts) were proposed and developed for field use.

In Europe, phenol and resorcinal formaldehydes, developed in the latter 1940s, came into use. During the next several years, ureaformaldehyde-based grouts giving high strength such as Halliburton's Herculox and Cyanamid's Cyanaloc were developed and marketed (about 1956). In 1957, Soletanche in France developed a single-shot silicate grout using ethyl acetate as the reagent. Other esters came into use in the following years.

Around 1960, Diamond Alkali Company entered the market with a single-shot silicate-based grout trade named SIROC, which offered high strength or low viscosity, each coupled with gel time control.

At about this time Terra Firma, a dried precatalyzed lignosulfonate, also entered the market.

Several years later Rayonier Incorporated marketed Terranier, a single-shot grout comprised of low-molecular-weight polyphenolic polymers (about 1963). More recently, Borden Inc. marketed Geoseal, a resin prepolymer (patent filed in 1968).

Developments in chemical grouting were also taking place in Asia. In Japan, an acrylamide grout was marketed in the early 1960s as Nitto SS, and the TACSS system, a polyurethane which uses groundwater as the reactant, was marketed several years later. In Europe, during the 1960s, refinements were made to the silicate systems, and in the late 1960s and early 1970s, acrylamide-based grouts appeared. Rocagil AL (Rhone-Poulenc Ind., France) is a mixture of an acrylic monomer and an aqueous dispersion resin, while Rocagil BT is primarily methylol acrylamide.

In the United States, the market was shared primarily by AM-9 and SIROC until 1978, with SIROC getting the lion's share. Proprietary grouting materials had and still have a small part of the market. In Japan, acrylamide grouts were banned in 1974 (five reported cases of water poisoning were linked to use of acrylamide on a sewer project), and several months later the ban was extended to include all chemical grouting materials except silicate-based grouts not containing toxic additives. These events were to have strong effects on grouting practice in the United States.

Since the early 1970s concern over environmental polution had been growing rapidly. In 1976 a federal agency (probably influenced by events in Japan) sponsored a study of acrylamide-based grouts used in the United States. Acrylamide is a neurotoxic material, and the first draft of the report (which was never published) recommended that acrylamide grouts be banned. In later revisions, the report recommended that regular medical supervision of personnel using acrylamide be made a condition for its use.

Concurrently with the acrylamide study, reports issued by other federal agencies suggested that DMAPN (the acrylamide accelerator) may be carcinogenic.

Implementation of the recommendations made in these reports became unnecessary, because early in 1978, the domestic manufacturer of acrylamide grout withdrew AM-9 from the market, and made its components unavailable to anyone who might wish to use them for grouts.

The loss of AM-9 as a construction tool was lamentable, but not catastrophic. The furor among grouters would have died down quickly except for one factor. Over the years since its introduction, a very specialized and sophisticated sewer sealing industry had grown around the use of AM-9 (see Chapter 12). Those involved in this industry began an immediate search for an AM-9 replacement.

This search quickly brought a Japanese equivalent to AM-9 to the United States. This product, originally known as Nitto SS became available early in 1979 as AV-100. European products were available for a short time on a trial basis only. They were not marketed commercially.

The search for new and less hazardous materials took longer to consumate. By the middle of 1979, Terragel became commercially available. This product was a concentrated solution of methylol–acrylamide. It was withdrawn from the market within a short time due to storage stability problems. In 1980 CR-250, a urethane product, was marketed specifically for sewer sealing applications. Improved and modified versions have since appeared. At the same time, Injectite 80, an acrylamide prepolymer (relatively nontoxic) became available also for sewer sealing applications. This product has not been marketed aggressively. Later in the same year, AC-400, a relatively nontoxic mixture of acrylates was marketed as a general replacement for acrylamides. Its properties are similar to those of acrylamide grouts, and AC-400 is regaining the market previously held by the acrylamides. Another acrylate grout, appeared on the market in 1985.

At present, virtually all construction grouting in the United States is done with silicates. This is not because other materials are not available. By way of contrast, phenoplasts, aminoplasts, chrome lignins, and acrylamides are all used in Europe. These products are well known to American grouting firms. However, in the United States, Terra Firma and Terranier (chrome lignin and phenoplast) fell by the wayside some years ago, primarily due to the toxic properties of the dichromate catalyst. Herculox and Cyanaloc (aminoplasts) had limited application to begin with because they require an acid environment. In addition, the formaldehyde component can cause chronic respiratory problems. Geoseal also contains formaldehyde.

Imported acrylamide dominated the sewer-sealing industry until the acrylate grouts appeared. These are gradually replacing acrylamide, as well as becoming the primary product for seepage control.

Although concern over environmental pollution and personnel health hazards has been an important factor in the limited use of specific chemical grouts (except for sodium silicate, all the chemical grouts are to some degree toxic, hazardous, or both), there is no ban against use of acrylamide-based grouts in the United States at the time this book goes to press. However, the health hazards to personnel using acrylamide in the field, coupled with the availability of good substitute products, will eventually phase out the use of acrylamide grouts.

1.5 FIELD PROBLEMS AMENABLE TO GROUTING

Grouting is but one of many methods of groundwater control used in construction. The choice of a method for any specific job almost always relates back to economics. Factors that contribute to economical selection include the size of the job, the job location, the operation time schedule, and the contractor's capabilities and experience. The problems described below (which were in these specific cases solved by grouting) should not be considered in general as yielding only to grouting techniques, or only to the specific materials that were used in these actual field jobs.

The purpose of grouting, as indicated in the definition of grout is to change the characteristics of the treated formation. These changes are either a decrease in permeability, an increase in strength, or both. Grouting the voids in a strongly cemented sandstone may have little effect on formation strength. Grouting the voids in sands may result in significant strength increase. The degree of strength increase required may determine the type of grout used. If only sufficient strength increase is needed to keep sand grains from changing relative positions (a process often termed "stabilization") a less expensive, relatively weak, grout may suffice. If strength increase is required to raise the safety factor against bearing failure, slope failure, or cave-ins, generally a very strong grout is dictated.

By contrast, water shut-off is generally not adequate unless it is total or nearly so. The choice of grouts for seepage control is selected on the basis of formation void size and effective grout penetrability.

The requirement of water shut-off and strength increase may be interrelated. The stability of an excavation face or a slope in fine-grained granular material is lowered by the flow of water through the face. Thus water shut-off must be accomplished in order to increase strength. Grout selection for such cases is also based on penetrability.

In many construction jobs plagued by seepage, there are no ancillary water problems, and the purpose of grouting is solely to reduce or shut off the flow of water. Typical cases include tunnels and shafts through sandstone or fissured sound rock, seepage through or around earth and concrete dams, and flow of water into underground structures of all kinds.

Figure 1.6 shows a small area along a river bank about a mile downstream of a power dam near Albany, New York. The source of the water in the upper left-hand corner of the picture is a channel leading back to the reservoir. Although the water loss is negligible, the possibility exists that continuing flow may enlarge the seepage

FIGURE 1.6 Seepage downstream from power dam.

channel and eventually lead to major problems. The small weir shown was constructed to provide a measure of the effectiveness of the grouting work. Procedures were to trace the channel as far back as practical from its outlet, using dyed water pumped into the ground, and then grout the channel to make it impermeable. An acrylamide grout was used on this project.

The major purpose of a grouting job may be to keep the soil in place. In Cleveland, Ohio, a new structure was designed between two existing buildings. Excavation for the new structure would be 11 ft below the foundation of one of the adjacent buildings into a loose, dry sand. Foundations for the existing structure were adequately supported as long as the soil under the footings did not move away. It was thus necessary to provide some means of keeping the sand in place to avoid settlement of the structure. Figure 1.7 shows the method used to stabilize the soil and keep it from moving during construction. This work was done with a silicate-based grout.

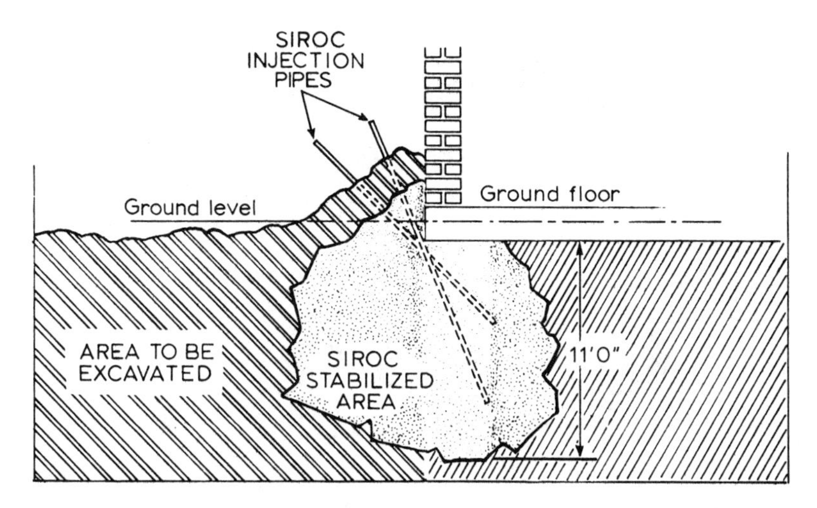

FIGURE 1.7 Stabilization of sand to prevent foundation movement. (Courtesy of Raymond Concrete Pile Division, New York.)

In some cases, the problem is simply one of increasing the formation strength. In Minneapolis, Minnesota, wood piling under a 165-ft-high brick chimney had deteriorated above the water table, and structural support of the foundation slab was disappearing. It was necessary to strengthen the foundation soil to the point where it could transfer the structural load from the foundation slab to the sound portion of the piles below the water table. Figure 1.8 shows the grouting procedures used to solidify and strengthen the soil beneath the chimney. A silicate-based grout was used on this project, with the outer calcium chloride curtain acting as a restraint to escape of grout placed inside the curtain.

There are also cases in which grouting serves a multiple purpose. The sewer tunnel shown in Fig. 1.9 was started in clay, and construction was proceeding smoothly until an unexpected stratum of fine, dry sand was encountered. The sand had to be stabilized to keep it from running into the tunnel, but it also needed sufficient shear strength to act as a support for the overburden, until steel support and lagging could be placed. Had the sand lens contained water, the grout would also have had to function as a seepage barrier.

Different materials and procedures are used in the field depending to a large extent on the purpose of grouting. They are discussed more fully in succeeding chapters.

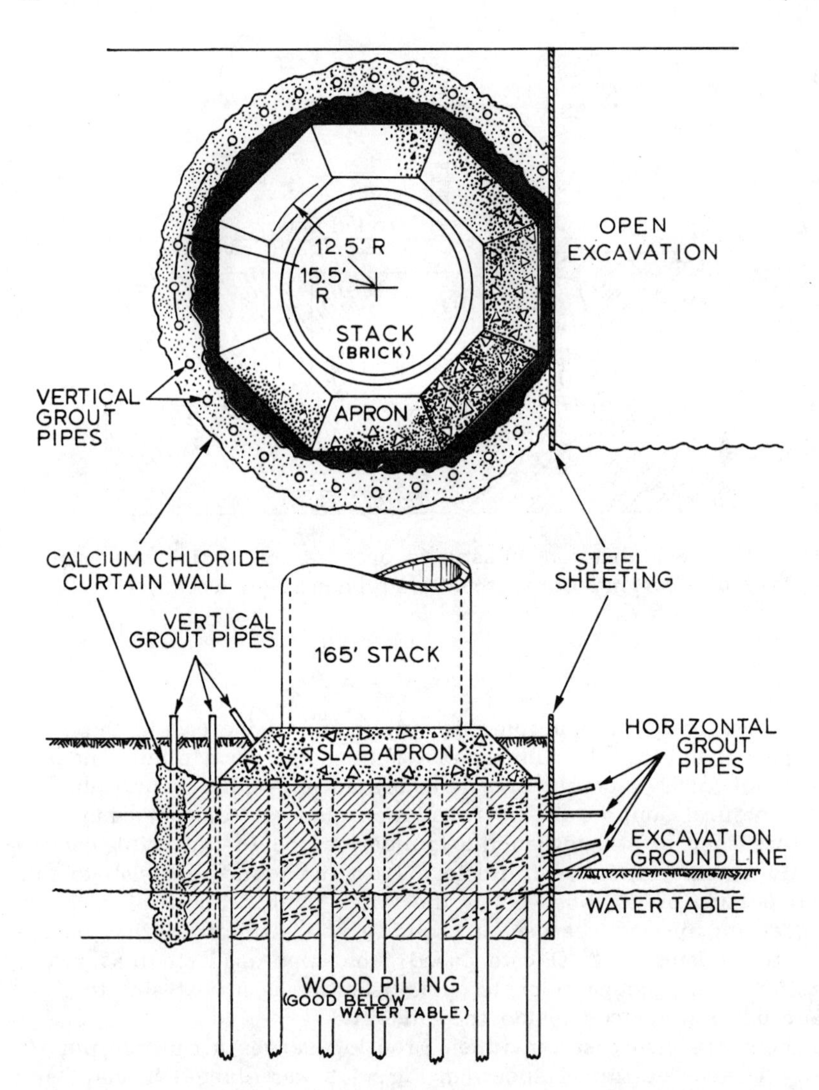

FIGURE 1.8 Strengthening of sand to increase bearing capacity. (Courtesy of Raymond Concrete Pile Division, New York.)

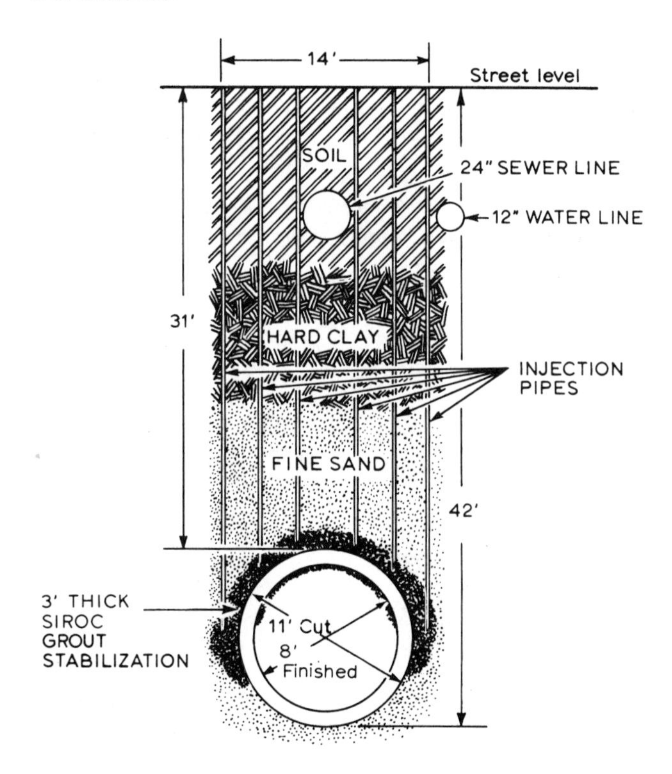

FIGURE 1.9 Stabilization and strengthening of sand. (Courtesy of Raymond Concrete Pile Division, New York.)

1.6 SUMMARY

Movement of water and/or soil into an excavation can interfere with construction schedules. Flow of water into a finished structure may interfere with effective use of the structure. Many field procedures are in regular use for control of water flow and/or increasing strength and/or stability of soil and rock formations. Grouting is one of these procedures.

Almost always the choice of method will be determined by economics, although economics for specific field emergencies may be influenced by such factors as availability of specialty contractors, equipment, and supplies. Grouting, in common with all other procedures, is not a cure-all. There are, however, many field situations for which grouting, when properly done, is the most cost effective method.

REFERENCES

1. Committee on Grouting, Preliminary glossary of terms related to grouting, *J. Geotech. Eng. Div. ASCE, 106*(GT7):803–805 (July 1980).

2. L. J. Goodman and R. H. Karol, *Theory and Practice of Foundation Engineering*, Macmillan, New York, 1968.

3. U.S. Department of the Interior, Bureau of Reclamation, *Policy Statement for Grouting*, Denver, Colorado, 1984.

2

The Grouting Medium

2.1 INTRODUCTION

Grouting is a procedure in which the grout replaces the natural fluids in the formation voids. For grouting to become a useful construction tool, displacement of the void fluids must take place at rates and pressures that make grouting economically competitive with other construction alternatives.

The ability of a formation to receive grout is called its groutability [1]. This characteristic is affected by many parameters. Often these parameters can be integrated to a reliable conclusion by conducting in situ pumping tests (see Chap. 7 for detailed discussion). When economics or other considerations preclude pumping tests, groutability of a formation must be estimated (or guessed at) from related characteristics. The most pertinent of these characteristics is the permeability.

The broad general classification of soils is an index of groutability. Thus, gravels and sands are generally groutable without difficulty, silts are usually difficult or impossible to grout, and clays cannot be grouted. The usual definitions of these materials in terms of grain size are shown in Fig. 2.1.

Classification of a soil is generally done by the driller taking the samples and is recorded on a boring log. The driller uses personal experience to estimate the percentage of various grain sizes in the granular portion of total soil sample and defines the soil in general terms as, for example, "sandy SILT, some clay." The capitalized word indicates the major component. Some classification systems give percentage ranges to words such as "little," "some," and "trace." The grouter using boring logs should become familiar with the specific classification system used.

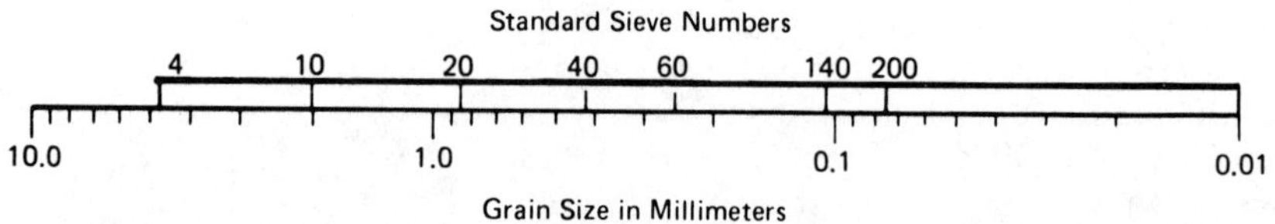

FIGURE 2.1 Classification by grain size.

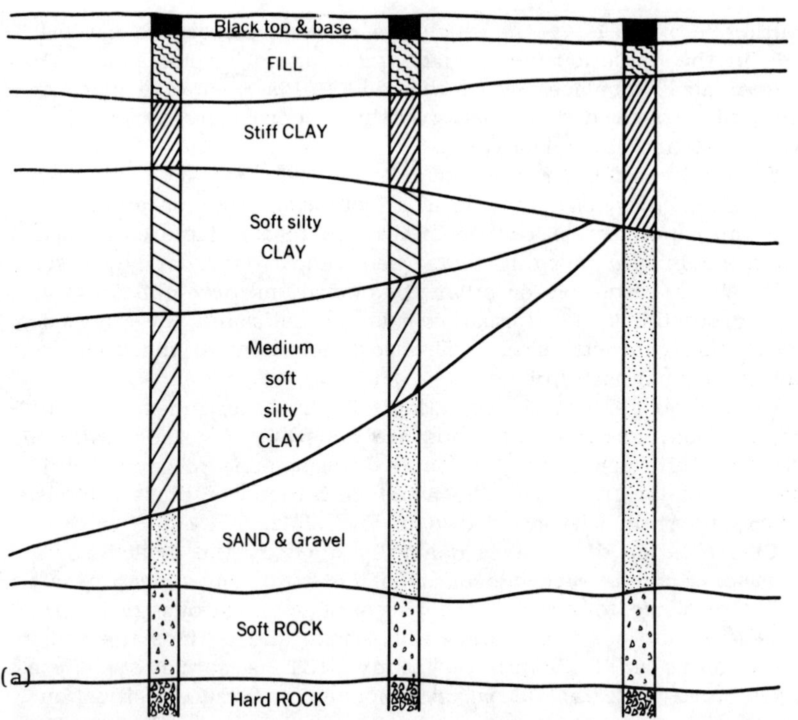

FIGURE 2.2 Erratic (a) and regular (b) soil profiles. [From R. H. Karol, *Soils and Soils Engineering*, Prentice Hall, New York, 1960 (out of print).]

Specific classification of soils can be made with the use of data from laboratory tests for grain size and consistency. The Unified Soil Classification System [2] is widely used in soil engineering reports to define foundation soils and their broad characteristics. Portions of the system are shown in Figs. 2.3 and 2.4. Of specific interest to the grouter are the columns headed "permeability" and requirements for seepage control."

When extrapolating boring logs for design purposes, however, it is first necessary to establish whether the deposits encountered in the bore hole are regular or erratic. Figure 2.2 shows such

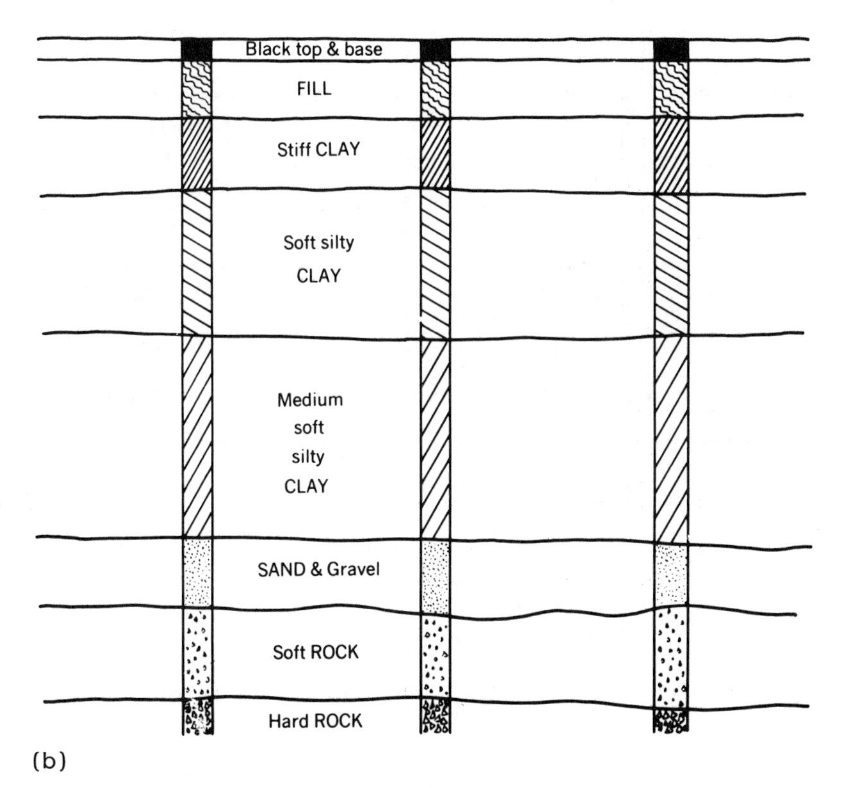

FIGURE 2.2 (Continued)

CHARACTERISTICS PERTINENT TO EMBANKMENTS AND FOUNDATIONS

Major Divisions (1)	(2)	Letter (3)	Symbol Hatching (4)	Symbol Color (5)	Name (6)	Value for Embankments (7)
COARSE GRAINED SOILS	GRAVEL AND GRAVELLY SOILS	GW		Red	Well-graded gravels or gravel-sand mixtures, little or no fines	Very stable, pervious shells of dikes and dams
		GP			Poorly-graded gravels or gravel-sand mixtures, little or no fines	Reasonably stable, pervious shells of dikes and dams
		GM		Yellow	Silty gravels, gravel-sand-silt mixtures	Reasonably stable, not particularly suited to shells, but may be used for impervious cores or blankets
		GC			Clayey gravels, gravel-sand-clay mixtures	Fairly stable, may be used for impervious core
	SAND AND SANDY SOILS	SW		Red	Well-graded sands or gravelly sands, little or no fines	Very stable, pervious sections, slope protection required
		SP			Poorly-graded sands or gravelly sands, little or no fines	Reasonably stable, may be used in dike section with flat slopes
		SM		Yellow	Silty sands, sand-silt mixtures	Fairly stable, not particularly suited to shells, but may be used for impervious cores or dikes
		SC			Clayey sands, sand-silt mixtures	Fairly stable, use for imperviou core for flood control structures
FINE GRAINED SOILS	SILTS AND CLAYS LL < 50	ML		Green	Inorganic silts and very fine sands, rock flour, silty or clayey fine sands or clayey silts with slight plasticity	Poor stability, may be used for embankments with proper control
		CL			Inorganic clays of low to medium plasticity, gravelly clays, sandy clays, silty clays, lean clays	Stable, impervious cores and blankets
		OL			Organic silts and organic silt-clays of low plasticity	Not suitable for embankments
	SILTS AND CLAYS LL > 50	MH		Blue	Inorganic silts, micaceous or diatomaceous fine sandy or silty soils, elastic silts	Poor stability, core of hydrauli fill dam, not desirable in rolle fill construction
		CH			Inorganic clays of high plasticity, fat clays	Fair stability with flat slopes, thin cores, blankets and dike sections
		OH			Organic clays of medium to high plasticity, organic silts	Not suitable for embankments
HIGHLY ORGANIC SOILS		Pt		Orange	Peat and other highly organic soils	Not used for construction

Notes:
1. Values in columns 7 and 11 are for guidance only. Design should be based on test results.
2. In column 9, the equipment listed will usually produce the desired densities with a reasonable number of passes when moisture conditions and thickness of lift are properly controlled.
3. Column 10, unit dry weights are for compacted soil at optimum moisture content for Standard AASHO

Taken from TM 3-357, The Unified Soil Classification System, Appendix A, Characteristics of Soil Groups Pertaining to Embankments and Foundations, by U. S. Army Engineer Waterways Experiment Station, March 1953.

FIGURE 2.3 Unified soil classification system. (From Ref. 2.)

Permeability Cm Per Sec (8)	Compaction Characteristics (9)	Std AASHO Max Unit Dry Weight Lb Per Cu Ft (10)	Value for Foundations (11)	Requirements for Seepage Control (12)
$k > 10^{-2}$	Good, tractor, rubber-tired, steel-wheeled roller	125-135	Good bearing value	Positive cutoff
$k > 10^{-2}$	Good, tractor, rubber-tired, steel-wheeled roller	115-125	Good bearing value	Positive cutoff
$k = 10^{-3}$ to 10^{-6}	Good, with close control, rubber-tired, sheepsfoot roller	120-135	Good bearing value	Toe trench to none
$k = 10^{-6}$ to 10^{-8}	Fair, rubber-tired, sheepsfoot roller	115-130	Good bearing value	None
$k > 10^{-3}$	Good, tractor	110-130	Good bearing value	Upstream blanket and toe drainage or wells
$k > 10^{-3}$	Good, tractor	100-120	Good to poor bearing value depending on density	Upstream blanket and toe drainage or wells
$k = 10^{-3}$ to 10^{-6}	Good, with close control, rubber-tired, sheepsfoot roller	110-125	Good to poor bearing value depending on density	Upstream blanket and toe drainage or wells
$k = 10^{-6}$ to 10^{-8}	Fair, sheepsfoot roller, rubber tired	105-125	Good to poor bearing value	None
$k = 10^{-3}$ to 10^{-6}	Good to poor, close control essential, rubber-tired roller, sheepsfoot roller	95-120	Very poor, suscepti-ble to liquefaction	Toe trench to none
$k = 10^{-6}$ to 10^{-8}	Fair to good, sheepsfoot roller, rubber tired	95-120	Good to poor bearing	None
$k = 10^{-4}$ to 10^{-6}	Fair to poor, sheepsfoot roller	80-100	Fair to poor bearing, may have excessive settlements	None
$k = 10^{-4}$ to 10^{-6}	Poor to very poor, sheepsfoot roller	70-95	Poor bearing	None
$k = 10^{-6}$ to 10^{-8}	Fair to poor, sheepsfoot roller	75-105	Fair to poor bearing	None
$k = 10^{-6}$ to 10^{-8}	Poor to very poor, sheepsfoot roller	65-100	Very poor bearing	None
	Compaction not practical		Remove from foundations	

FIGURE 2.3 (Continued)

UNIFIED SOIL CLASSIFICATION
(Including Identification and Description)

Major Divisions			Group Symbols	Typical Names	Field Identification Procedures (Excluding particles larger than 3 in. and basing fractions on estimated weights)		
1	2		3	4	5		
Coarse-grained Soils. More than half of material is larger than No. 200 sieve size.	Gravels. More than half of coarse fraction is larger than No. 4 sieve size.	Clean Gravels (Little or no fines)	GW	Well-graded gravels, gravel-sand mixtures, little or no fines.	Wide range in grain sizes and substantial amounts of all intermediate particle sizes.		
			GP	Poorly graded gravels or gravel-sand mixtures, little or no fines.	Predominantly one size or a range of sizes with some intermediate sizes missing.		
		Gravels with Fines (Appreciable amount of fines)	GM	Silty gravels, gravel-sand-silt mixture.	Nonplastic fines or fines with low plasticity (for identification procedures see ML below).		
			GC	Clayey gravels, gravel-sand-clay mixtures.	Plastic fines (for identification procedures see CL below).		
	Sands. More than half of coarse fraction is smaller than No. 4 sieve size.	Clean Sands (Little or no fines)	SW	Well-graded sands, gravelly sands, little or no fines.	Wide range in grain size and substantial amounts of all intermediate particle sizes.		
			SP	Poorly graded sands or gravelly sands, little or no fines.	Predominantly one size or a range of sizes with some intermediate sizes missing.		
		Sands with Fines (Appreciable amount of fines)	SM	Silty sands, sand-silt mixtures.	Nonplastic fines or fines with low plasticity (for identification procedures see ML below).		
	(For visual classification, the 1/4-in. size may be used as equivalent to the No. 4 sieve size)		SC	Clayey sands, sand-clay mixtures.	Plastic fines (for identification procedures see CL below).		
					Identification Procedures on Fraction Smaller than No. 40 Sieve Size		
					Dry Strength (Crushing characteristics)	Dilatancy (Reaction to shaking)	Toughness (Consistency near PL)
Fine-grained Soils. More than half of material is smaller than No. 200 sieve size.	Silts and Clays. Liquid limit is less than 50		ML	Inorganic silts and very fine sands, rock flour, silty or clayey fine sands or clayey silts with slight plasticity.	None to slight	Quick to slow	None
			CL	Inorganic clays of low to medium plasticity, gravelly clays, sandy clays, silty clays, lean clays.	Medium to high	None to very slow	Medium
			OL	Organic silts and organic silty clays of low plasticity.	Slight to medium	Slow	Slight
	Silts and Clays. Liquid limit is greater than 50		MH	Inorganic silts, micaceous or diatomaceous fine sandy or silty soils, elastic silts.	Slight to medium	Slow to none	Slight to medium
			CH	Inorganic clays of high plasticity, fat clays.	High to very high	None	High
			OH	Organic clays of medium to high plasticity, organic silts.	Medium to high	None to very slow	Slight to medium
Highly Organic Soils			Pt	Peat and other highly organic soils.	Readily identified by color, odor, spongy feel and frequently by fibrous texture.		

The No. 200 sieve size is about the smallest particle visible to the naked eye.

FIGURE 2.4 Unified soil classification system. (From Ref. 2.)

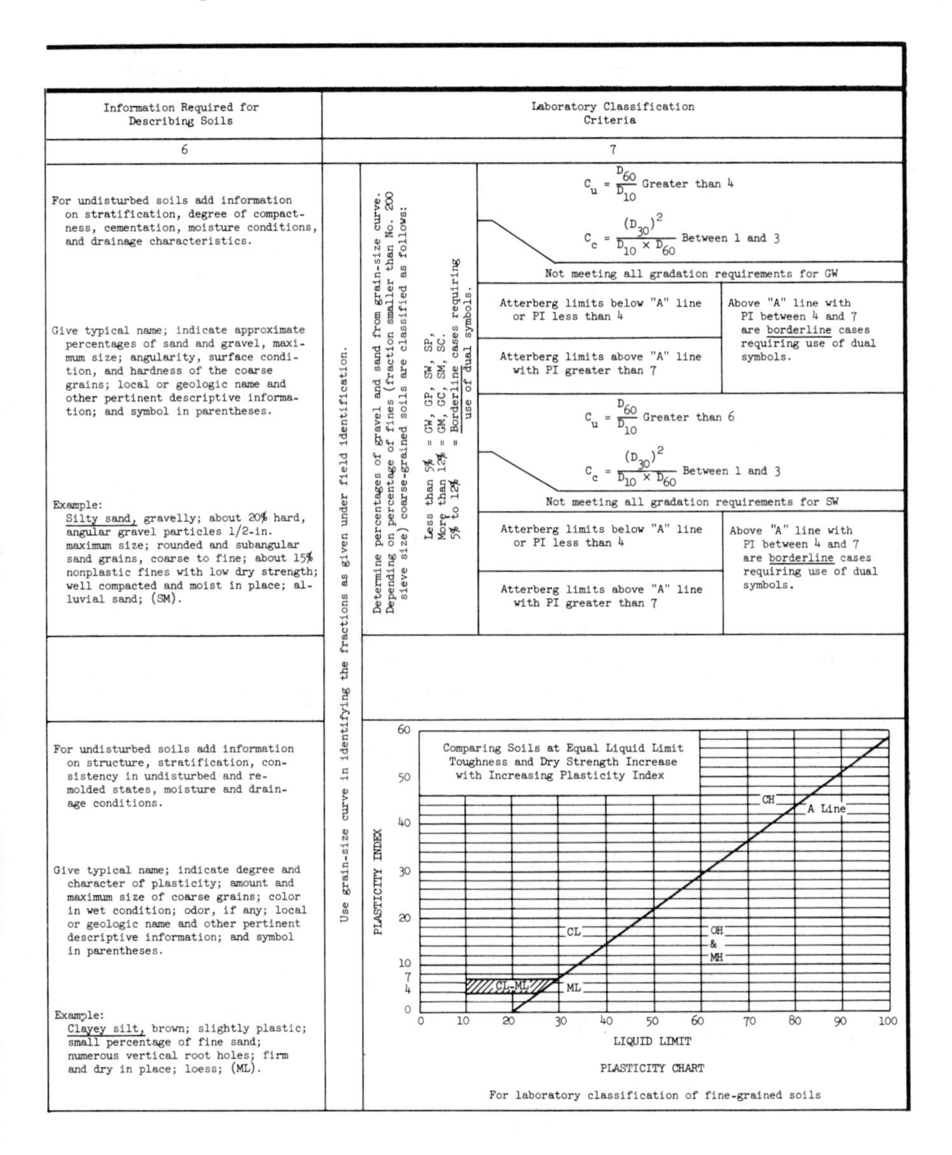

Information Required for Describing Soils	Laboratory Classification Criteria			
6	7			
For undisturbed soils add information on stratification, degree of compactness, cementation, moisture conditions, and drainage characteristics.	Use grain-size curve in identifying the fractions as given under field identification.	Determine percentages of gravel and sand from grain-size curve. Depending on percentage of fines (fraction smaller than No. 200 sieve size) coarse-grained soils are classified as follows: Less than 5% = GW, GP, SW, SP, More than 12% = GM, GC, SM, SC. 5% to 12% = Borderline cases requiring use of dual symbols.	$C_u = \frac{D_{60}}{D_{10}}$ Greater than 4 $C_c = \frac{(D_{30})^2}{D_{10} \times D_{60}}$ Between 1 and 3	
			Not meeting all gradation requirements for GW	
Give typical name; indicate approximate percentages of sand and gravel, maximum size; angularity, surface condition, and hardness of the coarse grains; local or geologic name and other pertinent descriptive information; and symbol in parentheses.			Atterberg limits below "A" line or PI less than 4	Above "A" line with PI between 4 and 7 are borderline cases requiring use of dual symbols.
			Atterberg limits above "A" line with PI greater than 7	
			$C_u = \frac{D_{60}}{D_{10}}$ Greater than 6 $C_c = \frac{(D_{30})^2}{D_{10} \times D_{60}}$ Between 1 and 3	
			Not meeting all gradation requirements for SW	
Example: Silty sand, gravelly; about 20% hard, angular gravel particles 1/2-in. maximum size; rounded and subangular sand grains, coarse to fine; about 15% nonplastic fines with low dry strength; well compacted and moist in place; alluvial sand; (SM).			Atterberg limits below "A" line or PI less than 4	Above "A" line with PI between 4 and 7 are borderline cases requiring use of dual symbols.
			Atterberg limits above "A" line with PI greater than 7	
For undisturbed soils add information on structure, stratification, consistency in undisturbed and remolded states, moisture and drainage conditions. Give typical name; indicate degree and character of plasticity; amount and maximum size of coarse grains; color in wet condition; odor, if any; local or geologic name and other pertinent descriptive information; and symbol in parentheses. Example: Clayey silt, brown; slightly plastic; small percentage of fine sand; numerous vertical root holes; firm and dry in place; loess; (ML).		PLASTICITY CHART (see figure text below) For laboratory classification of fine-grained soils		

FIGURE 2.4 (Continued)

deposits. Obviously, one bore hole cannot define erratic deposits, and should not be used by itself for design or planning purposes.

2.2 VOID RATIO AND POROSITY

Void ratio is defined as the quotient of the volume of voids divided by the volume of solids in a soil mass. Porosity is defined as the ratio of the volume of voids to the total volume. These relationships are shown pictorally in Fig. 2.5:

Void ratio: $$e = \frac{V_v}{V_s}$$

Porosity: $$n = \frac{V_v}{V}$$

From these definitions it can be seen that porosity is always less than unity, while void ratio can be more than 1. Void ratios for sand and silts are generally between 0.5 and 0.8. The relationship between void ratio and porosity is shown in Fig. 2.6.

The voids in a soil or rock can be filled with gas, liquid, or a combination of both. The most common materials in the voids are air and water. Saturated soils and rock formation have their voids

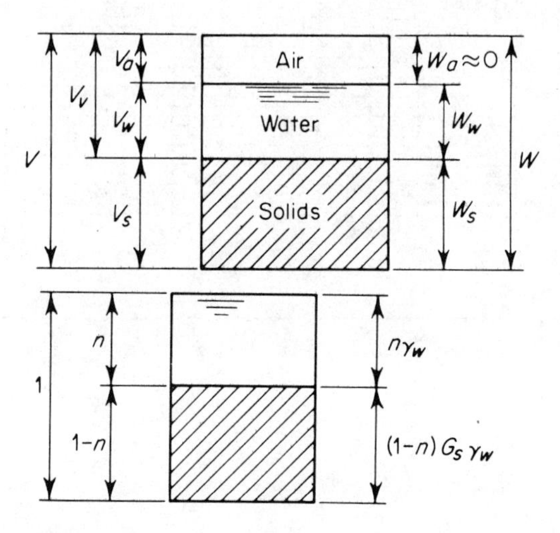

FIGURE 2.5 Representation of void ratio and porosity.

TABLE 2.1 Degree of Saturation (Relative Humidity)

Descriptive term	Percent saturation
Dry	0
Humid	1 to 25
Damp	25 to 50
Moist	50 to 75
Wet	75 to 99
Saturated	100

Source: Ref. 4.

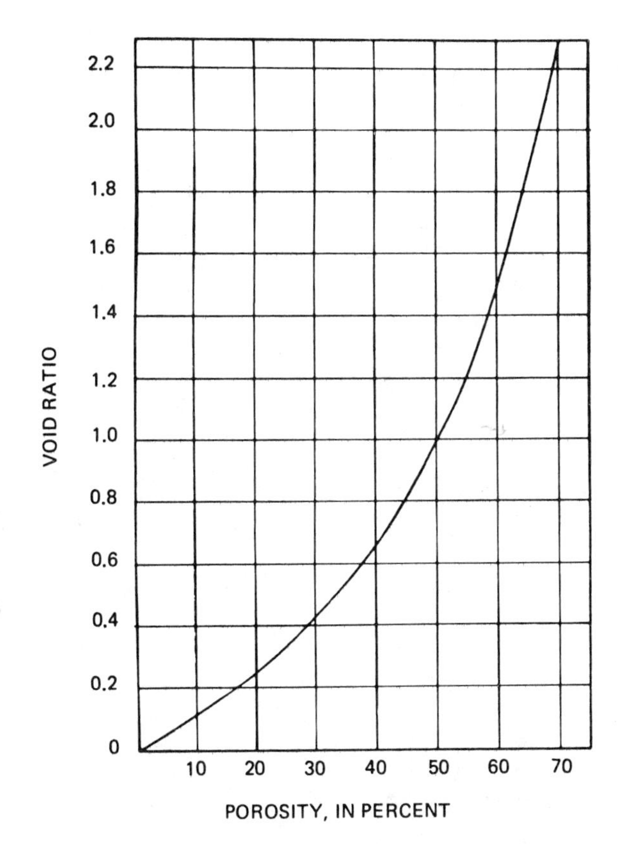

FIGURE 2.6 Relationship between void ratio and porosity.

completely filled with water. Dry formations contain only air in the voids. Between these two extremes the degree of saturation is defined as shown in Table 2.1.

2.3 DENSITY

The term density as used in soil mechanics is defined as the weight per unit volume and, unless otherwise specified, is assumed to be dry weight. The term is important in grouting because a soil in the loose state that may be readily grouted may be ungroutable in the dense state.

The relative density D_R or degree of density of a granular soil is a comparison of its in-place natural density with its loose and dense states. In terms of void ratio,

$$D_R = \frac{e_{max} - e}{e_{max} - e_{min}}$$

TABLE 2.2 Degree of Density of Sands

Descriptive term	Degree of density
Loose	0 to 1/3
Medium	1/3 to 2/3
Dense	2/3 to 1

Results of Standard Penetration Tests

Descriptive term	No. of blows per foot
Very loose	0 to 4
Loose	4 to 10
Medium	10 to 30
Dense	30 to 50
Very dense	50 and over

Source: Ref. 4.

where

e_{max} = void ratio at maximum density

e_{min} = void ratio at minimum density

e = natural void ratio

and descriptive terms for various states are shown in Table 2.2. This table also shows relationships between the descriptive terms and the standard penetration test (Penetration Test and Split-Barrel Sampling of Soils, ASTM designation: D 1586-67).

Granular soils of the same general classification (i.e., medium sands) may have a wide spread of grain sizes. The range of grain sizes is expressed by a factor called the uniformity coefficient $C_U = D_{60}/D_{10}$, where D_{60} and D_{10} are defined as the grain size of which 60% and 10% (respectively) of the soil is finer. The relationship between C_u and permeability is discussed in the following section.

2.4 PERMEABILITY

Flow of water through tubes and pipes is either laminar or turbulent, depending on the velocity of flow. Water particle paths are haphazard and irregular in turbulent flow and parallel to the container walls and each other in laminar flow. Below a limiting value, called the lower critical velocity, flow will always be laminar. The following equation, called the Reynolds number,

$$\frac{V_c D \gamma_w}{\mu g} = 2000$$

where

V_c = lower critical velocity

D = diameter of passage

γ_w = unit weight of water

μ = viscosity of water

g = acceleration of gravity

indicates that critical velocity is inversely proportional to the size of the passage. For the very small passages in silts, flow is never turbulent. When grain size exceeds 0.5 mm, turbulent flow can be expected. Burminster [3] has extended these criteria:

1. Gravel: Flow is always turbulent.
2. Sand: Flow is turbulent for gradients above 0.2 in the loose state and above 0.4 in the dense state.

Knowledge of the flow character is important to the grouter, since grout generally displaces groundwater in laminar flow and mixes with groundwater in turbulent flow. The degree or amount of mixing is also related to the ratio of grout injection rate at a point to groundwater flow past that point.

Permeability is the property of a soil which permits water to flow through its pores. Units are distance per time, commonly centimeters per second in soil mechanics. The equation used to determine permeability in soils is known as Darcy's law and is valid only for laminar flow:*

$$\frac{Q}{A} = v = ki$$

where

Q = total discharge

A = total soil cross section

v = velocity of flow

i = hydraulic gradient

k = (Darcy's) coefficient of permeability

The velocity determined by this equation is a superficial one, since it is based on the total cross-sectional area rather than the pore area. True velocities of flow can be obtained, if necessary, to determine whether flow is laminar or turbulent by using the true porosity value to modify A.

Permeability values for soils are generally determined by laboratory tests [4] (Test for Permeability of Granular Soils, ASTM designation: D 2434-68). Such tests on granular materials may be unreliable for grouting purposes, since it is virtually impossible to take a field sample of a granular soil without disturbing its natural density and stratification. Further, most natural soils, even when

*Recent research indicates that flow of viscous materials at high gradients deviates from that predicted by Darcy's law.

TABLE 2.3 Degree of Permeability

Descriptive term	k (cm/s)	Soils
High	10^{-1} and over	Gravel and coarse sand
Medium	10^{-1} to 10^{-3}	Sand and fine sand
Low	10^{-3} to 10^{-5}	Very fine sands
Very low	10^{-5} to 10^{-7}	Silts and clay
Impermeable	10^{-7} and less	Clay

Source: Ref. 4.

not visually stratified, have much greater permeabilities in the horizontal direction than in the vertical direction.

Field tests may also be performed to determine permeability. Such tests are discussed in Chap. 7.

The relationship between permeability and void ratio is empirical and for sands may be expressed as

$$\frac{k_1}{k_2} = \frac{e_1^2}{e_2^2}$$

TABLE 2.4 Approximate Soil Properties

Void ratio	Porosity	Soils	Grain size (mm)	Permeability (cm/s)
0.6 to 0.8	0.25 to 0.45	Gravel and coarse sand	0.5 and over	10^{-1} and over
0.6 to 0.8	0.25 to 0.45	Medium and fine sand	0.1 to 0.5	10^{-1} to 10^{-3}
0.6 to 0.9	0.25 to 0.5	Very fine sand	0.05 to 0.1	10^{-3} to 10^{-5}
0.6 up	0.25 up	Silts	0.5 and less	10^{-5} to 10^{-7}
0.6 up	0.25 up	Clays	0.05 and less	10^{-7} and less

TABLE 2.5 Porosity, Void Ratio, and Unit Weight of Typical Soils in Natural State

Description	Porosity, n (%)	Void ratio, e	Water contents $w_{sat.}$ (%)	Unit weight			
				g/cm^3		lb/ft^3	
				γ_d	$\gamma_{sat.}$	γ_d	$\gamma_{sat.}$
1. Uniform sand, loose	46	0.85	32	1.43	1.89	90	118
2. Uniform sand, dense	34	0.51	19	1.75	2.09	109	130
3. Mixed-grained sand, loose	40	0.67	25	1.59	1.99	99	124
4. Mixed-grained sand, dense	30	0.43	16	1.86	2.15	116	135
5. Glacial till, very mixed grained	20	0.25	9	2.12	2.32	132	145
6. Soft glacial clay	55	1.2	45	–	1.77	–	110
7. Stiff glacial clay	37	0.6	22	–	2.07	–	129
8. Soft slightly organic clay	66	1.9	70	–	1.58	–	98
9. Soft very organic clay	75	3.0	110	–	1.43	–	89
10. Soft bentonite	84	5.2	194	–	1.27	–	80

Note: $w_{sat.}$ = water content when saturated in percent of dry weight

γ_d = unit weight in dry state

$\gamma_{sat.}$ = unit weight in saturated state

n = porosity = $e/(1 + e)$

e = void ratio = $n/(1 + n)$

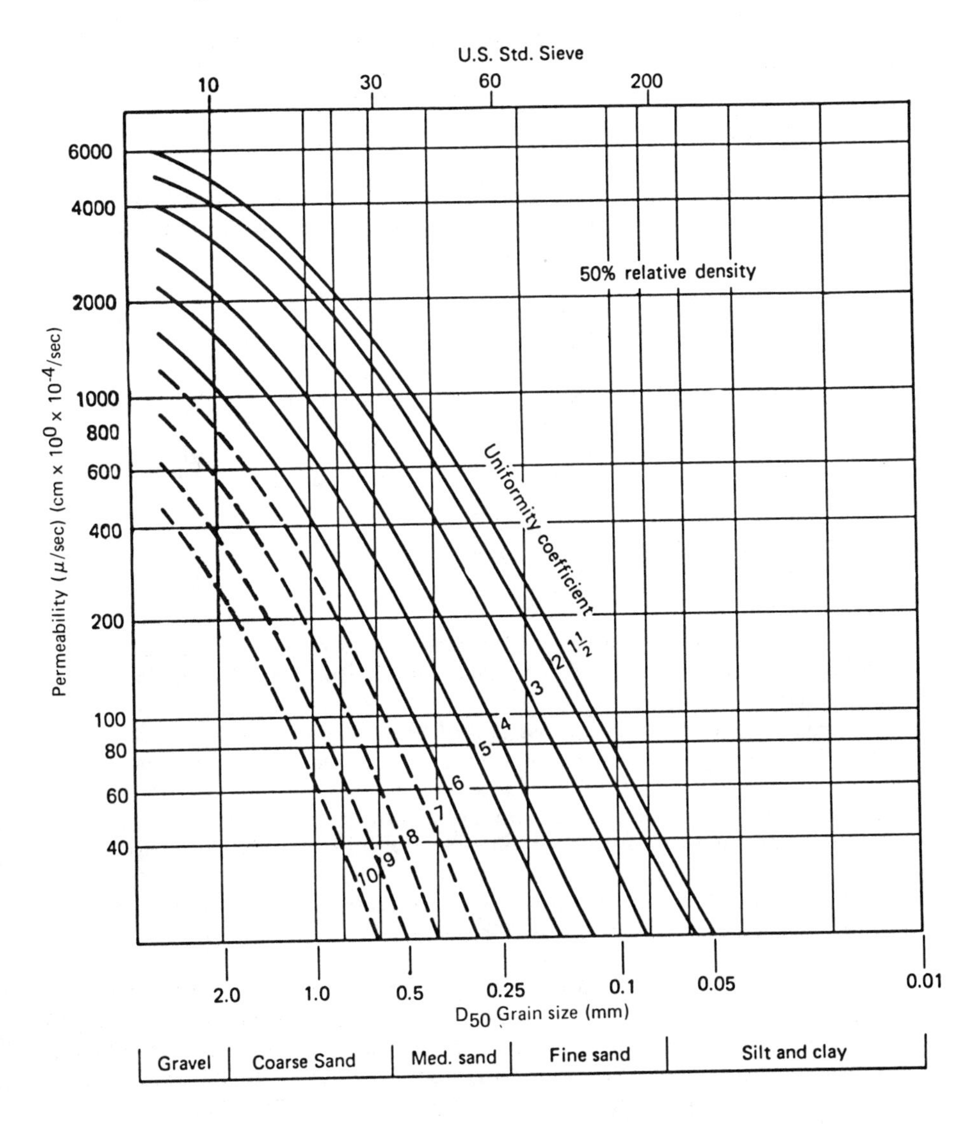

The uniformity coefficient (C_u) is useful in describing a soil. It is defined as the ratio of the D_{60} size of the soil (the size than which 60% of the soil is finer) to the D_{10}:

$$C_u = \frac{D_{60}}{D_{10}}$$

FIGURE 2.7 Empirical relationship between the grain size, uniformity coefficient, and the permeability.

TABLE 2.6 The General Relationship Between Permeability and Groutability

$k = 10^{-6}$ or less	Ungroutable
$k = 10^{-5}$ to 10^{-6}	Groutable with difficulty by grouts with under 5 cP viscosity and ungroutable at higher viscosities
$k = 10^{-3}$ to 10^{-5}	Groutable by low-viscosity grouts but with difficulty when μ is more than 10 cP
$k = 10^{-1}$ to 10^{-3}	Groutable with all commonly used chemical grouts
$k = 10^{-1}$ or more	Use suspended solids grout or chemical grout with a filler

The degree of permeability is defined as shown in Table 2.3. Tables 2.4 and 2.5 show approximate soil properties for various typical soils.

Granular soils with uniform grain size will also have fairly uniform pore size. As the spread of grain sizes increase, some of the smaller grains may fit into the larger voids. Thus, as the uniformity coefficient increases, the permeability will decrease. Empirical relationships between these two factors are shown in Figure 2.7.

Permeability is the soil property most closely related to groutability. The general relationships between these two factors are shown in Table 2.6.

Permeability of soils may be expressed in many related units. Most common in the United States is cm/sec (centimeters per second). Conversion to other units such as ft per day will cause no problems. In Europe, however, the term "Lugeon" is used, which cannot be converted directly (empirical correlations can be found in the literature). One Lugeon is defined as a water loss of one liter per minute per meter of hole at 10 atmospheres. This is approximately equivalent to 0.0005 gallons per minute per foot of hole at one psi, or a permeability of 10 ft per year in an AX size drill hole.

2.5 ROCK FORMATIONS

The porosity and permeability of massive rock formations are both very low, compared to soil formations, and actual values do not

TABLE 2.7 Porosity Ranges for General Rock Classification

Rock	Porosity in (usual)	Percent (max.)
Sandstone	15	35
Limestone	10	30
Shale	10	25
Granite	2	5

Note: These values should not be used for design purposes, and do not reflect large isolated fissures or caverns.

lend themselves to generalization for different types of rock. It is possible, by using pumping data and fracture spacings measured in many drill holes, to estimate with reasonable accuracy the porosity and permeability of a massive rock formation [5]. Such data would be important in estimating grout volume and pressure requirements for dam foundations. When excavations must be made through massive rock, as for tunnels and shafts, the rate of water inflow is often an adequate guide to groutability.

Sandstone and siltstone often have porosities and permeabilities of the same order of magnitude as granular materials and are treated in much the same way. Porosity ranges for general rock classification are shown in Table 2.7. Limestone and other rock masses containing large internal openings (which may or may not be interconnected) do not lend themselves to preplanning of a grouting operation.

The extent and width of cracks and fissures in concrete can be assessed only by their surface manifestations. Such assessments are often unreliable. For this reason, the grouting of cracks in concrete is often a trial-and-error process in which each grouting attempt is also a pumping test. (Procedures are discussed in Chap. 8.)

2.6 SUMMARY

The ability of a porous formation or structure to accept grout is a function of its permeability (the volume of grout that can be placed per unit volume of formation is a function of porosity). For granular materials, the relationships between a formation's grain size and density can be used to estimate permeability. Such values are accurate enough for preliminary feasibility estimates.

In granular soils with less than 20% silt, it is reasonable to assume that chemical grouting will be effective. For other conditions, including fissured and fractured rock and concrete, additional data beyond grain size distribution is needed to estimate groutability.

REFERENCES

1. Committee on Grouting, Preliminary glossary of terms related to grouting. *J. Geotech. Eng. Div. ASCA, 106*(GT7): 803–815 (July 1980).

2. The Unified Soil Classification System, *Characteristics of Soil Groups Pertaining to Embankments and Foundations*. TM3-357, Appendix A, U.S. Army Engineer Waterways Experiment Station, March 1953.

3. Donald M. Burmister, *Soil Mechanics*, Vol. I, Columbia University, New York, 1950, p. 112.

4. R. H. Karol, *Soils and Soil Engineering*, Prentice-Hall, Englewood Cliffs, N.J., 1960, pp. 46–47.

5. D. T. Snow, Rock fracture spacings, openings and porosities, *J. Soil Mech. Foundation Div. ASCE, 94*(SM 1): 73–91 (Jan. 1968).

3

Grouting Theory

3.1 INTRODUCTION

The flow of grout through a formation is governed by the properties of the formation and the grout, as well as the hydraulic gradient. While much has been written about the theory of flow through porous media, soils are so heterogeneous in nature that it is difficult to apply such theories and with the exception of Darcy's law, little application has been made to grouting practice. Nonetheless, there are theoretical considerations that can be studied advantageously.

3.2 BASIC CONSIDERATIONS

In theory, a liquid can be pumped into any porous formation, provided no pressure or time constraints exist. In practice, chemical grouts must be placed at pressures consistent with good engineering practice and at rates that make use of chemical grouts economically feasible. In the discussions that follow, it is assumed that those practical considerations are met for all conditions. It is also assumed, unless otherwise stated, that grout penetration into a formation is by permeation, not fracturing.

When grout is injected through the open end of a small pipe placed below the water table in a uniform, isotropic granular soil, the flow of grout is radial. Thus, under certain conditions, the grout–groundwater interface becomes and remains spherical. The required conditions include interface stability and hinge upon specific relationships between viscosities and flow rates. Flow rates in turn, are dependent on pumping pressure for given formation permeability and grout viscosity.

3.3 STABILITY OF INTERFACE

Groundwater is generally moving, although movement is usually very slow except in the vicinity of an open discharge. However, many construction projects provide such open discharges by excavation for shafts, tunnels, etc. When grouting is done to shut off water entering the excavation, placement of grout takes place in moving groundwater, and mixing with groundwater can occur. If groundwater is moving rapidly enough to cause turbulent flow, mixing to some degree is sure to occur. If groundwater flow is laminar, mixing will still occur if the rate of water moving past the grout injection point* approaches the rate of grout placement.

If groundwater is static, or if the rate of grout placement is large compared to the flow of groundwater, the interface between grout and groundwater will be stable† during grouting when the grout viscosity exceeds that of the groundwater. Experience verifies that this is true even when the grout viscosity approaches that of water (as it does for acrylamide-based grouts, with viscosities as low as 1.2 cps). When the interface is stable, its shape under uniform isotropic conditions is spherical.

The lower boundary of the grout–groundwater interface is inherently unstable [1]. The grouted mass (for grouts denser than groundwater) tends to sink under the action of gravity. If the downward movement is slower than the expansion of the grout–groundwater interface, the sinking has little effect on the shape and location of the grouted mass. However, if pumping stops and the grout remains in place as a liquid, it will continue to migrate downward until it sets.

A grout with a viscosity lower than that of water will always have an unstable interface with groundwater. Instead of a spherical interface, the grout will penetrate the groundwater with a number of intrusive fingers and lenses. This factor is used to advantage

---

*This is a rather nebulous number to compute, since it involves not only knowing the actual speed of the groundwater but assuming a reasonable cross-sectional area of groundwater flow. Nonetheless, it is important to be aware of the possibility of grout dilution when pumping at slow rates.

†Stable in the sense that mixing does not occur to any important degree.

in the two-shot system (the Joosten process, for example, sodium silicate and calcium chloride) by injecting the most viscous material first. In such systems, the intruding fingers and lenses interlock to form a more or less continuous gel matrix, without displacing all the formation water from the voids. In contrast, an expanding stable interface displaces virtually all the pore water.

3.4 FLOW THROUGH SOIL VOIDS

Under uniform, isotropic conditions with laminar flow of a Newtonian fluid (such as water and the low viscosity chemical grouts) the theoretical flow rate [2] equals:

$$Q = \left[\frac{4\pi kH}{\mu \; \frac{1}{r} + \frac{1}{R} + \frac{1}{R}} \right]$$

where

Q = flow rate

k = permeability

H = hydraulic gradient (head)

μ = viscosity of liquid

r = radius of spherical injection source

R = radius of liquid penetration

The value of R is often dictated by the design of the grouting operation. The equation above can thus be used as a coarse check on whether the parameters Q and H fall within cost effectiveness and safety. For a cylindrical injection source of length L and diameter D, r may be taken as $1/2 \sqrt{LD}$. Also on a coarse level, the indication is that grout acceptance will vary directly with the pressure, and inversely with the viscosity.

The open passages through soils and many rock formations consist of tortuous winding routes of variable cross-sectional size and shape made up of interconnected adjacent voids. Analysis of flow through such individual passages is unproductive. If we straighten the route and unify the cross-sectional size and shape, analysis becomes meaningful, although not necessarily applicable. Poiseuille's law does this for a long tube of small diameter through which a liquid is flowing slowly:

$$V = \frac{\pi p r^4 t}{8L\mu}$$

where

V = volume of liquid escaping in time t

t = time for V

p = pressure differential between tube ends

r = tube diameter

L = tube length

μ = viscosity of liquid

(The notation η is often used for viscosity and has been changed to μ here for consistency.)

Qualitatively, the following general conclusions can (again) be drawn in applying Poiseuille's law to flow through soils: For any given pressure, the rate at which the formation will accept grout will vary directly with the pressure and inversely with the viscosity. This relationship appears to become invalid at very high gradients and viscosities, but is of little interest to chemical grouters in those ranges. Further, the acceptance rate is directly proportional to the fourth power of the average void size (and therefore also of the average grain size for granular materials).

3.5 EFFECT OF PUMPING RATE ON GROUT FLOW

In practical terms, grout should always be pumped as rapidly as possible in order to minimize job costs. Structural safety and sometimes equipment limitations will always impose a pressure ceiling, which in turn may limit the rate at which grout is placed.

When pumping into a uniform, isotropic granular soil, as long as the movement of the interface exceeds the rate of sinking caused by gravity, pumping rate has no effect on the flow of grout through the soil (other than affecting the rate of interface movement). However, few soils are uniform and isotropic, and even those which appear to be so are almost always considerably more permeable in the horizontal direction than in the vertical. This is often due to layering caused by size separation of waterborne materials or by seasonal deposits of differing grain sizes. Grout pumped into such formations, particularly if they are varved, will tend to penetrate the coarser strata as parallel sheets. From these sheets, penetration vertically into the less pervious zones will occur. At slow pumping rates, the penetration phenomena occur simultaneously, and the grouted soil mass is shaped like a flattened spheroid. At rapid pumping rates the horizontal advance in coarse strata may be so rapid as to leave much of the finer strata ungrouted. Similar reasoning applies to grouting in fissured rock and points out the usual necessity for sequential grouting in the same zone.

If the pumping rate is constant, the rate of expansion of the grout–groundwater interface is constantly decreasing. Thus, for large volumes of grout pumped at a constant rate, the grout flow at the periphery is laminar. If pumping very slowly through a large open-ended grout pipe it is possible to have laminar flow throughout the entire grout zone. However, with the pipe sizes and pumping rates usually required for field work, the flow rate at the end of the grout pipe is great enough to create turbulence. Thus, a zone of turbulence (analogous to quick conditions in granular soils) will exist at the end of the grout pipe. In uniform soils the shape of this zone will be roughly spherical, and its size is a function of the pumping rate, increasing as the pumping rate increases.

The creation of a turbulent zone effectively enlarges the pipe size and thereby facilitates the placement of grout. As long as the zone of turbulence is a small portion of the grouted zone, it is a desirable condition. The relationship between the sizes of the two zones is a function of both the pumping rate and the total grout volume. To avoid turbulence throughout the total grouted zone,* when placing small volumes of grout, it may be necessary to pump at rates slower than those appropriate for larger volumes.

Head loss will occur as the grout moves through the voids in a granular soil or fissured rock. The longer the flow paths and the finer the pore paths, the greater the head loss. Thus, under some field conditions it may be necessary to increase the pumping pressure in order to keep grout moving into the formation. If the pumping pressure rises above the fracturing pressure, fracturing may occur. Often, this is undesirable and counterproductive. When such situations arise, good practice calls for additional grout holes at smaller spacing.

3.6 EFFECT OF PUMPING PRESSURE ON GROUT FLOW

The rate of pumping grout under steady-state conditions into a formation is directly proportional to the pumping pressure. In

*Turbulence leads to mixing of grout and groundwater and is therefore undesirable if it occurs throughout the entire grouted zone.

practice, steady-state conditions are transitory, and while pressures must always be increased to increase pumping rate, they must often be increased merely to maintain a constant pumping rate.

The act of pumping liquid into a formation creates a zone of high pressure potential. Because of this, the entering fluid flows away from this zone in all directions, locally changing the normal groundwater flow pattern. At low pumping pressure, the excess pressure potential dissipates at short distances from the injection point, and it may be possible to establish new equilibrium conditions with steady-state inflow.

The greater the pumping pressure, the farther the distance for effective pressure dissipation and the longer the time for establishment of steady-state flow conditions. During the time interval between pressure application and steady-state flow, pumping pressure increases if the pumping rate is held constant. When pumping at low pressures into more pervious materials such as coarse sand and fine gravel, new equilibrium conditions establish themselves quickly. When pumping at high pressures and into less pervious materials such as fine sand, silts, and siltstone, new equilibrium conditions are generally not reached during the grout pumping time (i.e., either pumping pressure must be slowly increased to maintain the pumping rate or pumping rate must be decreased to keep pressures from becoming excessive). Of course, if grout begins to set up in the formation while pumping continues, this reduces the formation permeability and either decreases the flow or increases the pressure or does both.

Limitations on allowable pumping pressures should be established by the job owner's engineer. Such allowable pressures should recognize the fact that when working at short gel times, only a very small quantity of grout is in the liquid state within the formation at any one instant. For example, when grouting near a dam with cement grout or with large volumes of chemical grout at long gel times, it may be necessary to assume that a major portion of the dam base may be subjected to the uplift forces at the grouting pressure. On the other hand, an insignificant portion of the dam base would be subjected to fluid pressure when grouting with chemicals at short gel times. Thus, the allowable grouting pressures can be much higher than if cement were used or long gel times with chemical grouts.

When only safety against formation uplift need be considered, the following criteria are helpful: In granular (cohesionless) soils such as sands and silts, pressures of 1 psi (pound per square inch) per foot of distance from unsupported face or per foot of depth from ground surface gives a theoretical safety factor close to 2 against uplift and about 1 against fracturing. (Since failure cannot occur within a small enclosed soil mass, the actual safety factor is almost always greater than the theoretical value.) When grouting cohesionless strata underlying clayey (cohesive) soils, uplift (if it occurs) will generally occur at strata interfaces. Thus, cohesive strength of clayey soils does not necessarily permit higher grouting pressures than could be used in sands.

When grouting in sands and silts with short gel times or with very small grout volumes, particularly in strata overlain by clays, grouting pressures may be safety increased above the values based on cover alone. Such pressures cannot be computed without extensive laboratory test data since they are related to the plastic characteristics of the soils. However, it is reasonable to grout at pressures equal to the full overburden under such conditions (this will generally be close to 2 psi per foot of depth).

When grouting in cemented sand or siltstone or in fractured rock, allowable grouting pressures may be significantly higher than those which can be used in granular deposits. Such pressures should be determined by experience or by tests on representative field samples. General criteria for allowable pressures in such materials are not available as yet.

True grouting pressures are those at the elevation of the grouted stratum. A pressure gage at this location reads the true pressure. Gages at the pump do not include the elevation head, nor can they account for friction losses in the piping system.

3.7 SUMMARY

Darcy's law and Poiseuille's law both deal with the flow of fluids through porous media. Because of the haphazard arrangement of pore sizes and passages in geologic formations, these theoretical equations cannot be directly applied to grouting problems. They do, however, give order-of-magnitude data about the effects of changing the controllable variables.

Pumping rates and pressures are interdependent, and one or both are easily controlled during a grouting operation. The upper limits for each are determined by specific job conditions.

REFERENCES

1. R. A. Scott, Fundamental considerations governing the penetrability of grouts and their ultimate resistance to displacement. In *Grouts and Drilling Muds in Engineering Practice*, Butterworth's, London, 1963, pp. 6–7.

2. J. F. Raffle and D. A. Greenwood, The relationship between the rheological characteristics of grouts and their capacity to permeate soils. *Proc. 5th International Conference on Soil Mechanics and Foundation Engineering*, Vol. 2, pp. 789–793, London, 1961.

4

Chemical Grouts

4.1 INTRODUCTION

For many years, the term chemical grout was synonymous with sodium silicate and the Joosten process (see Sec. 1.4). However, sodium silicate is the basis for many chemical grout formulations in addition to the silicate-chloride two-shot injection. Further, the last three decades have witnessed the introduction and use of many materials totally unrelated to the silicates. Today, commercial chemical grouts* cover a wide range of materials, properties, and costs and give the grouter the opportunity to match a grout with specific job requirements, as well as to treat problems that could not be solved with high viscosity silicates.

4.2 GROUT PROPERTIES

The specific mechanical properties of each grout that are important factors in the selection of a grout for a specific job include mechanical permanence, penetrability, and strength. Similarly, the chemical properties include chemical permanence, gel time control, sensitivity, and toxicity, and the economic factors include availability and cost.

---

*A commercial chemical grout is defined as one which has been used on actual field projects and whose components can be purchased on the open market. Proprietary materials, particularly those whose components and properties are not publicly divulged, are not considered commercial grouts and are not discussed in this book.

Permanence

All grouts that contain water not chemically bound to the grout molecules are subject to mechanical deterioration if subjected to alternately freeze-thaw and/or wet-dry cycles. The rate at which such deterioration occurs varies with the amount of free water available in the grout as well as with the degree of drying or freezing. Laboratory tests on small sand samples grouted with a gel having a high free water content indicated that significant deterioration began only after the completion of four to six complete cycles. The test conditions were probably more severe than field conditions would be, since total drying seldom occurs close to the water table and freezing in temperate zones seldom extends more than 5 to 10 ft below the ground surface. Nonetheless, it should be kept in mind that mechanical deterioration of grouts can occur under certain conditions.*

Chemical deterioration of grouts can occur if the grouts react with the soil or groundwater to form soluble reaction products, if the grout itself is soluble in groundwater, or if the reaction products which form the grout are inherently unstable. Generally, materials with any of these unfavorable characteristics would not be proposed or used as a grout. However, there may be locations in which unusual concentrations of strong reactants are present in groundwater† and may have deleterious effects on grouts which are otherwise considered permanent.

With the exception of sodium silicate grout gelled with a bicarbonate (such as sodium bicarbonate), all chemical grouts discussed later are generally considered to be permanent materials.

Penetrability

The comparative ability of grouts to penetrate a formation is mainly a function of their relative viscosities. Although there is truth to the contention that lubricating fluids pass through pipes with less friction loss than nonlubricating fluids, the difference is small and cannot make up for significant viscosity differences.‡ In general,

---

*For example, complete wet-dry cycles may occur near a poorly insulated underground steam pipe.

†In the vicinity of a chemical plant, for example.

‡The effectiveness of pumping silicates into a formation to increase the take of the following cement grout is more probably due to fracturing the formation or to preventing the smaller voids from filtering the cement particles than to making the voids more "slippery."

viscosity alone should be used as the guide to relative penetrability of chemical grouts.

In promotional, advertising, and even technical literature there are anomalies in the data representing viscosities of grout solutions. Since grout viscosity is one of the more important factors in regulating the movement of grout injected into a porous formation, it is desirable to define the terms clearly.

Viscosity of a substance is defined as follows [1]: "The tangential force per unit area of either of two horizontal planes at unit distance apart, one of which is fixed, while the other moves with unit velocity, the space being filled with the substance." Viscosity is expressed in dyne-seconds per square centimeter or poises.* The centipoise (cP) is equal to 0.01 poise (P).

When a solid is subjected to a shearing force, the solid (simultaneously with the application of force) deforms, and internal stresses develop until a condition of static equilibrium is reached. Within the elastic limit of a substance, these internal stresses are proportional to the induced shearing strains (deformation). The ability of a material to reach static equilibrium, rather than deform continuously, is due to a property called shear strength.

Fluids do not possess shear strength as such. Fluids do offer resistance to deformation, due to internal molecular friction. However, under the influence of a shearing force, deformation will continue indefinitely. The property called viscosity is actually a measure of the internal friction mobilized against shearing forces.

Viscosities of fluids are generally not measured directly. Instead, a parameter dependent on viscosity is measured and a predetermined relationship used to arrive at an actual value. There are many different commercial viscosimeters available, utilizing such principles as flow through an orifice or resistance offered to a known torque. In general, viscosimeters cover only a limited range of values. Before using one for measuring the viscosity of chemical grouts, it is necessary to determine that the instrument functions properly in the required range. For measuring a fluid whose viscosity approaches that of water, a viscosimeter that can measure accurately at 1 to 5 cP is required. Any instrument can be readily checked in this range by using it to measure the viscosity of water at room temperature. Unless the instrument gives repeated readings of 1.0 within a few percent, it should not be relied upon. Methods for determining grout viscosity have been studied and standardized. See ASTM O-4016.

The viscosities of four different grouts, as measured by a Stormer viscosimeter, are shown in Fig. 4.1. Viscosities, of course,

---

*Related terms are *kinematic viscosity*, the ratio of viscosity to density, and *fluidity*, the reciprocal of viscosity.

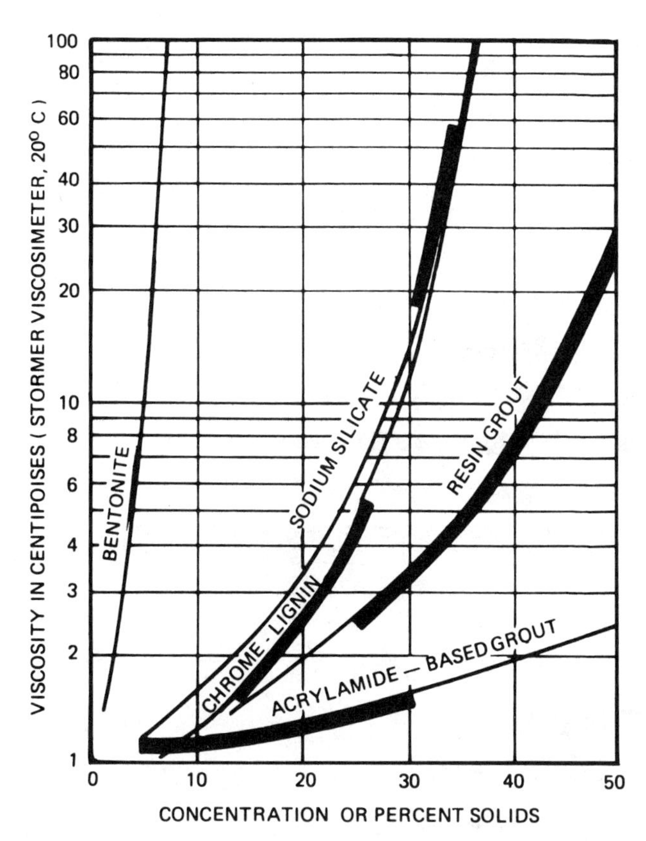

Figure 4.1 Viscosities of several chemical grouts. Heavy lines indicate the solution concentration normally used for field work.

vary with percent (dissolved) solids, and the chart is presented in that fashion. The usable viscosities of the various materials depend on the minimum desirable field concentration of solids. Thus, while it is obviously possible to work with a 20% sodium silicate solution in the Joosten process (a viscosity of between 3 and 4 cP), a gel would not form at that low a concentration, and it would be misleading to claim a 4 cP viscosity. In a similar fashion, other silicate formulations can be used to give either low viscosity of high strength, and it is misleading to list those values simultaneously as if they were the properties of the same fluid.

The penetrability of various chemical grouts is shown in relation to soil grain size in Figs. 4.2a and b. Other methods of ground-

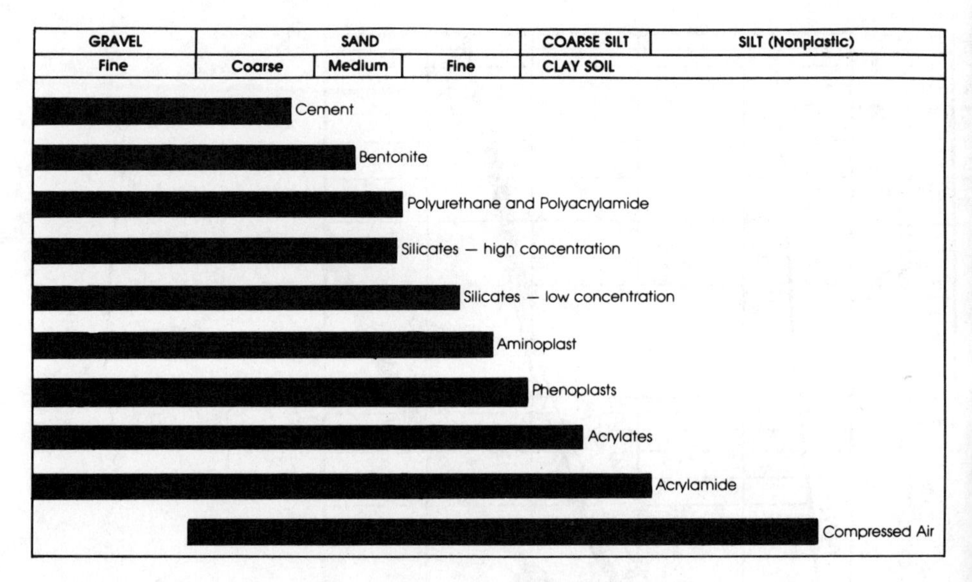

(a)

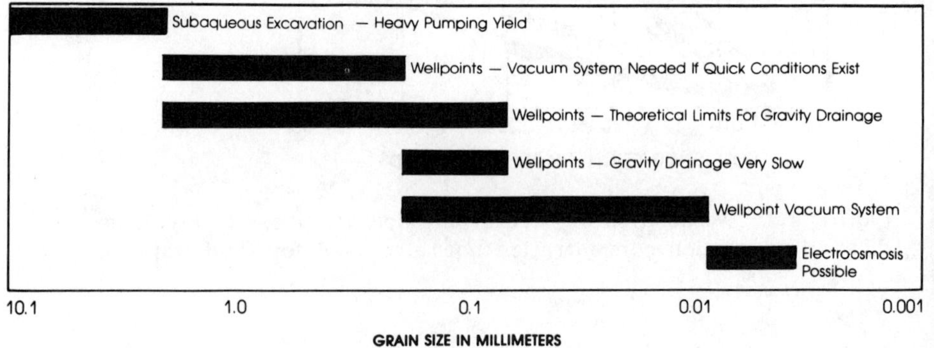

(b)

FIGURE 4.2 (a) Penetrability of various grouts. (b) Effective range of groundwater control measures. (c) Indicative range of grout treatments. (From Ref. 15.) (d) Grout penetrability. (From Ref. 16.) (e) Groutability of soils by various solution grouts. (From Ref. 17.)

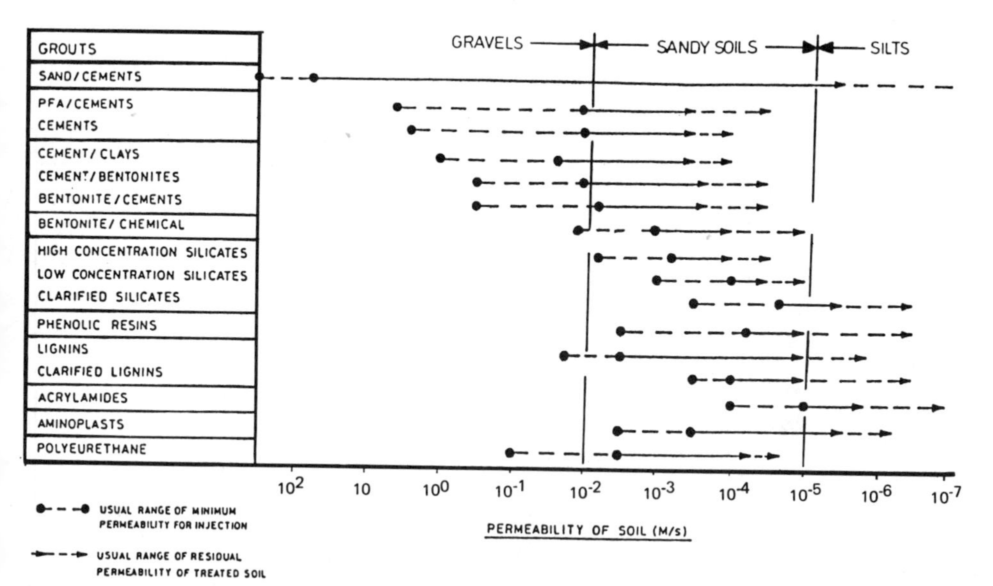
GROUTS
SAND/CEMENTS
PFA/CEMENTS
CEMENTS
CEMENT/CLAYS
CEMENT/BENTONITES
BENTONITE/CEMENTS
BENTONITE/CHEMICAL
HIGH CONCENTRATION SILICATES
LOW CONCENTRATION SILICATES
CLARIFIED SILICATES
PHENOLIC RESINS
LIGNINS
CLARIFIED LIGNINS
ACRYLAMIDES
AMINOPLASTS
POLYEURETHANE
GRAVELS
SANDY SOILS
SILTS
10^2 10 10^0 10^-1 10^-2 10^-3 10^-4 10^-5 10^-6 10^-7
PERMEABILITY OF SOIL (M/s)
USUAL RANGE OF MINIMUM PERMEABILITY FOR INJECTION
USUAL RANGE OF RESIDUAL PERMEABILITY OF TREATED SOIL
N B 1) THE LOWER THE INITIAL PERMEABILITY THE HIGHER THE RESIDUAL PERMEABILITY
2) PERMEABILITY RANGES INDICATED ARE FOR SUPERFICIAL SOILS ONLY

(c)

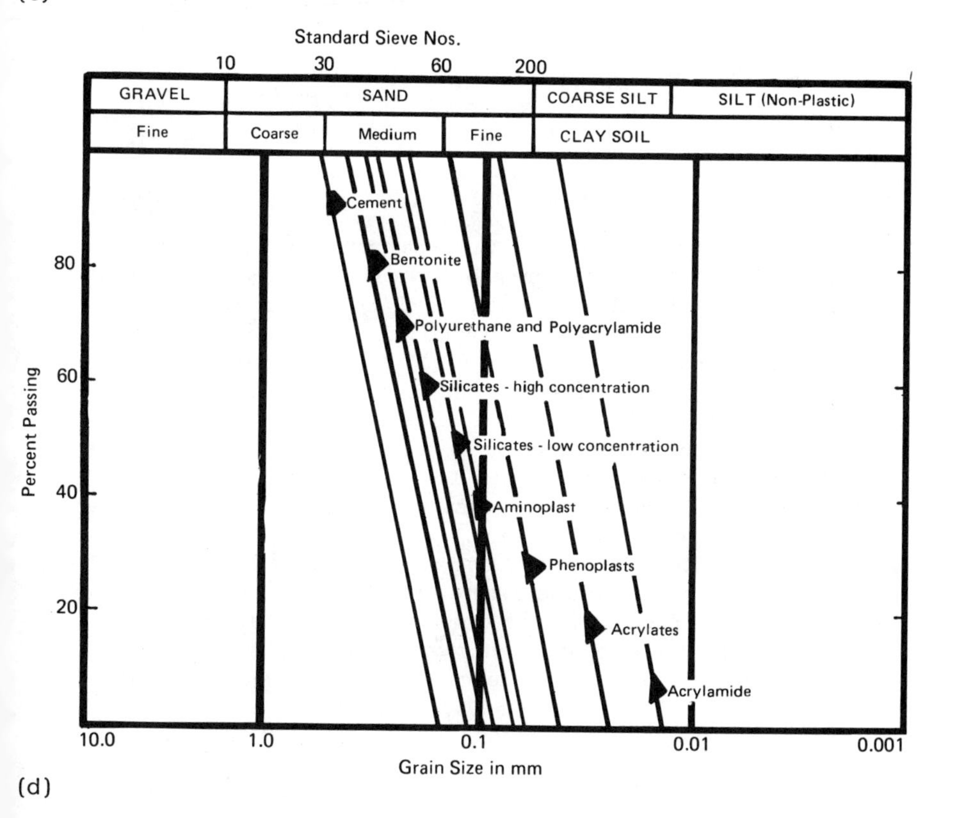
Standard Sieve Nos.
10 30 60 200
GRAVEL
SAND
COARSE SILT
SILT (Non-Plastic)
Fine
Coarse
Medium
Fine
CLAY SOIL
Cement
Bentonite
Polyurethane and Polyacrylamide
Silicates - high concentration
Silicates - low concentration
Aminoplast
Phenoplasts
Acrylates
Acrylamide
80
60
40
20
Percent Passing
10.0 1.0 0.1 0.01 0.001
Grain Size in mm

(d)

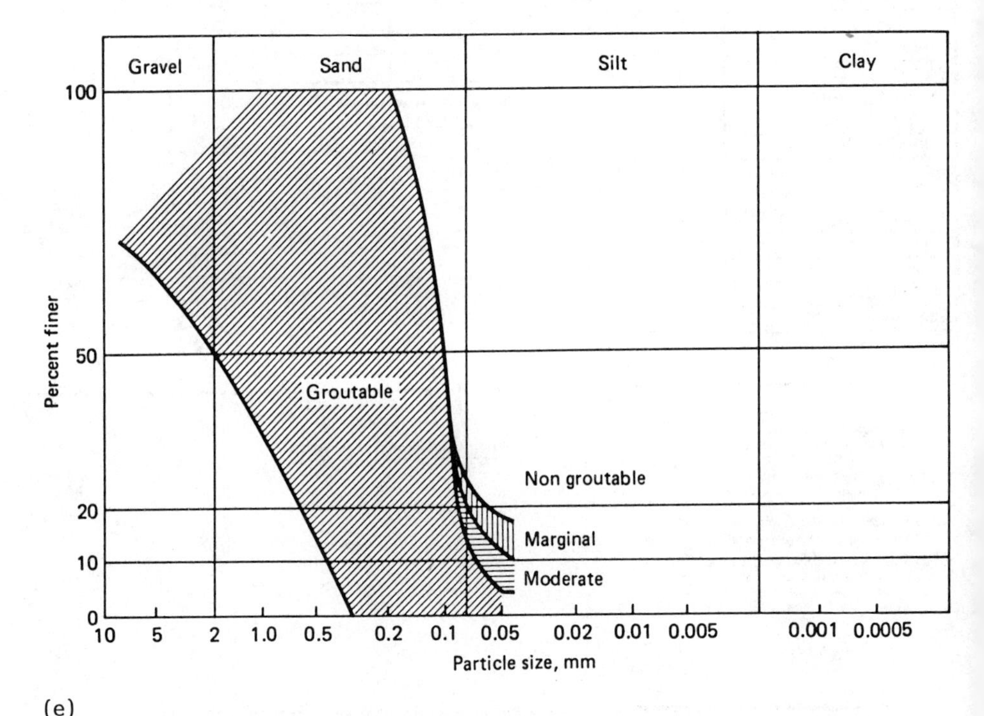

(e)

FIGURE 4.2 (Continued)

water control are also shown, so that comparison is also possible with Fig. 1.5. In terms of permeability a conservative criterion is that grouts with viscosities less than 2 cP such as acrylamide-based materials can usually be pumped without trouble into soils with permeabilities as low as 10^{-4} cm/s. At 5 cP, grouts such as chrome-lignins and phenoplasts may be limited to soils with permeabilities higher than 10^{-3} cm/s. At 10 cP, silicate-based formulations may not penetrate soils below $k = 10^{-2}$ cm/s. All grouts may have trouble penetrating soils when the silt fraction exceeds 20% of the total. (This is based on the assumption that the soil is not fractured by the grouting pressure.) It is also important to consider that permeability of a loosely compacted granular material may be 1 or more orders of magnitude greater than that of the same soil in the dense state. These considerations have led to proposed new methods of presentation of data on grout penetrability. Figure 4.2 (c, d, and e) show three of these new approaches (Refs. 15–17).

Strength

Chemical grouts (except for epoxies and polyesters) have little strength compared to cement, and the actual strength of a solid mass of chemical grout is of academic interest only. However, the strength of soil formation stabilized with chemical grout is of very practical interest.

Grouts that completely displace the fluid in the soil pores form a continuous but open and nonuniform latticework that binds the grains together. In so doing, the grout increases the resistance to relative motion between soil grains and thus adds shear strength to the soil mass. Grouts that form as fingers and lenses within a soil mass also add shear strength, depending partially on the degree to which the lenses and fingers interlock. Until recently, standard methods of testing and reporting the strength of stabilized sands did not exist, and the literature is full of data which can be readily interpreted. (Since 1979, however, ASTM has been working on standards related to grouting. Several have now been published, including D-4219, which deals with unconfined compression testing of grouted soil samples. Strength data published after 1983 should be more reliable.)

Grouting with chemicals is often done primarily to add strength to a formation. Other major applications of chemical grouts are to prevent anticipated groundwater problems or alleviate existing ones. Most applications, regardless of purpose, result in placing grout below the water table, and desiccation of the grouted mass will never occur. Thus, the most significant strength factor is the *wet* strength of the stabilized soil formed and remaining immersed in a saturated formation.

By contrast, much of the literature devoted to strength data reports *dry* strength: strength of stabilized soils in which some or all of the contained water has been removed by air or oven drying. When water held by a grout is lost through desiccation, the grout matrix shrinks. Shrinkage of the grout between soil grains sets up forces analogous to capillary tensile forces but often much stronger. This increases the resistance to relative motion between grains and adds shear strength to the soil mass. Thus, desiccated grouted soils will show strengths equal to or higher than wet strengths, often higher by as much as 10 times. Dry values of strength are meaningless for grouts that are placed and remain under the water table, and all references to strength in this book, unless otherwise stated, refer to wet strength.

The strength of a fully permeated grouted soil depends on the specific grout used but also on other factors. Chief among these are the density, the average grain size, and the grain size distribution of the soil. In general, strength increases with increasing density and with decreasing effective grain size. Well-graded soils give higher strengths than narrowly graded soils with the same effective grain size. Because of these variables, general discussion of the anticipated strength of grouted soils should present ranges of values rather than a specific number.

The shear strength of soils is composed of two components: (1) the mechanical resistance of the individual soil particles to sliding and rolling over each other (termed frictional resistance and measured by the angle of internal friction) and (2) the resistance of individual particles to relative motion because of the attraction between particles (termed cohesion). Pure clays have virtually no frictional resistance and a friction angle of zero. Clay particles, however, are very small, and the attraction between particles is large compared to the particle size. Clays can develop appreciable cohesion. On the other hand, the particles of granular soils are so large in comparison to molecular attractive forces that for practical purposes such soils have zero cohesion. However, the particles of granular soils interlock and develop significant frictional resistance, with friction angles of the order of 35°. Many natural soils are mixtures of fine and coarse particles and exhibit both friction and cohesion.

Soils that have appreciable cohesion are generally too impermeable to accept grout. Thus, grouting is done primarily in soils with zero cohesion, and the effect of filling the voids with gel is equivalent to adding a cohesive component to the soil shear strength. If the grout by its lubricity also makes it easier for the soil grains to move past each other, there will also be a reduction in the soil friction angle.

It is common practice to define the shear strength of a soil on coordinate axes, where y represents shear forces [2]. The strengths of granular materials are defined by the straight sloping line passing through the origin and making an angle ϕ (the friction angle) with the x axis, as shown in Fig. 4.3a. This line is called the failure line, and all stresses which fall below this line do not cause shear failure. It can be seen from the diagram that the shear strength of a soil increases as the normal stress increases.

The dashed line in Fig. 4.3a illustrates the change in strength due to grouting—the addition of cohesive forces and a possible decrease in the friction angle. If shear strengths are compared at points such as A and B, it is seen that at small loads the effect of grouting is to give a significant increase in strength but that at large loads the increase due to grouting is insignificant. (In fact, if grouting reduces the friction angle, it is possible that at high normal stresses the shear strength of the soil will be reduced by grouting.) Each specific job must be analyzed separately to determine if grouting can indeed provide a significant strength increase.

The idealized sketch in Fig. 4.3a would require traixial testing in order to define with accuracy the values of cohesion and friction angle. Figure 4.3b shows actual test data illustrating one case in which grouting decreased the friction angle (ϕ).

Almost all strength values reported in the literature for grouted soils are the results of unconfined compression tests (samples loaded in compression along one axis only). However, some of those tests are on cube-shape samples, and many others are on samples whose height (long axis) is less than twice the minimum cross-sectional dimension. All such tests give higher than true strength values, because there is not sufficient height for a failure plane to develop without intersecting the loading blocks. It is also probable that many tests are run at strain rates too rapid for determining true static strength. Those tests also give results higher than true strength values. More uniform test data can now be obtained by following the recent ASTM standards.

Unconfined compression tests measure the strength of the test sample at zero lateral pressures. For in situ soils, this occurs only very close to ground surface. Most grouting work places the grout below the ground surface, and even at shallow depths significant lateral pressures develop. Therefore, the triaxial test [2] is a better replication in the laboratory of actual field stress conditions. Whenever it is important to know the actual strength of a specific grouted formation, triaxial tests are a better choice of test procedure. The unconfined compression test is still very useful for comparative purposes, for example, checking the effects of additives to a grout or comparing different grouts.

Although adequate testing has not been performed on all grouts, unless contrary evidence can be produced, it should be assumed that chemically grouted soils will be subject to creep. Stress-controlled unconfined compression tests on acrylamide-based grouts, silicate-based grouts, and lignin-based grouts have shown that sustained loads well below the unconfined compressive strength can cause creep, leading to failure in hours, days, or even months, depending on the ratio of applied load to unconfined compressive strength. (See Refs. 7–9, Chap. 14, and Appendix F.)

Unconfined compression tests of stabilized soils give data which can be used to determine the effective cohesion due to the grout (the vertical axis intercept on Fig. 4.3a). Short-term tests, whether stress or strain controlled, give data shown typically in Fig. 4.3c. Such test results do not reflect the influence of creep. Tests to determine creep strength consist of applying a constant load to the stabilized soil and recording deformation versus time. Typical test

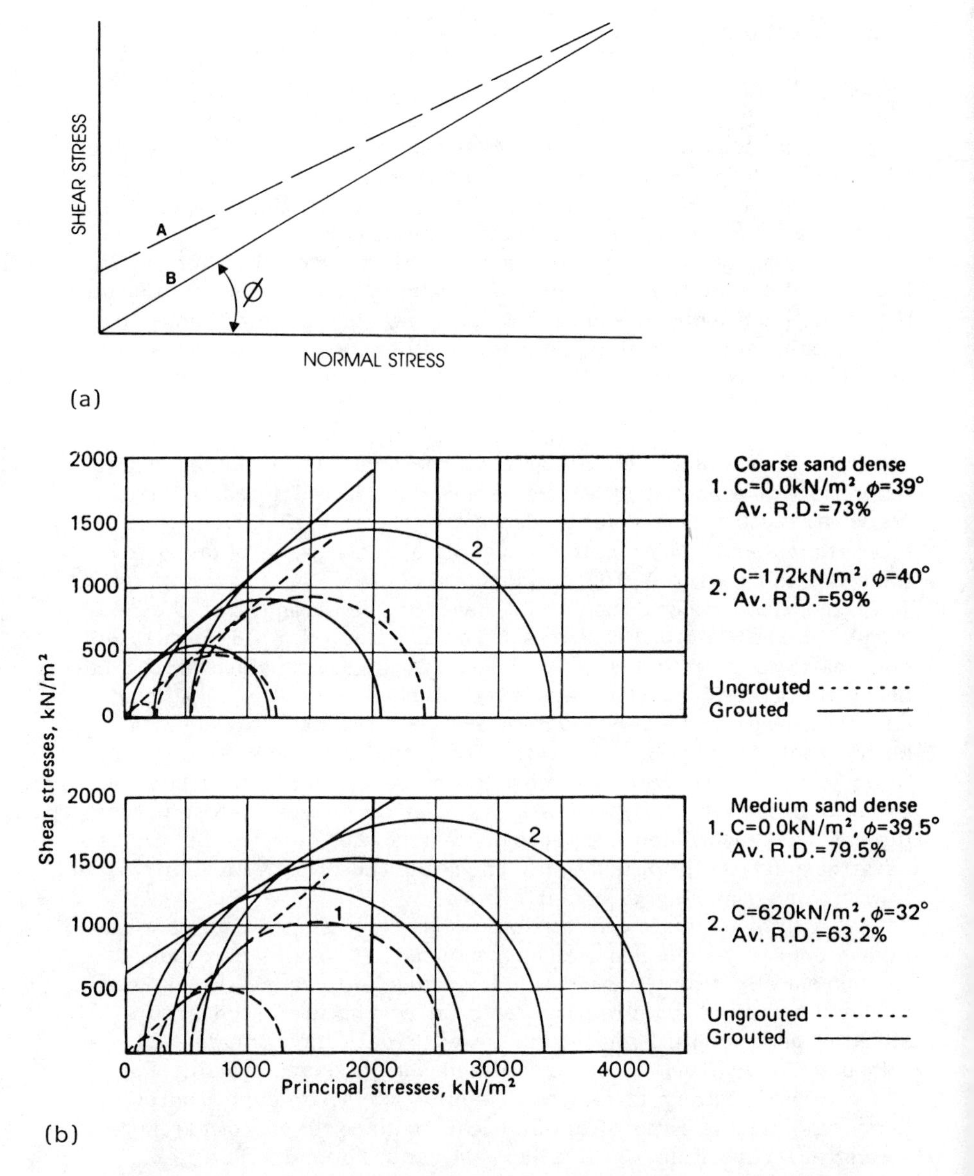

FIGURE 4.3 (a) Failure line for granular soils. (b) Drained triaxial test results for silicate-grouted coarse and medium sands. (From Ref. 15.) (c) Typical stress–strain curve from unconfined compression test on chemically grouted sand. (d) Compression versus time data for creep test on chemically grouted sand, under constant load. (e) Failure time versus percent of unconfined compression failure load. +, Unconfined tests; triaxial tests with S_3 = 25% of S_1; 1, no failure plant shutdown turned off air supply; 2, no failure—plant shutdown; 3, data at November 25, 1959—test still continuing.

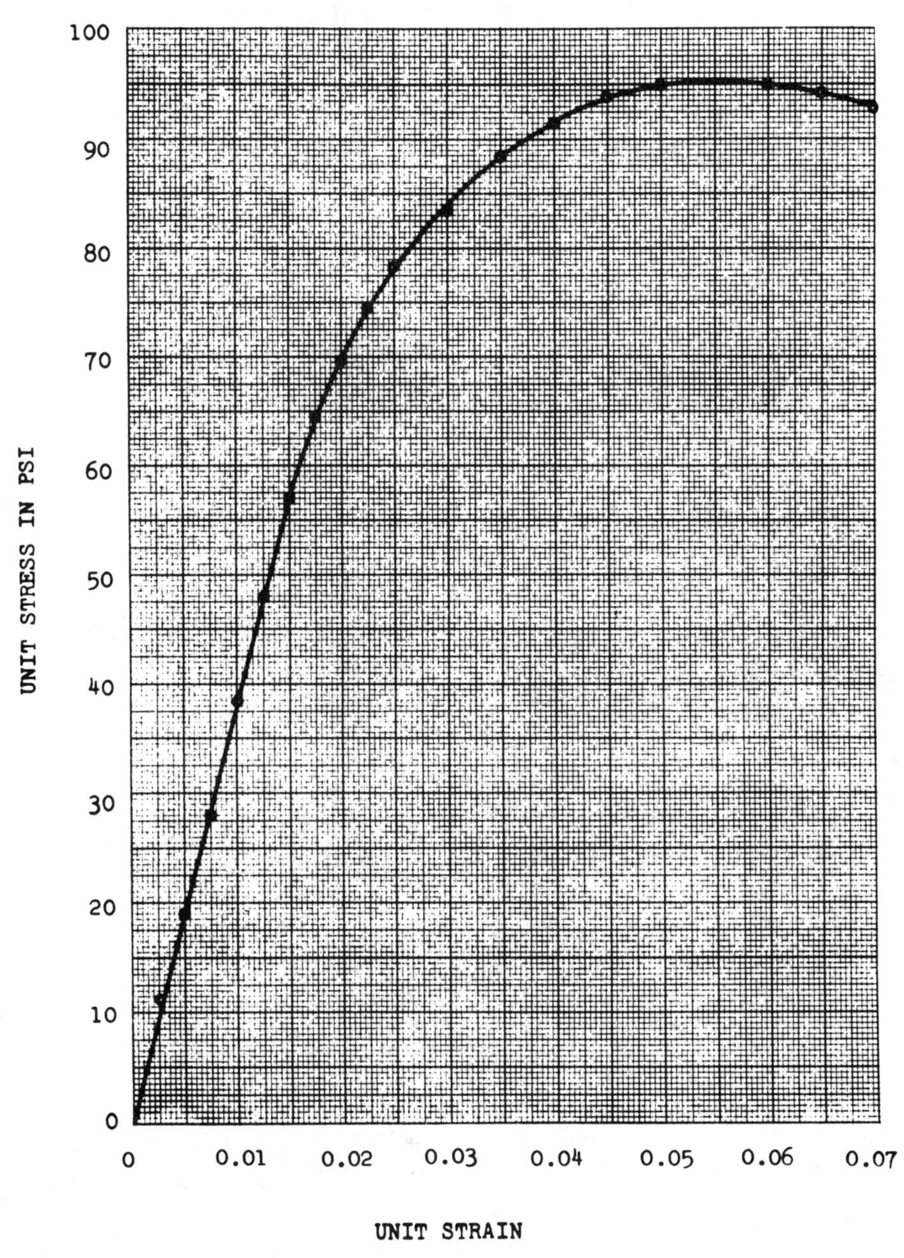

(c)

FIGURE 4.3 (Continued)

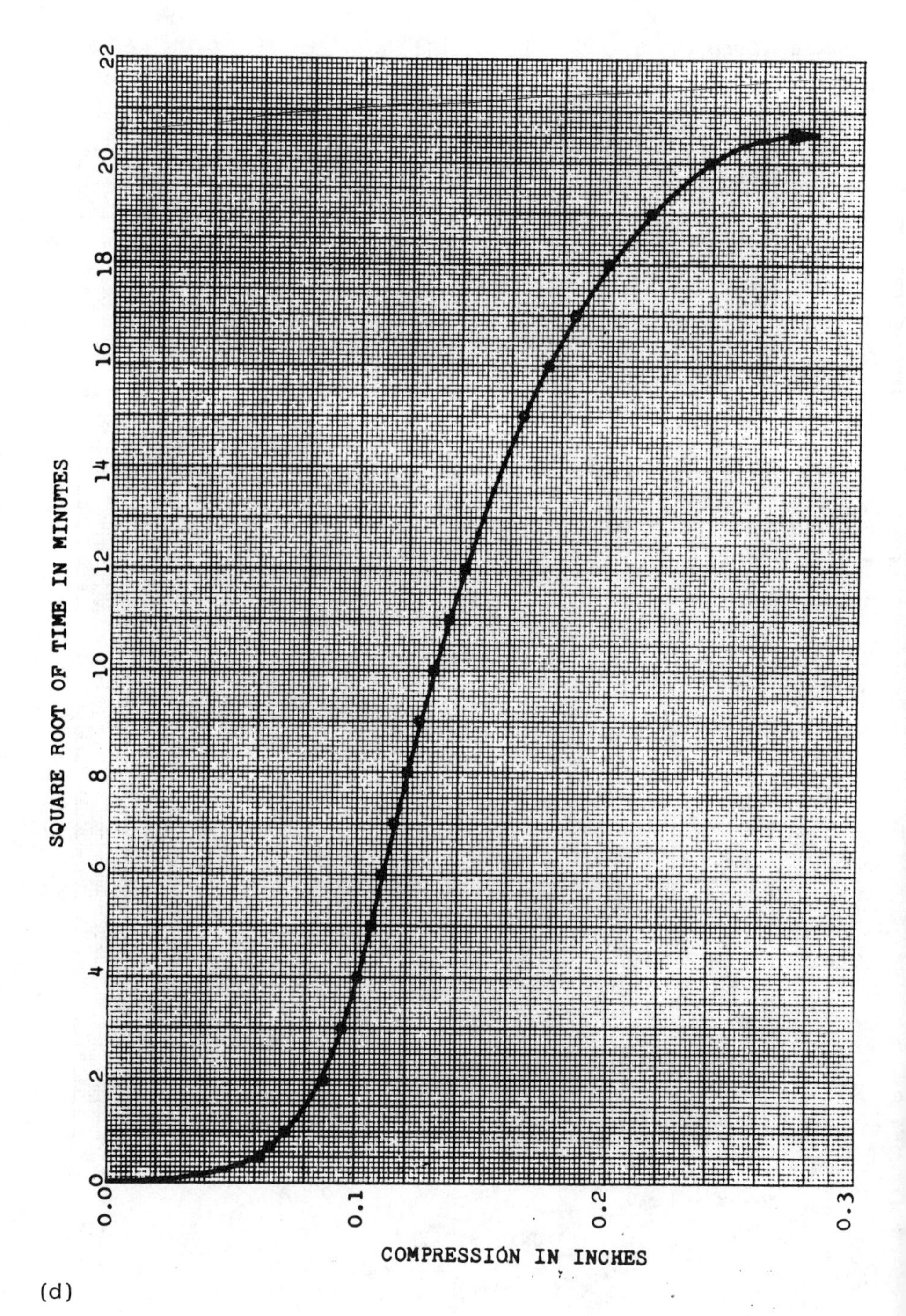

(d)

FIGURE 4.3 (Continued)

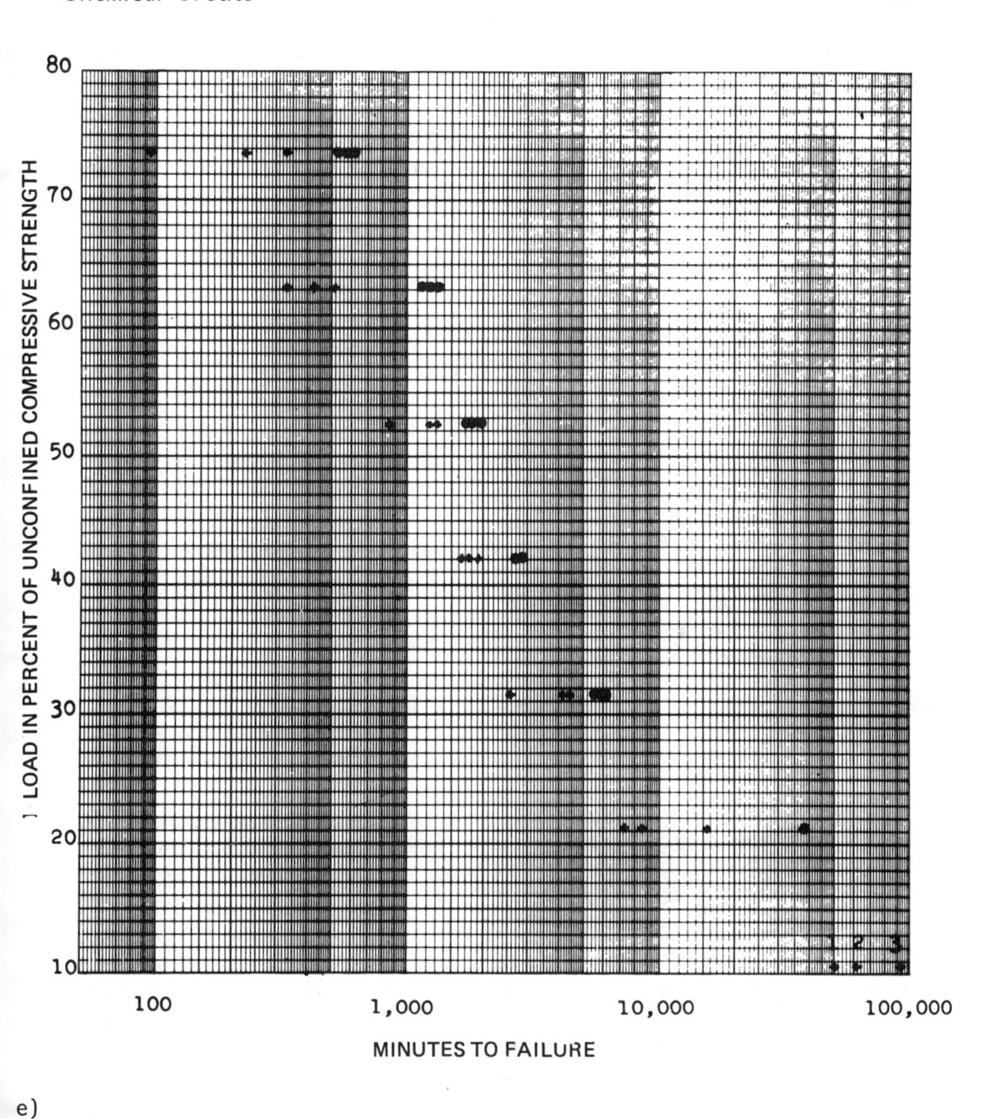

e)

FIGURE 4.3 (Continued)

data are shown in Fig. 4.3d, which indicates failure occurred after 6 hours of loading. At loads approaching the ultimate values obtained from short-term tests, stabilized soils fail rather quickly due to creep. As the loading decreases, it takes longer and longer for the samples to fail. At some value of the applied loads, failure will not occur. Data to find the nonfailing value take much effort to accumulate and have been published only for acrylamide and silicate grouts [3-5]. Figure 4.3e shows these data for both unconfined compression and triaxial creep tests.

It is possible to delineate a creep endurance limit below which failure will not occur regardless of the load duration. Limited data suggest that the endurance limit is about 25% of the unconfined compressive strength for the materials tested when applied to field conditions where stabilized soil masses are unsupported on one face, as in open cuts, tunnels, and shafts. For field conditions where lateral support exists for the grouted mass, triaxial tests are more suitable. Limited data indicates that, using "at-rest" lateral pressures (see Ref. 2, p. 104), the triaxial creep endurance limit approaches half of the unconfined compressive strength. These data, while approximate and in need of further study, indicate the values of the safety factor to be used when grouted soil strength is a design consideration.

Gel Time Control

The ability to change and control the setting time of a grout can be an important factor in the successful completion of field projects. Excepting the Joosten process, there is always a time lag from the mixing of the chemical components to the formation of a gel. If all other factors are held constant, this time lag (referred to as the gel time or induction period) is a function of the concentration of activator, inhibitor, and catalyst in the formulation.* With most grouts, the gel time can be changed by varying the concentration of one or more of those three components. If a wide range of gel times can be obtained and accurately repeated, gel time control is called good or excellent. If only a narrow range of gel time is possible and repeatability is difficult, gel time control is rated fair or poor.

Certain chemical grouts, after catalyzation, maintain a constant viscosity and at the completion of the induction period turn from liquid to solid almost instantaneously. Other grouts maintain

---

*For most grouting materials, the gel time is relatively independent of grout concentration.

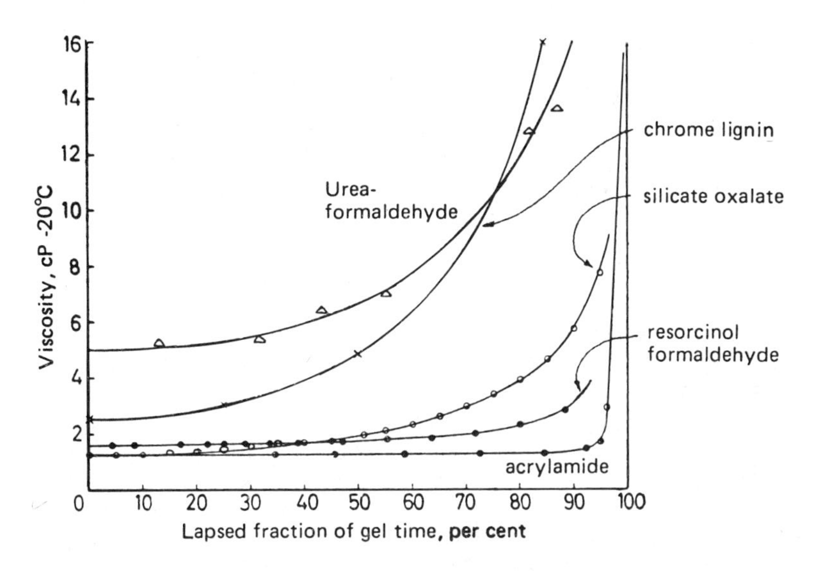

FIGURE 4.4 Growth of viscosity in period before gelation. (From Ref. 15.)

constant viscosity for much less than the total induction period. Still others increase in viscosity from the moment of catalysis to gel formation, as shown in Fig. 4.4. For the last two categories, it may not be possible to pump grout into a formation during the full induction period. It is also difficult to judge when gelation occurs and to gage accurately the effects on gel time of external variables.

Sensitivity

For any given grout formulation, the gel time is a function of temperature. Chemical reactions slow down as the temperature drops, and for many materials the effect is to approximately double the gel time for every 10°F drop in temperature.* In addition to temperature effects, when a grout is used in the field, many other factors may affect the gel time. Primary among these is dilution with groundwater which (in addition to changing the solution temperature) will dilute the concentration of activator, inhibitor, and catalyst. Of course excessive dilution may bring the grout itself below a concentration where gel will form. Groundwater may also carry

*Silicates are an exception. See Sec. 4.4.

dissolved salts which affect the gel time either by chemical activity or by change in pH. Gel times may also be affected by contact with undissolved solids, such as the materials used in pumps, valves, and piping. The degree to which a grout is sensitive to such factors is termed its sensitivity. Materials which are highly sensitive generally cause difficulty in handling in the field.

Toxicity

Health and occupational hazards have come under increasing scrutiny in the last decade, resulting in several instances of banning of grout components by federal agencies. Among the products used in the past, and still in used, are ones classed as neurotoxic, cancer causing, toxic, corrosive, highly irritating, etc. While there have been a number of incidents of people adversely affected by exposure to grouts, a study of each case involving field projects for which data have come to the author's attention indicates that the problems could have been readily avoided by the use of common sense and fol following the manufacturer's recommendation for handling.

Almost all of the reported incidents of grout-related health problems have involved persons handling the materials, not the general public. Efforts to contain the toxicity problems have been made largely by the distributors of the products, and largely aimed at the users. Manufacturers, on the other hand, have responded by either withdrawing products from the market, or developing less toxic replacements. Many of the manufacturing decisions are dictated by real or anticipated liability concerns, which may bear little relationship to actual hazards of properly handled materials.

Hazardous chemicals are normally tested by manufacturers in accordance with standardized procedures, and are often described by a value for LD_{50}. This is the dosage expressed in milligrams per kilogram of body weight per day at which 50% of the experimental animals die. United States Federal agencies rate products as follows:

Very toxic LD_{50} = 5 to 50
Moderately toxic LD_{50} = 50 to 500
Very slightly toxic LD_{50} = 500 to 5000

When LD_{50} values are known, they will be listed in later sections as various grouts are discussed.

All chemical grouts should be handled with care in the field, with safety and cleanliness equal to or better than the manufactorer's recommendations.

Economic Factors

Cost is of course a very important factor in selecting the construction method best suited to solving a specific field problem. When grouting is compared with other methods such as well pointing or slurry trenching, chances are that the cost comparisons will be realistic. When comparing two different grouts, however, the mistake is is often made of comparing raw materials costs rather than in-place costs. Chemical grouts are available commercially at prices ranging from 50 cents to $40 per gallon. However, the ease of placing and the effectiveness of these materials vary, and the cost of placement is almost always a major part of the in-place cost. Thus, while raw material costs may vary over a 80 to 1 ratio, in-place costs generally range from 5 to 1 to about equal. Selecting the most suitable material for a specific job will generally overcome the possible lower material cost of other products.

4.3 THE IDEAL CHEMICAL GROUT

If the goals of research to develop a new chemical grout were to be listed, they would state that the basic materials should be as follows:

1. A powder readily soluble in water (this eliminates the expense of transporting a solvent, and water is the least expensive solvent)
2. Inexpensive and derived from chemicals in abundant supply
3. Stable at all anticipated storage conditions
4. Nontoxic
5. Noncorrosive
6. Nonexplosive

and the grout solution should be:

7. A low-viscosity solution, preferably that of water
8. Stable under all normal temperatures
9. Nontoxic, noncorrosive, nonexplosive
10. Catalyzed with common, inexpansive chemicals, meeting the criteria of 8 and 9
11. Insensitive to salts normally found in groundwater
12. Of stable pH on the positive side (so that it may be used in conjunction with cement)
13. Readily controlled for varying gel times
14. Able to withstand appreciable dilution with groundwater

and the end-product should be:

15. Permanent gel
16. Unaffected by chemicals normally found in groundwater
17. Nontoxic, noncorrosive, nonexplosive
18. High strength

Of course, no such material exists. However, every criterion listed can be found in one or more commercially available materials. It is important, therefore, to determine which grout properties are critical to a specific project in order to have a sound basis for selecting a grout.

4.4 COMMERCIAL CHEMICAL GROUTS

Attempts to develop classification systems for chemical grouts started in the early 1960s. They were generally based on mechanical properties. There are now sufficient diverse products commercially available to warrant classification and discussion by chemical groupings. One of the recent groupings is from a 1973 U.S. Government publication [4] which on the last page lists 19 available chemical grouts. A more detailed listing was prepared for the Federal Highway Administration in 1977 [5]. Data from this report have been used with modification and amplification in this section.

If a count were made of all the chemical grout formulations which have passed through patent offices in the past 30 years, it would reach into the hundreds. Although most of those formulations have commercial possibilities, very few have been developed commercially. The majority have been eliminated by the normal industrial selection process. In the United States, from about 1970 onward, two specific materials (silicates and acrylamide-based* grouts) shared a market percentage estimated at between 85% and 90%, with the remaining portion of the market divided among only six other products.

It should be expected that new products will appear on the market from time to time. Those in current use will not be readily displaced unless the new products have significant advantages. Thus, it is reasonable to expect that in the future only a limited number of materials will have significant commercial use and that these will behave much the same as those in current use, except for improvements in such areas as toxicity, control, and cost. For this reason, much of what is covered in this chapter and in the sections describing field technology should remain applicable to new products.

The most widely used chemical grouts are aqueous solutions, and these materials will be covered in detail. Other types of grouts will be listed only briefly in this section.

---

*More recently, acrylates have been replacing acrylamides.

Specific grouting materials discussed in succeeding paragraphs are divided into the following chemical families:

1. Sodium silicate formulations
2. Acrylics
3. Lignosulfites–Lignosulfonates
4. Phenoplasts
5. Aminoplasts
6. Other materials

Sodium Silicate Formulations

Gels can be formed from many silica derivatives. For example, fluorosilicates can be precipitated by hydroxides, and silicon esters precipitate in the presence of an alkaline solution. Organic derivatives (such as methyl silicate) can be gelled by polybasic acids (such as citric acid) in various intervals of time after mixing. The silicate derivatives form the largest single group of related grouting materials. However, the alkali silicates, and in particular sodium silicate, is the only one used to any extent for chemical grouting.

Sodium silicate, n—$SiO_2 \cdot Na_2O$, is commercially available as an aqueous (colloidal) solution.* The silica/alkali ratio n is important in that ranges of 3 to 4 yield gels with adhesive properties, particularly suitable for grouting (beyond a ratio of 4, the silicate becomes unstable). Table 4.1 shows some of the related properties of commercially available† sodium silicate solutions.

When sodium silicate solution and a concentrated solution of appropriate salt are mixed, the reaction forming a gel is virtually instantaneous. The earliest successful field process is credited to Joosten (the field procedure still bears his name) and consisted of injecting a concentrated solution of sodium silicate at about 38° Be into the ground immediately followed by an injection of calcium chloride‡ solution at about 35° Be. The reaction in the ground is expressed as

$$SiO_2 \cdot Na_2O + CaCl_2 + H_2O \longrightarrow Ca(OH)_2 + SiO_2 \cdot 2NaCl$$

*Industrially, this product has many uses including adhesives, catalysts, deflocculants, detergent bleaches, etc.

†These data are average values for European manufacturers. American products are similar.

‡Other chlorides, such as that of magnesium, may also be used.

TABLE 4.1 Silicate Compounds Commonly Used in Grouting

Baume degrees	Weight ratio, SiO_2/Na_2O	Viscosity (cP)	% SiO_2	% Na_2O	% H_2O
30-31	3.9-4.0	40-50	22.7	5.7	71.6
30-31	3.87	40-50	23.1	6.0	70.9
35-37	3.3-3.4	60-70	25.8	7.7	66.5
36-38	3.4	50-100	26.5	7.8	65.7
38-40	3.3-3.4	160-200	27.7	8.3	64.0
40-42	3.15-3.25	200-260	28.7	9.0	62.3

Source: Ref. 5.

As shown in Sec. 3.3, the grouting sequence results in a unstable interface, and if the silicate solution has not been moved away (by groundwater or gravitational forces), it is penetrated by thin fingers and lenses of chloride solution. The resulting gel contains calcium hydroxide, silica, and sodium chloride. Because the reaction is so rapid, not all the solutions can reach contact, but the unstable intrerface generally ensures that sufficient contact occurs to provide a continuous gel network through stabilized soil.

The Joosten process, properly used, results in a strong gel that can give unconfined compressive strengths above 500 psi. However, the utility of the process is limited by the high viscosity of the solutions and the need for many closely spaced grout holes. Also, the nature of the reaction prohibits complete reaction of the two liquids. For many years research and development were aimed at a silicate-based grout with gel time control to eliminate the inherent disadvantages of the two-shot system. As those research efforts met with success, the Joosten system was phased out as a construction tool. Today it is virtually a method of the past.

Silica is a weak acid, and sodium silicate is therefore basic. Silicate will be precipitated as a gel by neutralization. Thus, a dilute sodium silicate solution mixed with certain acids or acid salts will form a gel after a time interval related to chemical concentrations. The best known combination is the sodium silicate–bicarbonate mixture. The key word is *dilute*, and in essence a reaction that would be instantaneous in concentrated solutions is delayed by diluting the sodium silicate with water and also limiting the amount of reactant by dilution with water. The dilution yields a grout of very low viscosity but also of very low strength.* Control of gel time is obviously not precise. Such grouts are considered temporary, although they do have application for stabilization of tunnel faces.

There are reported case histories of these so-called "temporary" grouts having performed satisfactorily for many years. The special circumstances leading to such longevity have not been established. These projects are related mainly to the use of aluminate salts, for example, the discussion in Sec. 9.2.

The search for a single-shot permanent silicate grout with high strength eventually led to the use of organic compounds as catalysts. These react with the water–silicate mixture to form an acid or an acid salt, which causes precipitation of the silica. Setting time is controlled by the rate of acid formation and therefore by the quality of organic compound. Since the sodium silicate need not be diluted, strong gels can be formed.

In the late 1950s, a French patent [6] proposed the use of ethyl acetate with a detergent, and a U.S. patent [7] proposed the use of formamide. Refinements of these two patents have formed the basis for most of the silicate formulations used in the United States and Europe at present.

For many years since its introduction in 1960, a product trade named SIROC was the most widely used silicate formulation in the United States. Figure 4.5, taken from the manufacturer's data (Diamond Alkali Company), shows the various chemicals used. In the SIROC system, the silicate and reactant are always used, generally with one of the accelerators or with cement. The silicate (gel base) is a 38% solution of solids in water, having a specific gravity of 1.4 (41.4° Be) and a viscosity of 195 cP at 68°F. (Viscosity of silicate solutions is highly temperature dependent, as shown in Table 4.2 for the product used in SIROC. Viscosity also varies with the silica/alkali ratio, approximately doubling as the ratio goes from 3 to 4. At any fixed ratio, the viscosity is quite low for Baume's of 30 to 35 but increases at an increasing rate from 35 to 40.) This solution is diluted with additional water for field use and ranges from 10% to 70% of the total grout volume. Lower values are used only for temporary work. Field grouts usually are in the 30% to 40% range. (It is common field practice to consider the silicate product, as purchased from the manufacturer, as a 100% solution even though it is in reality about a 38% colloidal suspension. Thus, when a field concentration is referred to as 50%, it really means that the manufacturer's product has been diluted with an equal volume of water.)

---

*In all silicate-based grouts the strength and viscosity are functions of the sodium silicate content and are directly related.

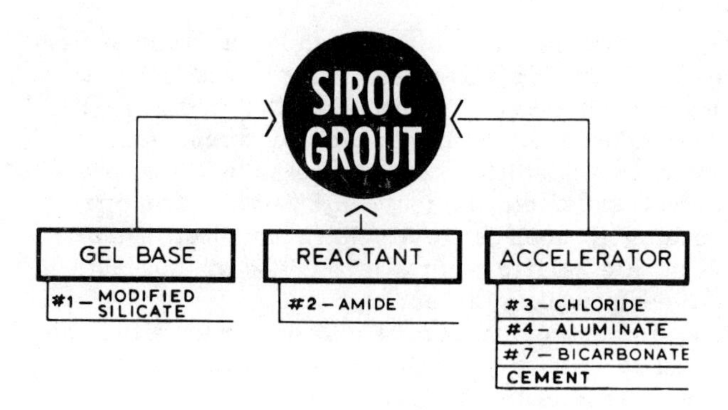

FIGURE 4.5 SIROC family of chemicals. (Courtesy of Diamond Shamrock Corp., Painesville, Ohio, formerly Diamond Alkali Company, Chicago.)

The variation of strength with silicate concentration is shown in Fig. 4.6, taken from the manufacturer's data. (The same data plotted versus viscosity of the grout solution are shown in Fig. 4.12.) These data are approximately representative of all silicate formulations having gel time control.

Gel time of the grout solution can be controlled by varying the percentages of either the reactant or the accelerator, or both. Figure 4.7 shows the gel time range possible for 30% SIROC and

TABLE 4.2 Variation of Physical Properties with Temperature

Temperature (°F)	Density conversion			Viscosity (cP)
	Sp. gr.	°Be	lb/gal	
40	1.409	42.1	11.75	440
50	1.406	41.9	11.72	344
60	1.402	41.6	11.68	253
68	1.399	41.4	11.66	195
70	1.398	41.3+	11.65	179
80	1.306	41.2	11.63	138

Source: Reference 5.

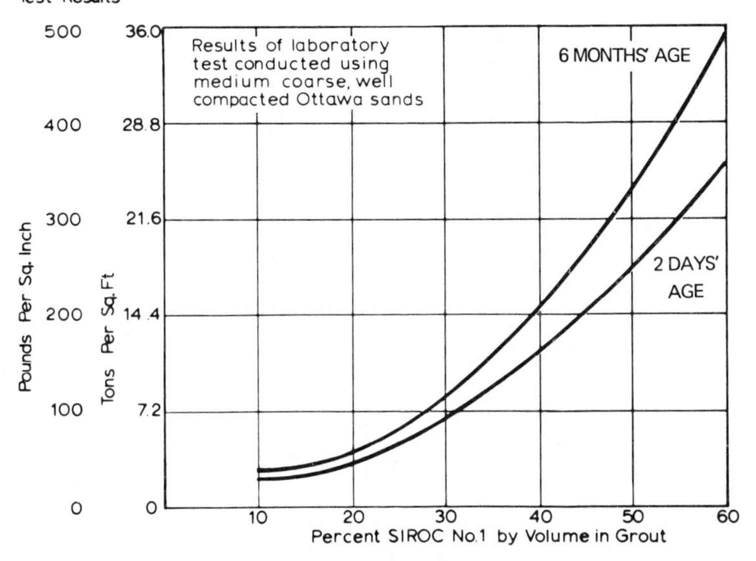

FIGURE 4.6 Strength versus silicate concentration. (Courtesy of Diamond Shamrock Corporation, Painesville, Ohio.)

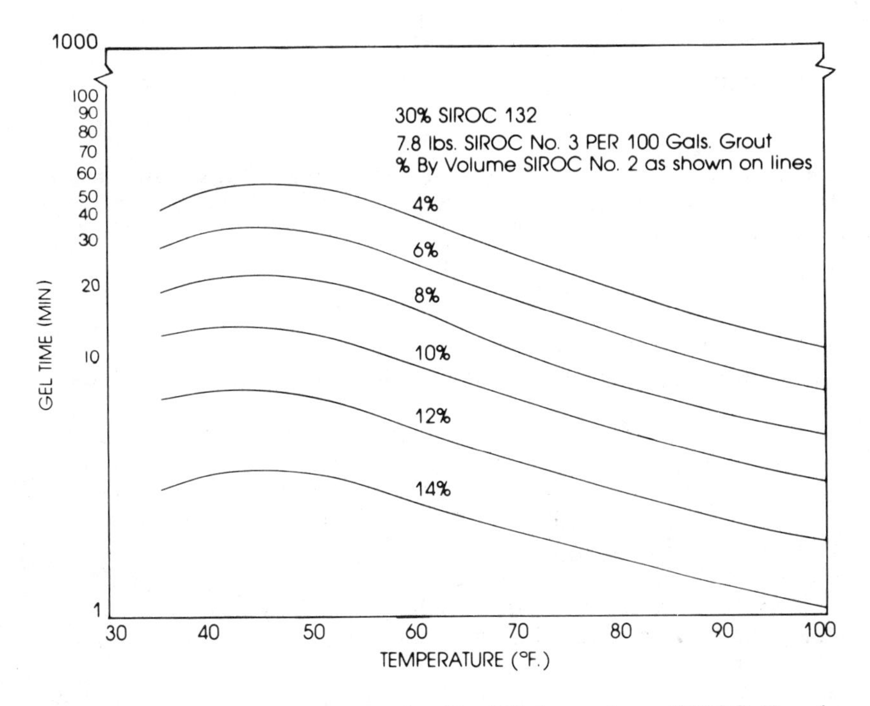

FIGURE 4.7 Gel time control with 30% by volume SIROC No. 1, using Formamide percentages as shown.

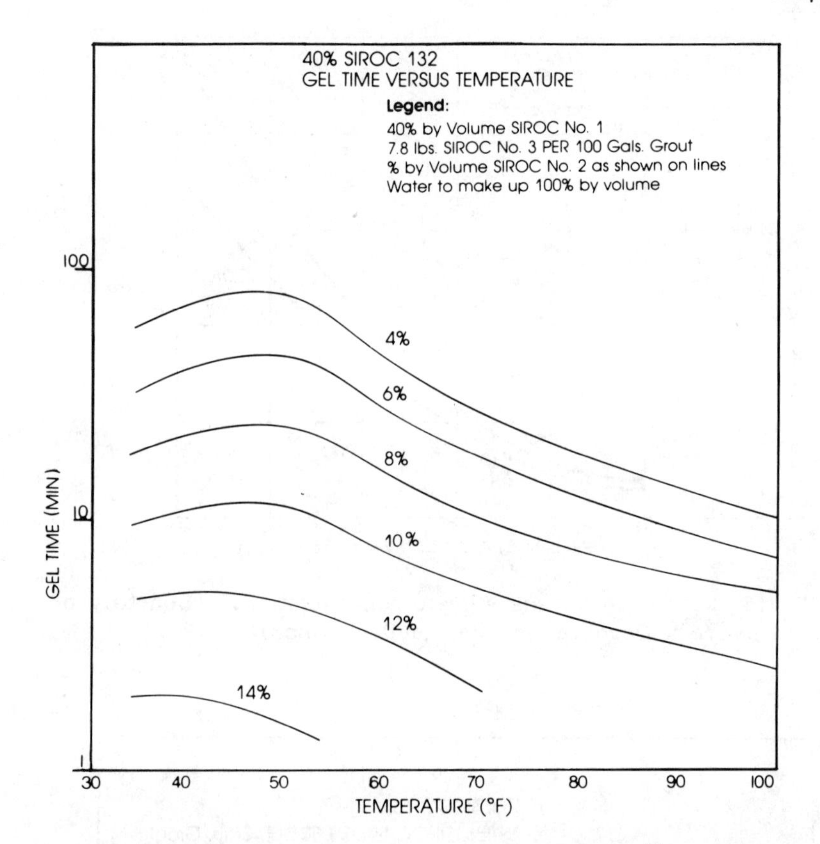

FIGURE 4.8 Gel time control with SIROC, using Formamide percentage as shown. (Courtesy of Diamond Shamrock Corp., Painesville, Ohio.)

calcium chloride using formamide reactant. Figure 4.8 shows the range of gel times possible for a 40% SIROC solution using the same catalyst system.

Within the past several years considerable field work has been done using sodium silicate solution and organic reactants without accelerators. Much of this relates to vehicular tunnels in the Washington, D.C. area and is well documented [8]. Contractors' field charts demonstrate gel times of the grout in Fig. 4.9.

In Europe, most of the silicate grouts use organic reactants only. One manufacturer (Rhone Poulenc, France) describes his product (called Hardener 600) as a "mixture of open chain compound

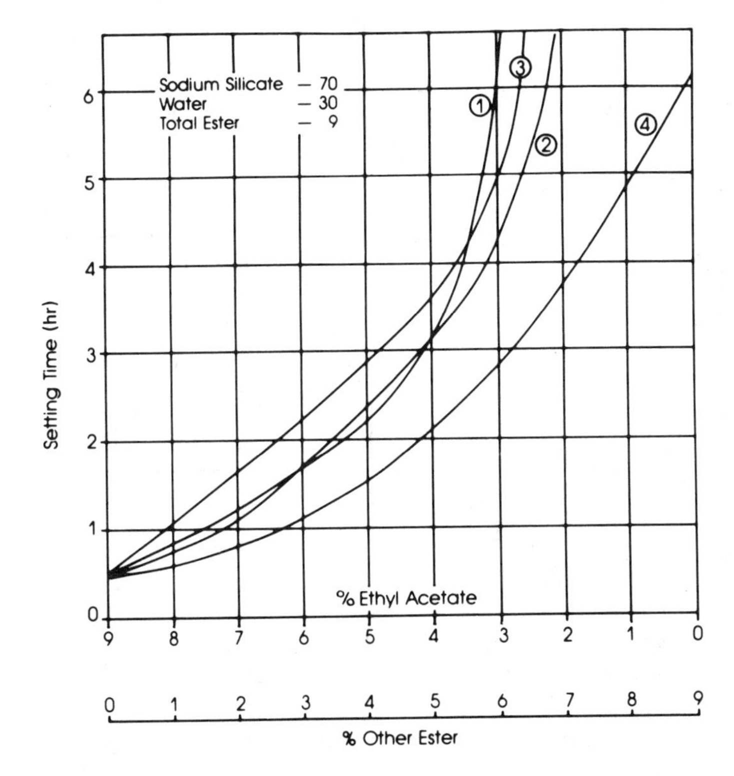

FIGURE 4.9 Properties of silicate grout using organic reactant. 1, butyl acetate; 2, isopropyl acetate; 3, isobutyl acetate; 4, ethyl succinate.

diacid esters." Figures 4.10 and 4.11 show technical and application data from the manufacturer's literature.

The strength that a silicate gel imparts to a stabilized soil is primarily a function of silicate content. For consolidation of strata where laboratory UC values of 100 psi or more are desired, silicate contents required are such that the solution viscosity is in the 10 cP and higher range. For waterproofing of soils where less strength is needed, solution viscosity will be in the 3 to 10 cP range.* (These are initial viscosities. During the first three-quarters of

*Recent research urges caution when using silicates for waterproofing [18].

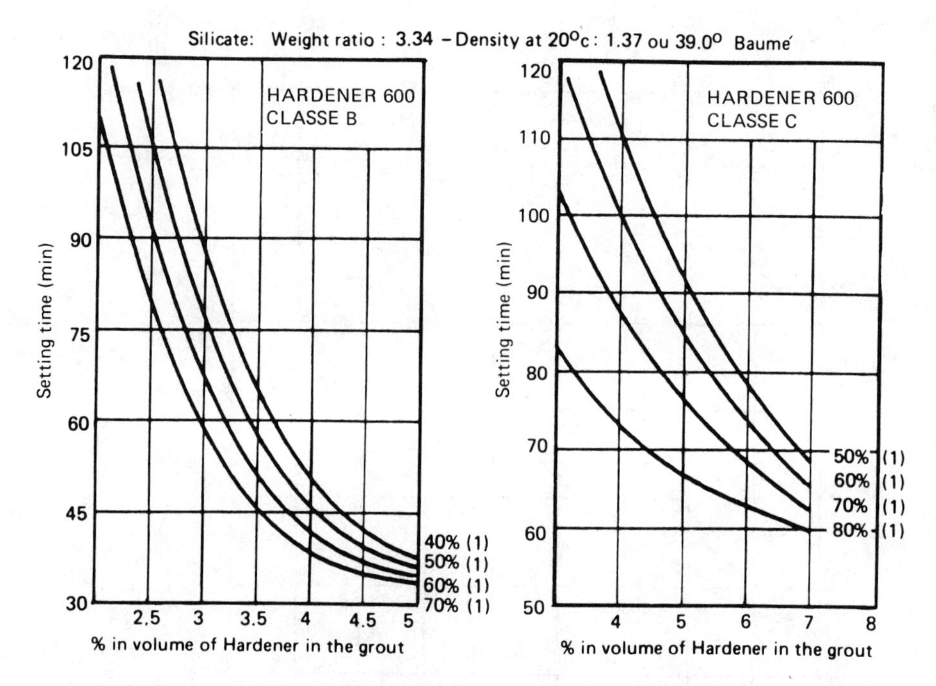

FIGURE 4.10 Properties of silicate grouts in European practice. (Courtesy of Rhodia, Inc., Monmouth Junction, New Jersey.)

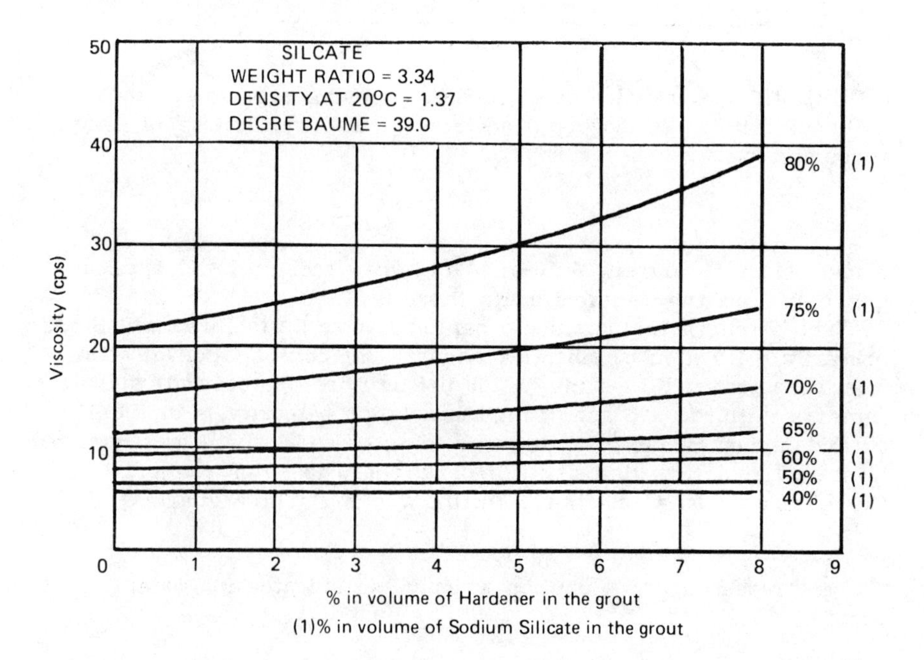

FIGURE 4.11 Properties of silicate grouts in European practice. (Courtesy of Rhodia, Inc., Monmouth Junction, New Jersey.)

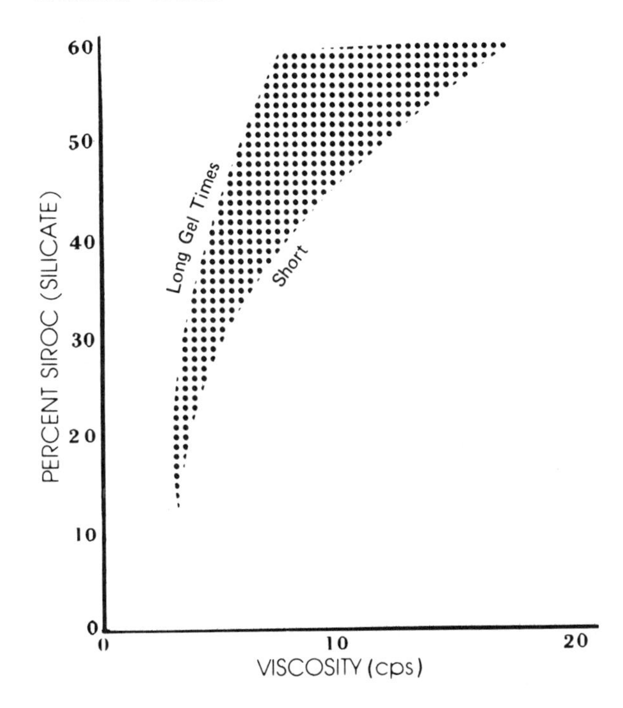

FIGURE 4.12 Relationship between percent SIROC and viscosity.

the induction period, the viscosity approximately doubles, as shown in Fig. 4.4.) Figures 4.12 and 4.13 illustrate some of the relationships. Other factors also affect the strength when silicate content is constant.

One study [9] has shown that the short-term strength of the pure gel is inversely proportional to setting time. Some of this relationship must carry over to grouted soil and probably to long-term strength. Since setting time is related to catalyst concentration, a relationship is thus established that strength increases with increasing catalyst concentration. Strength will also vary with different catalysts, all other factors being equal.

The strengths obtained in laboratory UC tests of grouted soils also depend on the curing conditions and the testing rate. Since higher strain rates also give higher strength, some standard rate must be chosen so that tests from different sources are comparable. [Values obtained from such tests should be considered as comparative indices. If true strength values are needed for design purposes, they should be obtained by stress-controlled triaxial tests that define the (creep) endurance limit. (See Appendix F.)]

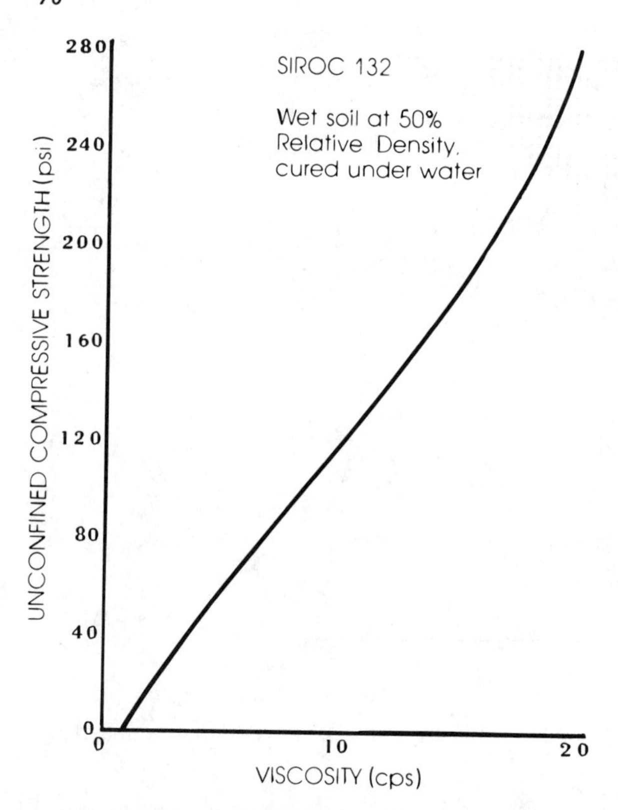

FIGURE 4.13 Relationship between viscosity and strength.

ASTM standards have been published that define procedures for grouted sample preparation and for Unconfined Compression tests to determine strength indices (D-4320). Use of these standards should permit valid comparisons of strength data from different sources. (As this manuscript is being finalized, an ASTM standard on triaxial creep testing of grouted soil samples is in the draft stage. This standard should be completed by the near future.)

Dry cure of grouted soils gives numbers much higher than wet cure, and they tend to be misleading. Wet cure of a small stabilized mass in a large quantity of water may also be misleading in the other direction, because of dissolution (discussed later in this section). For short-term testing, samples should be left in their fabrication forms until just prior to testing. For long-term tests, samples should be placed in a container of slightly larger diameter and the annular space filled with saturated dense granular soil (see ASTM standard D-4320).

Although most of the silicate formulations are considered permanent materials, the end product is subjected to two phenomena which often tend to cause doubt about permanence. A newly made silica gel will, upon standing, exude water and shrink. This phenomenon is called syneresis and occurs at a decreasing rate. The total water loss is related to the gel properties, generally decreasing with increasing silicate content and shorter setting times. Syneresis also takes place in the voids of stabilized soil masses. In a soil whose voids were completely filled with new gel, the shrinkage accompanying syneresis results in an increase in residual permeability after several weeks. In coarse-grained soils this may partially negate the initial effectiveness of water shutoff. As soils become progressively finer, the practical effects of syneresis become smaller. For medium and fine sands, the effects are generally considered negligible.

The time dependency of the syneresis phenomenon is shown in Fig. 4.14a, for a 40% silicate solution catalysed with 15% glyoxal. Similar amounts of syneresis occur with other catalyst systems, as shown in Fig. 14.4b. Several well documented field projects (Chap. 14, Refs. 7–9) indicate that final field permeability of a granular formation grouted with silicates will not be better than 10^{-4} to 10^{-5} cm/sec. This is probably due to the syneresis phenomenon.

Stabilized sand cylinders stored under water for a period of time show loss of strength which varies from negligible to total (complete falling apart of soil grains) depending on the grout chemistry. The dissolution of the gel is caused by the unreacted portion of the soda, which reverses the process which formed the polysilisic acid. (For this to occur, there must be ionic movement, as would occur if the gel were underwater.) High soda contents generally result from low reactant concentrations associated with long setting times. Soda content high enough to completely dissolve the silica gel is generally due to error or improper use of materials. However, most gels will contain some soda, and therefore gels made in the field underwater will always show some loss of strength. This is a phenomenon limited in extent to the contact zone between grouted soil and groundwater and for the normal scale of ground injection does not have significant effect on the overall project. In contrast, on a laboratory scale, the phenomenon may readily encompass the entire grouted sample.

Recent studies indicate that, under some conditions of underground water flow and pressure, silicate grouts may exhibit less than total stability. Krizek and Madden [18], after cautioning that their data is based on very high gradients (100) conclude:

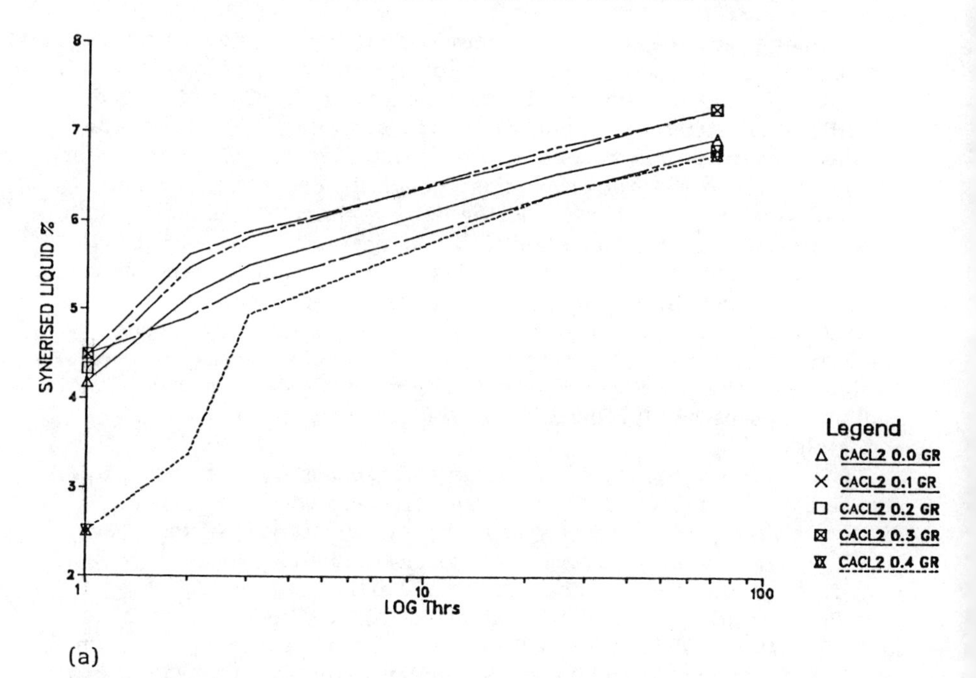

FIGURE 4.14 (a) Rate of synergesis for sodium silicate stabilized with glyoxal. (From Ref. 19.) (b) Syneresis rates of silicate grouts. (From Ref. 18.)

2. Specimens injected with the sodium silicate grouts underwent large variations in permeability during the early stages of testing, but, once the permeability stabilized, it remained relatively constant for the remainder of the test. The value at which the permeability stabilized appears to be dependent on the permeability of the ungrouted soil and, to a lesser extent, on the chemical characteristics of the grout. In general, it appears that, at gradients between 50 and 100, the maximum long-term reduction in the permeability of a soil due to the injection of these grouts is one to two orders of magnitude.

3. The amount of rate of grout elutriation appear to be dependent on the strength of the gel and the amount of syneresis experienced in the grout. The silicate grouts require several days to achieve their maximum strengths

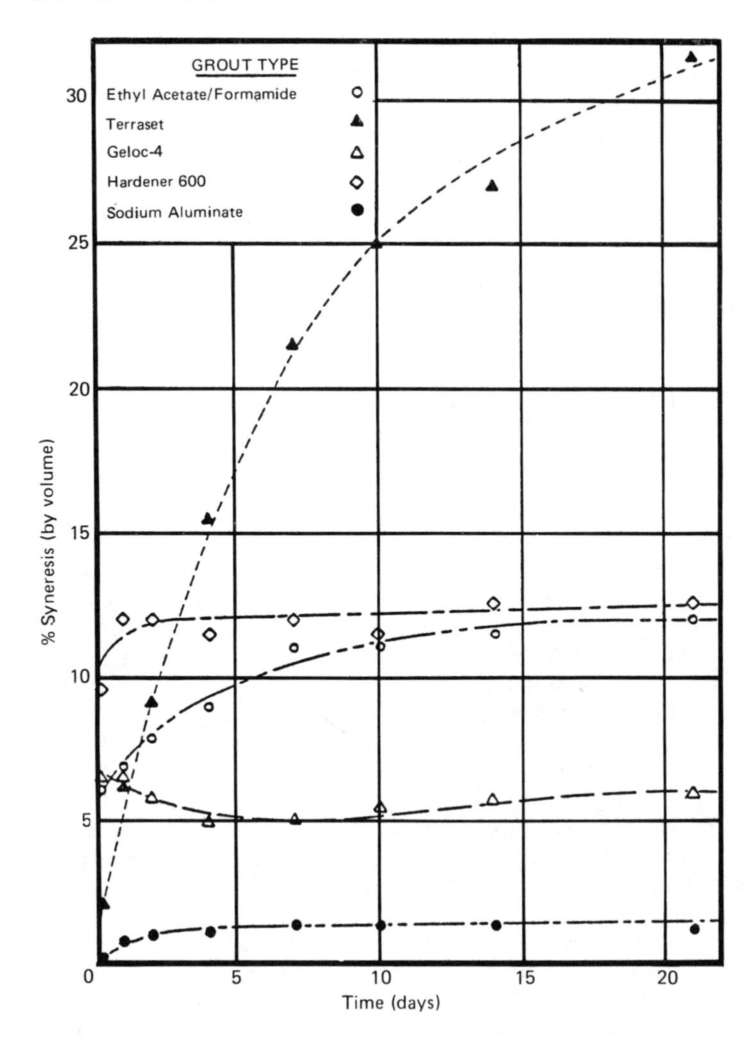

(b)

FIGURE 4.14 (Continued)

and exhibit reductions of as much as 25% of their original volume due to syneresis. Specimens injected with silicate grouts and cured for less than one day experienced rapid, and usually complete, elutriation due to lack of strength. In older specimens the elutriation was a gradual process, and the rate at which the permeability increased was apparently accelerated by increases in the degree of syneresis.

4. For the silicate grouted specimens in which most of the grout was eroded, that grout which remained appeared to be concentrated at the contact points between the soil grains.

5. Extreme caution is recommended when considering the silicate grouts tested in this program for use in situation where they will be subjected to high gradients.

If nothing else, this data suggest caution in recommending the use of silicate grouts for waterproofing and seepage control.

The phenomenon of creep in grouted soil masses has been recognized since the late 1950s (the earliest studies were made by the author with acrylamide grouts in 1957. Results were published shortly afterward in an inter-company report, and are shown in the section on acrylic formulations). More recently details of the creep phenomenon in silicate-grouted soils were reported (Refs. 7–9 of Chapter 14, and Refs. 5, 8, and 12 of this chapter). Some of the relationships between strength and times are shown in Fig. 4.15.

Silicate grouts, depending upon grout and catalyst concentrations, may take from several days to as much as 4 weeks to reach maximum strength. Creep studies should be correlated to specific field loading schedules in order to give realistic results.

Of course, SIROC is not the only domestic silicate grout in use. GELOC is proprietary formulation using esters as catalysts. Data regarding strength and gel time control is similar to those shown in Figs. 4.9, 4.10, and 4.11. Terraset 55-03 is a product of Celtite Corp. (Cleveland, Ohio). Data are shown in Fig. 4.16. In Japan, Glyoxal is the catalyst generally in use for silicates. This product is also available domestically. The relationship between catalyst concentration and gel time is shown in Fig. 4.17. Figure 4.18 shows strength data versus time for sand samples grouted with silicate and Glyoxal.

One of the peculiar characteristics of silicate grouts (and one not noted for any other grouting material) is a reversal in the temperature–time relationship at low temperatures. Normally, it is anticipated that gel times would get long, as temperatures decrease. This is so for all other grouts. With silicates, however, when temperature of the grout solution drops below 15°C or 50°F, gel

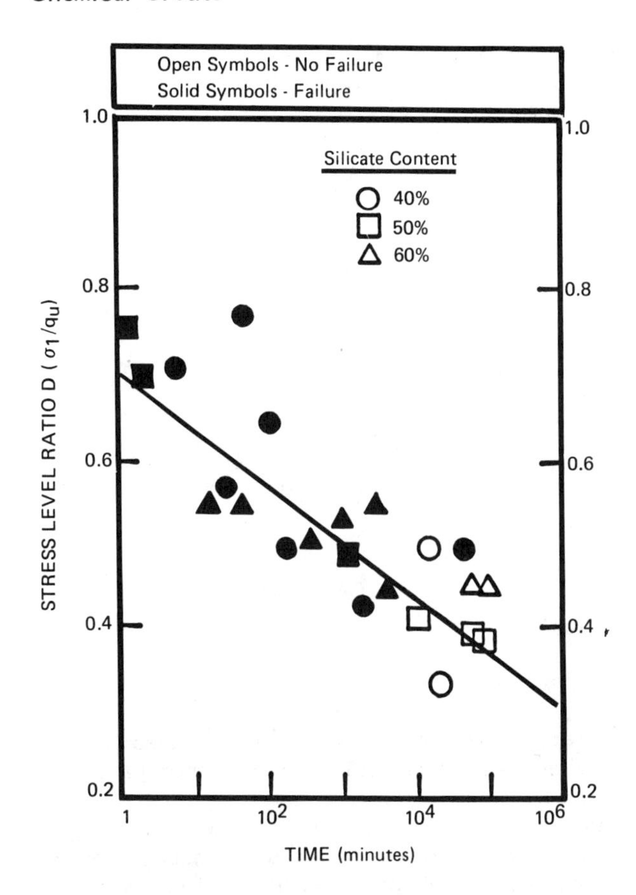

(a)

FIGURE 4.15 (a) Creep endurance limits for silicate grouts. (From Ref. 22.) (b) Unconfined creep results for 50% silicate, moist-cured specimens. (From G. Wayne Clough, "Silicate stabilized sands," *J. Geotech. Eng. Div.*, ASCE, Jan. 1979, p. 75.)

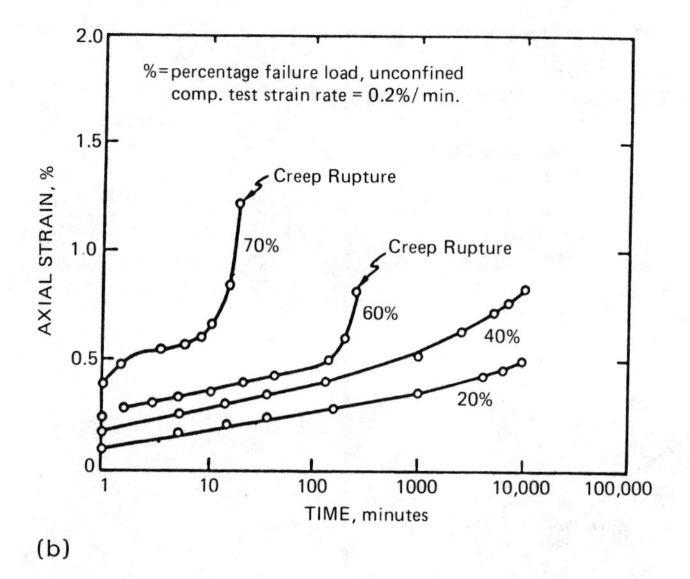

FIGURE 4.15 (Continued)

times start getting shorter, as shown in Fig. 4.19, and also in Figs. 4.7 and 4.8. In field work this could cause unanticipated flash sets when working at temperatures near freezing.

Portland cement can be used with silicates, and acts like a catalyst (Figs. 4.16 and 4.20). However, the use of normal cements adds solids, which reduces the penetrability of the grout suspension. This negative aspect is overcome through the use of special products such as Microfine 500 discussed in Appendix A.

Sodium silicate solution is generally considered totally nontoxic and free of health hazards and environmental effects. Sodium salts may be exuded from silicate gels. They might be classed as environmental hazards in special circumstances. Some of the organics used for reactants may have toxic, corrosive, and/or environmental effects. Manufacturers' recommendations for handling those products should be followed.

Acrylamide Grouts

The rapid expansion and development of the markets for organic chemicals and products in the 1940s led to the discovery in 1951 of a product that became the first of the new chemical grouts. This was a mixture of organic monomers, which could be polymerized at

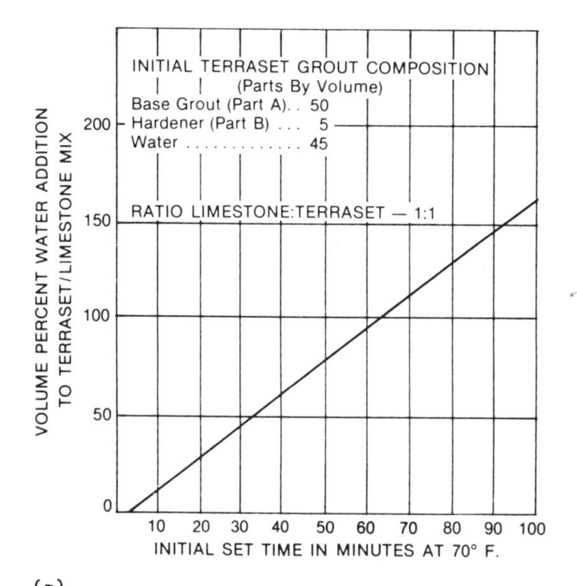

(a)

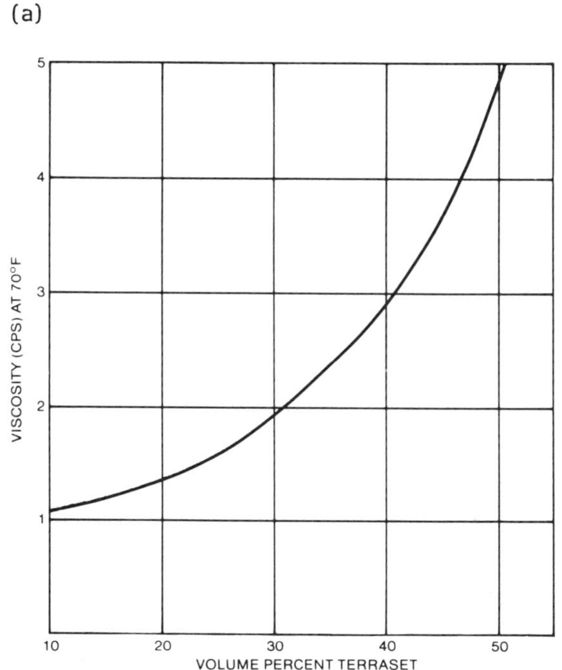

(b)

FIGURE 4.16 Celtite 55-03 terraset chemical grout. (a) Initial set times at terraset/limestone grout, (b) 55-03 terraset, variation in gel time with change of accelerator, and (c) 55-03 terraset variation at gel time with percent terraset at normal temperature, and (d) initial set times at terraset/cement grouts using Type 1 Portland cement. (Courtesy of Celtite, Inc., Cleveland, Ohio.)

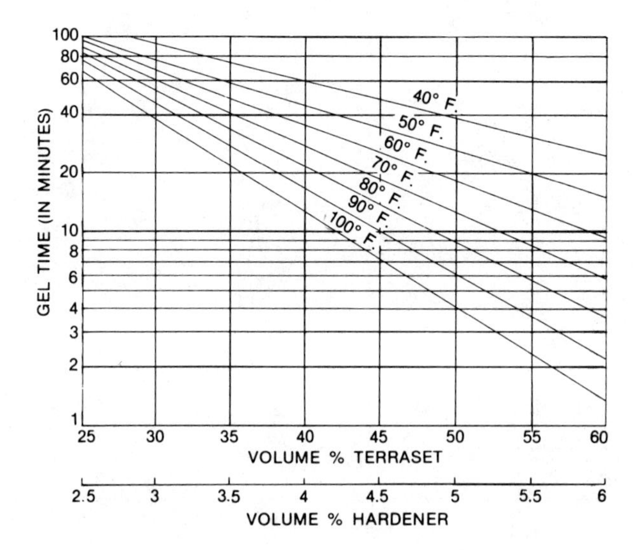

(c)

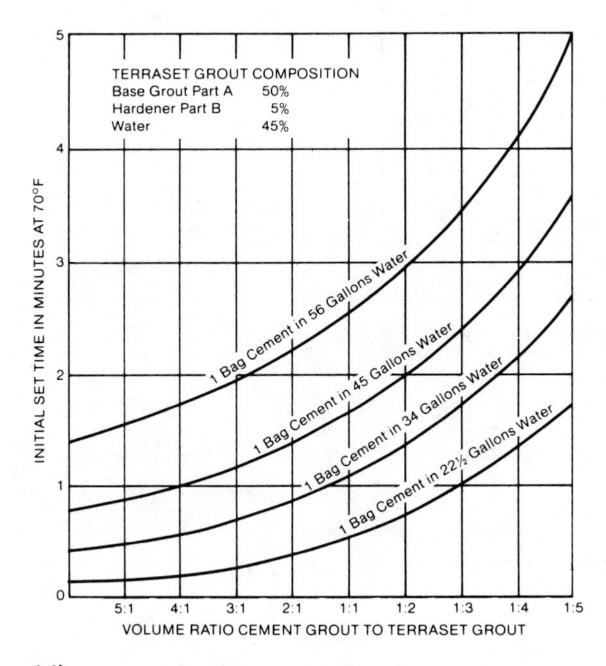

(d)

FIGURE 4.16 (Continued)

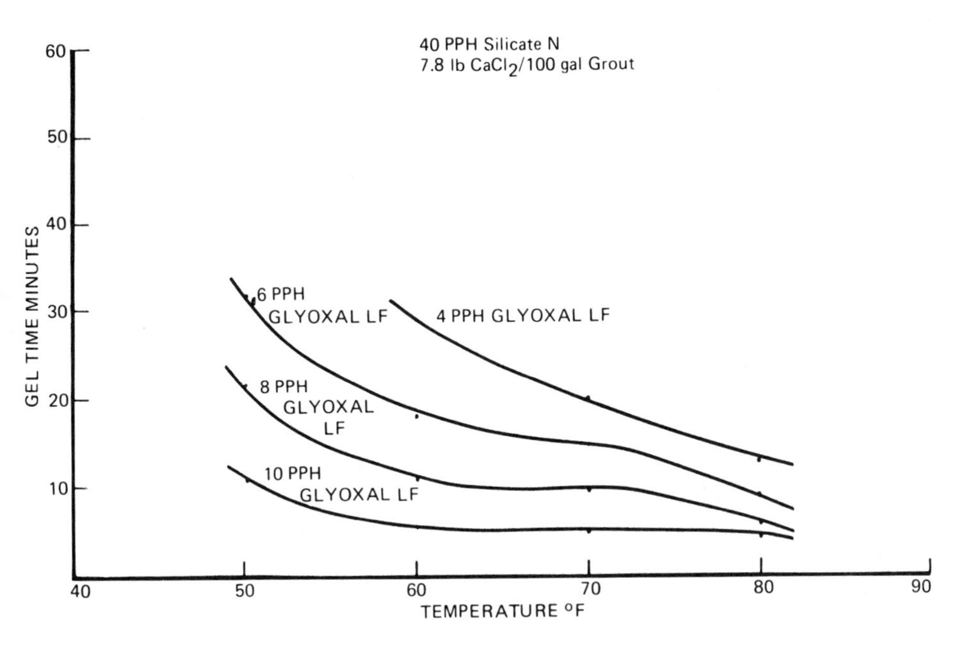

FIGURE 4.17 Influence of temperature and Glyoxal concentration on gel time. (Courtesy of American Cyanamid, Wayne, New Jersey.)

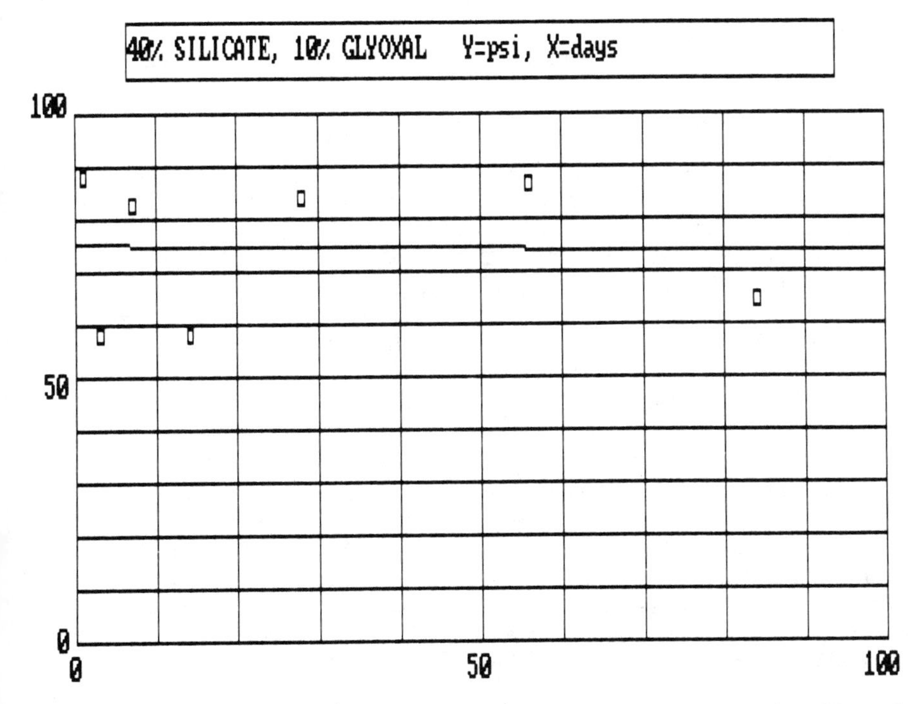

FIGURE 4.18 Aging effects for sodium silicate stabilized with Glyoxal.

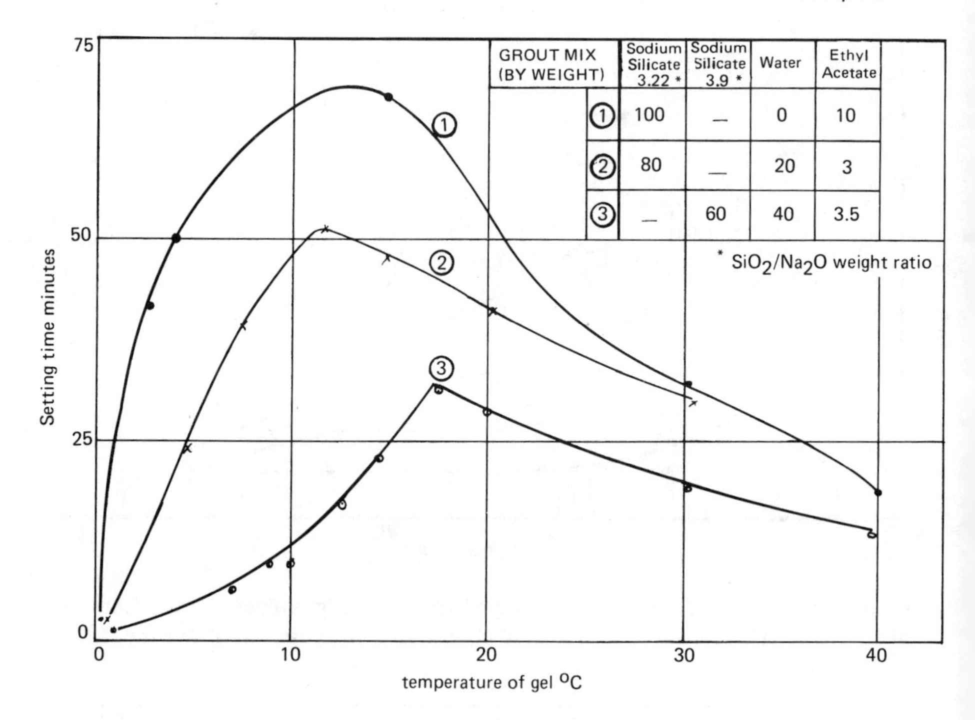

FIGURE 4.19 Effect of temperature on setting time of silica gels. (From Ref. 5.)

ambient temperatures, with setting time a direct function of catalyst percentage. The solution had a viscosity and a density close to those of water, and both properties remained virtually constant during the induction period. The change from liquid to solid was almost instantaneous for the shorter gel times. The manufacturer (American Cyanamid Company, Wayne, New Jersey) was quick to recognize the potential of the product as a grout and continued the search for an optimum mixture in their laboratories and with the Massachusetts Institute of Technology. In 1953, a product called AM-955 (so labeled because it was a mixture of dry powders of which 95% was acrylamide and 5% methylene-bis-acrylamide), later shortened to AM-9, was made in a pilot plant in sufficient quantities to begin field evaluation. Initial field successes led to the establishment by the manufacturer of an AM-9 Field Service Group in 1955 and an Engineering Chemicals Research Center in 1957. Both organizations devoted full time to the development of the product and techniques and equipment for field application. These efforts were highly

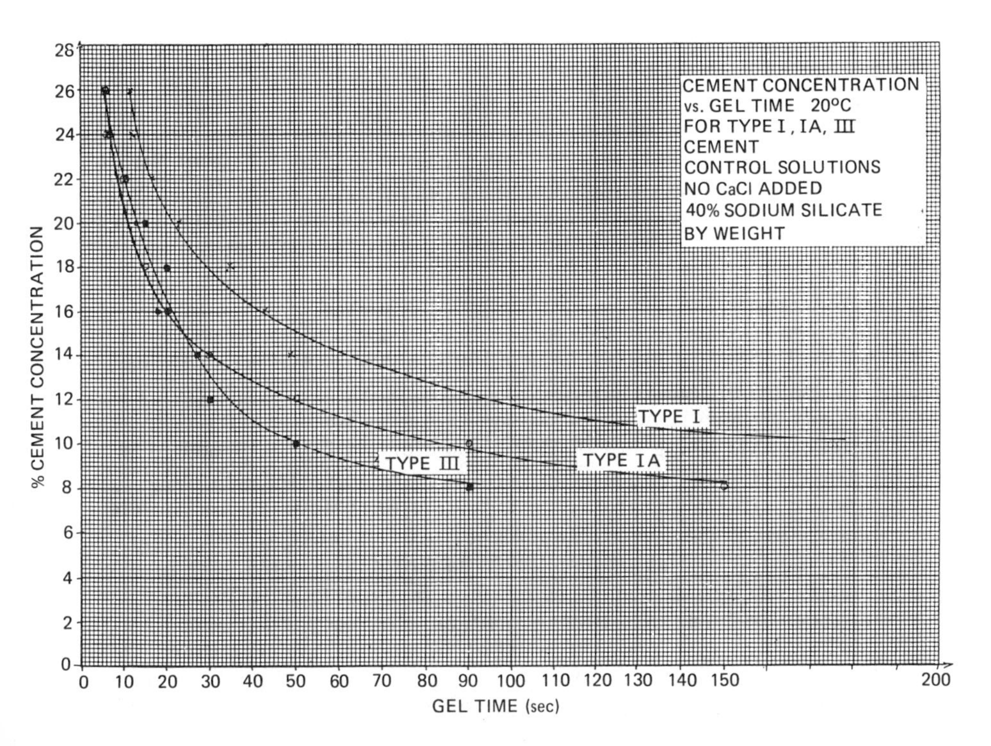

FIGURE 4.20 Cement as a catalyst for sodium silicate. Cement concentration versus gel time at 20°C for Types I, IA, and III cement control solutions. No $CaCl_2$ added, 40% sodium silicate by weight. (Unpublished research at Rutgers University, New Brunswick, New Jersey, 1985, by Xiao Tianyuan, visiting scholar from The People's Republic of China.)

successful and were the major spur to the development and marketing of chemical grouts by other manufacturers. (Two of these new products are also acrylamide based and were introduced between 1960 and 1970. Acrylamide-based grouts continue to appear, the latest in the early eighties, tradenamed Injectite-80. This is a low-molecular-weight soluble polyacrylamide (which eliminates the toxicity at the expense of a large increase in viscosity), developed primarily for the sewer sealing industry. It has not been marketed aggresively, due to the almost simultaneous appearance of the acrylate grouts.

Acrylamide-based materials come closest in terms of performance to meeting the specifications for an ideal grout. They penetrate more readily, maintain constant viscosity during the induction period, and have better gel time control and adequate strength for most applications. They are however, more costly per pound (or per gallon) than silicates, and acrylamide is neurotoxic.

The hazards of acrylamide grouting are detailed in a memorandum from NIOSH (National Institute for Occupational Safety and Health) appended to a memorandum dated September 19, 1985, from the United States Environmental Protection Agency to its Water Management Division directors. The following paragraph is exerpted:

> Acrylamide monomer, which is frequently used as a grouting material to prevent the infiltration of ground water into sewer lines, is toxic to the body's nervous system. At least 56 reported cases of poisoning have occurred in workers exposed to acrylamide. Thirteen of these cases occurred in individuals using acrylamide for waterproofing and soil stabilization at a construction site. Possible long term health effects from chronic low level exposures to acrylamide are also of concern. Two recent studies have shown that acrylamide can cause cancer in laboratory animals. It is currently not known if acrylamide is capable of causing cancer in humans. Human contact with acrylamide can occur through dermal exposure, inhalation, or ingestion. Skin contact with acrylamide monomer is especially hazardous because acrylamide is readily absorbed through the skin.

Although small, single doses of acrylamide to experimental animals (the acute oral LD_{50} for rodents is approximately 200 mg/kg) are not particularly hazardous, the product possesses a high degree of cumulative toxicity. Repeated and prolonged intake of small quantities by experimental animals results in disturbance of certain functions of the central nervous system. The effect is manifested by muscular weakness and disorders of equilibrium and locomotion.

If exposure to acrylamide is terminated when signs of poisoning are first observed, complete recovery may be expected to occur within a relatively short period of time. However, if poisoning is severe, or if exposure is allowed to continue after evidence of poisoning has developed, recovery may require a longer period of time.

In gel form, acrylamide-based grout contains very little free acrylamide. The gel produced when a 10% solution of grout is properly catalyzed, either in the laboratory or in the field, contains less than 0.02% acrylamide. Dilution of acrylamide to this degree is considered to render it nonhazardous.

It is known that many microorganisms found in soil and natural waters assimilate ungelled acrylamide so that hazards of the unreacted material persist only a relatively short period of time. These organisms do not affect the gel itself. However, an application near a water supply which may be used for drinking or recreational purposes should be undertaken only when conditions indicate that no appreciable quantity of acrylamide will find its way into water. Under most usual circumstances, significant contamination will be unlikely. In this connection, it is important that excessive dilution of catalyzed solutions be prevented until the gelling reaction has had sufficient time to proceed to completion.

Despite the known hazards, millions of pounds of acrylamide grouts have been placed over the past several decades, and these products continue in use domestically and overseas. When specific basic safety and safe handling precautions are faithfully observed, toxicity hazards can be reduced to negligible values.

Until early 1978, three acrylamide grouts and one acrylamide-based grout were commercially available. At that time the announced, without explanation or amplification, that the manufacture of AM-9 had been stopped and that neither the product nor its components would be sold for grouting purposes (see Sec. 1.4).

By early 1979 the Japanese product Nitto SS (product of the Nitto Chemical Industry Co., marketed in the U.S. as AV-100) and the French product Rocagil BT (Rhone Poulenc, France) were both available in the United States. Nitto SS has essentially the same toxicity level as AM-9 (it was banned in Japan in 1974, following a careless application near a well which led to several cases of acrylamide poisoning). It is marketed as a dry powder. Rocagil BT contains methanol acrylamide, and its toxicity is about half that of Nitto SS. Rocagil BT is marketed as a 40% solution in water. There is also Rocagil 1295 grout (a mixture of acrylamide and methanol acrylamide), as well as a discontinued Japanese material, Sumisoil, which is in essence a copy of AM-9. None of the Rocagil acrylamides are currently available domestically. Two products formerly marketed in the United States, Q-Seal and PWG, were actually distributors' trade names for AM-9.

All the acrylamide-based grouts have similar characteristics and properties and are treated together in the following discussion.

The acrylamide-based grouts consist of a mixture of two organic monomers: acrylamide (or methanol acrylamide, methacrylamide, etc.), which is generally 95% of the mixture and will polymerize into long molecular chains, and 5% cross-linking agent such as methylene-bis-acrylamide, which binds the acrylamide chains together. The stiffness of the grout can be varied by changing the 95:5 ratio. If 97:3 is used, a sticky, very elastic transparent gel of low strength is obtained. If 90:10 is used, a harder, stiffer, opaque white gel is obtained. Between the limits of the values given above, there is some difference in the UC value for stabilized soils, and the slope of the stress-strain curve becomes steeper and the UC value higher as the cross-linking agent increases. More importantly, as the cross-linking agent decreases, the gel will absorb water from its wet environment, and it will expand and become weaker. All commercial products are sold as a mixture of fixed ratio, and the user must purchase the components separately if he or she wishes to modify the ratio.

Grout solutions up to 20% solids have viscosities well under 2 cP. [Ref. 5 indicates that the viscosity doubles with the addition of catalyst and activator. This could well be related to the specific activator used TEA (triethanolamine), since tests did not show a viscosity change when DMAPN (dimethylaminopropionitrile) was added.] Such solutions, when properly catalyzed, will, after a length of time which depends on catalyst concentration, change almost instantly to a solid irreversible gel (see Fig. 4.4).

The minimum concentration of grout from which a gel will form is temperature dependent, as shown in Fig. 4.21. The percent monomer selected for field use must take into account the possibility of dilution with groundwater below the values at which gels will form. Most field work is done at about 10%. When pumping into flowing water, concentrations as high as 15% to 20% may be used. When large grout volumes are placed at short gel times, 7% to 8% may be used, and in special cases it may be reasonable to work at 5%.

Gels that form from grouts normally used for field work contain 8% to 12% solids. These solids, in the gel, are in the form of long molecular chains, randomly cross-linked to each other. The rest of the grout, 88% to 92% by weight, consists of water molecules mechanically trapped in the "brush-heap" structure of the gel. If submersed in water or under saturated soils or placed in an environment of 100% relative humidity, the water in the gel remains in place, and the gel undergoes no volume change. However, gel placed in an arid environment will lose water and can shrink to the volume occupied by its solids, about 10% of the original gel volume. Mechanical pressure can also force water from the gel. (Gels subjected to a typical soil consolidation test show permeabilities of about 10^{-10} cm/s.) Hydraulic pressure, on the other hand, will force water into the gel and cause it to expand. When the forces causing volume change are removed, in the presence of a water source the gel will return to its original volume. Thus, gels can reabsorb the 90% water they have lost and expand to fill the voids in the formation in which they were placed. Shrinkage cracks which may have formed in the gel

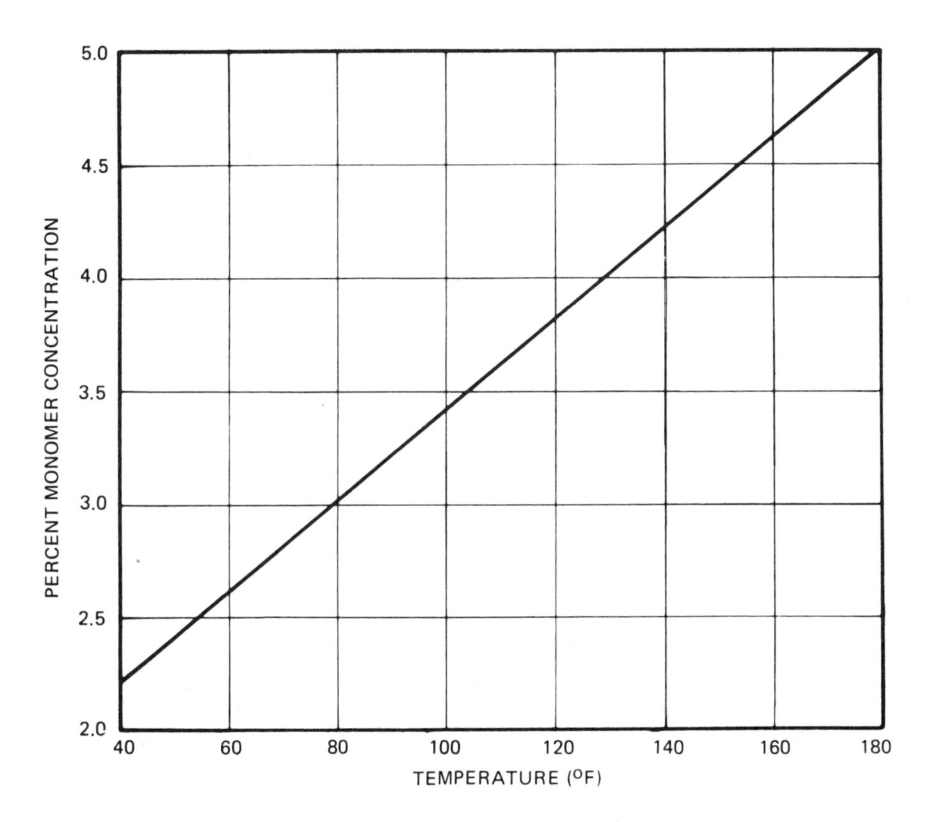

FIGURE 4.21 Minimum monomer concentration for gel formation (acrylamide grout).

will reseal but not heal, so the permeability of the formation may be somewhat greater and the strength somewhat less than prior to desiccation.

Ions can migrate through the water in the gel. Soluble salts that are added to a grout solution for various reasons may be expected to leach out in time. Similarly, gels made with pure water can be expected to absorb salts from a saline environment.

Acrylamide grouts at ambient temperatures are catalyzed with a two-component redox system. One part, the initiator or catalyst, can be a peroxide or a persalt. Ammonium persulfate (AP), a powder, is most commonly used. The second part, the accelerator or activator, is an organic such as nitrilo-tris-propionamide (NTP), dimethylaminiopropionitrile (DMAPN), and triethanolamine (TEA).

All three have disadvantages. DMAPN, a liquid, is best from a control point of view but is considered a health hazard. NTP, a powder, has limited solubility in water, particularly at low temperatures. TEA, a liquid, is somewhat metal sensitive. At the present time virtually all U.S. applications use TEA. There are also materials which act as inhibitors and can be used reliably to control gel time. Potassium ferricyanide, KFe, is most often used.

Gel time is independent of monomer concentration but is directly dependent on the temperature and concentration of the catalyst, activator, and inhibitor. Although the gel time can be changed by varying any one component, there is an optimum ratio between catalyst and activator for the most effective use of each. The charts in Fig. 4.22a to 4.22d show that AP and DMAPN should be used in about equal amounts (by weight). Figures 4.23 and 4.24 show typical gel time charts for Nitto SS and Rocagil BT. Short gel times, down to 5 and 10 s (such short gel times are occasionally required in the field), are obtained by omitting the KFe and increasing both catalyst and activator. For work at temperatures above 100°F, AP is generally used without an activator. For very long gel times (12 h, for example) AP is used with KFe. Whenever an activator is not used, a buffer such as disodium phosphate is required to maintain the pH of the solution around 8. If the pH of the grout goes below 7, which can happen if excess AP is used to shorten the gel time, or if a buffer is omitted, gel time control becomes very erratic.

Sunlight can cause catalysis of grout solutions to which an activator has been added. Such solutions should be kept covered in daylight. Entrained air from overmixing can reduce catalyst concentration and prolong gel times.

A catalyzed grout solution stays at constant viscosity until just before gelation. At that time the temperature begins to rise and the grout gels. For a 20 m gel time, the time from start of gelation to completion is less than a minute in the laboratory. The reaction is exothermic, and the temperature rise for 100 cc of 10% grout at room temperature can be 30 to 40°F. For 30% and 40% grouts, the temperature rise will go above the boiling point, and steam formation will shatter the gel. The exotherm contributes to the speed with which gelation is completed. When grouting soils, the temperature rise is much less than for a solid gel, and in the ground the change from liquid to solid is not as rapid as in samples of pure grout.

Uncatalyzed grout solutions have a pH of 4.5 to 5. Catalyzed solutions have a pH of about 8. Gel time control is predicated on maintaining the pH between 8 and 11. Higher pH will shorten gel times, while pH on the acid side will lengthen gel times or prohibit gel formation entirely. Generally, groundwater and ground formations

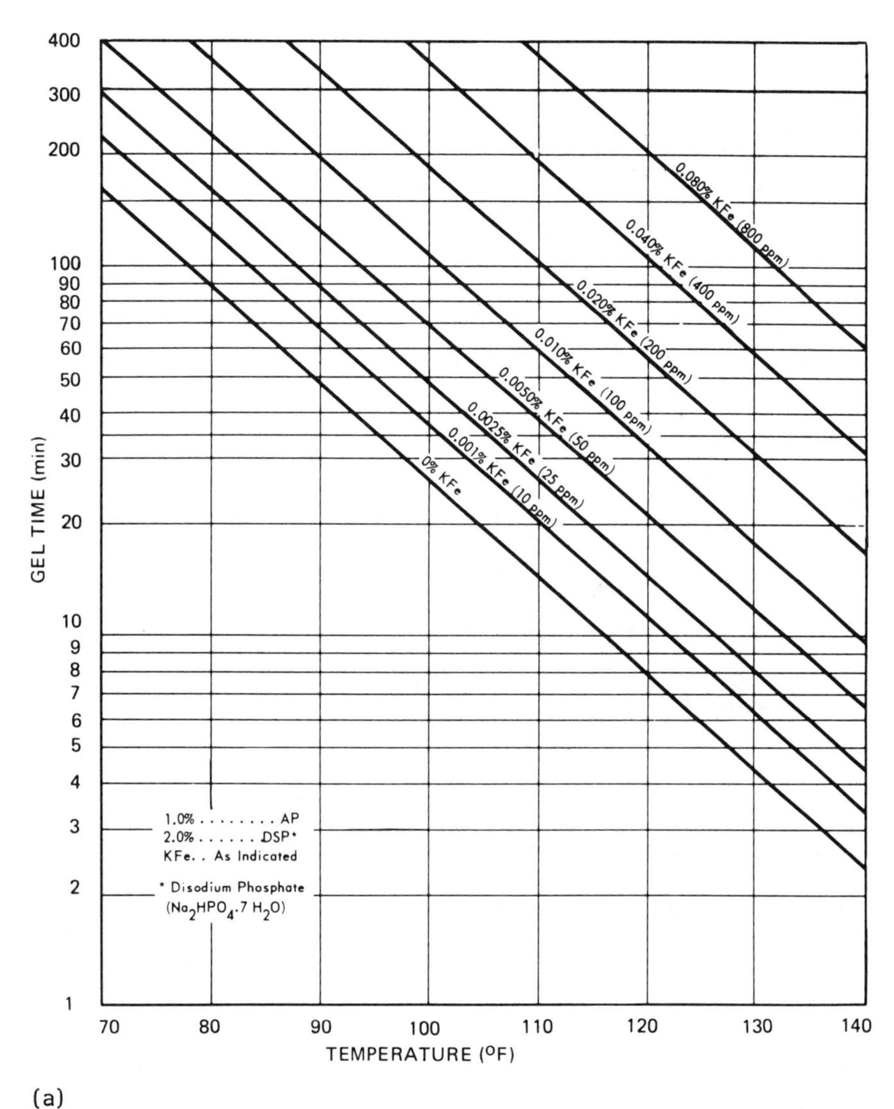

(a)

FIGURE 4.22 Varying catalyst and activator concentrations as they affect gel time of 10% acrylamide-based grout.

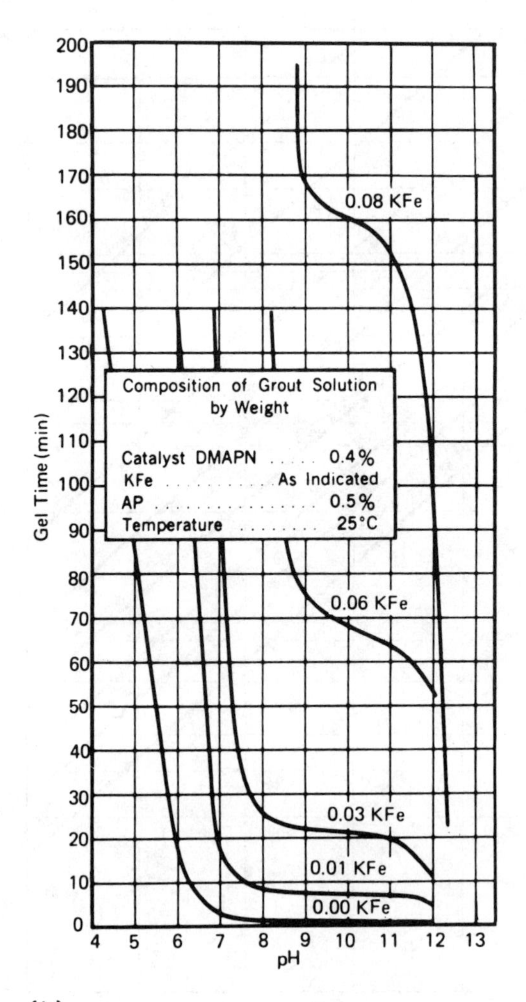

(b)

FIGURE 4.22 (Continued)

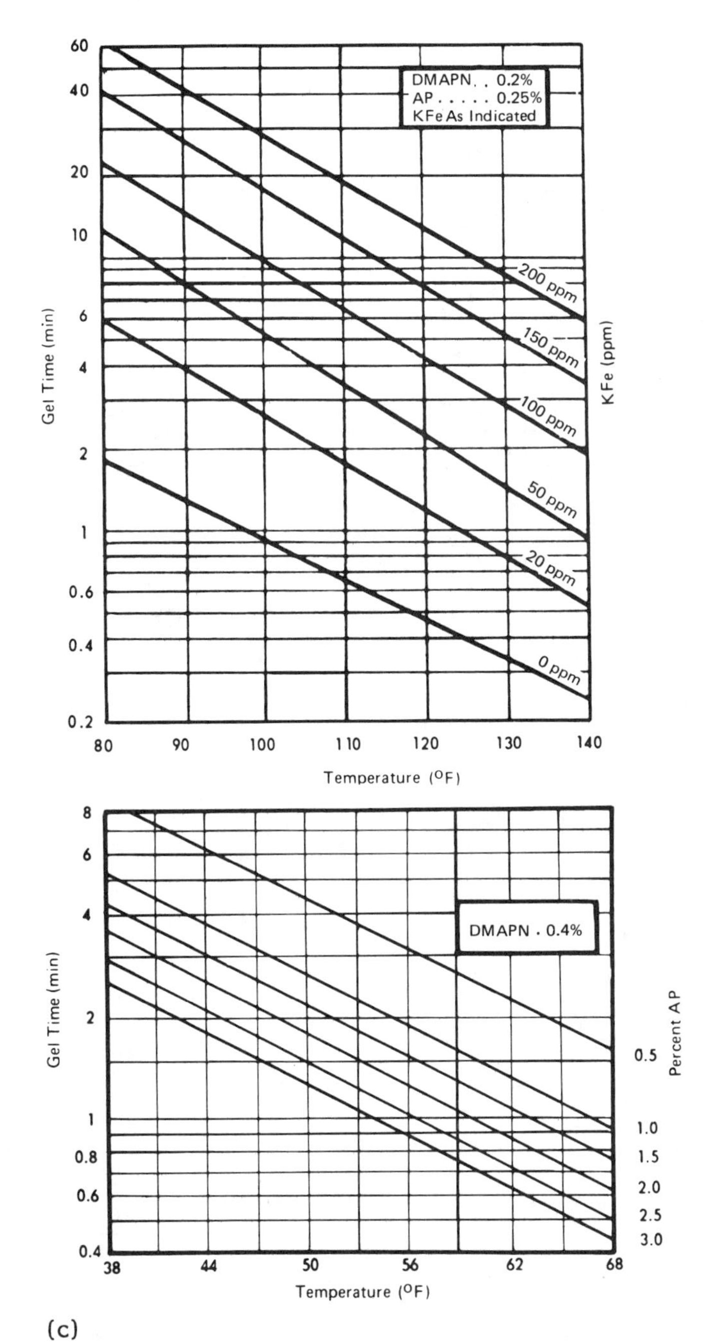

(c)

FIGURE 4.22 (Continued)

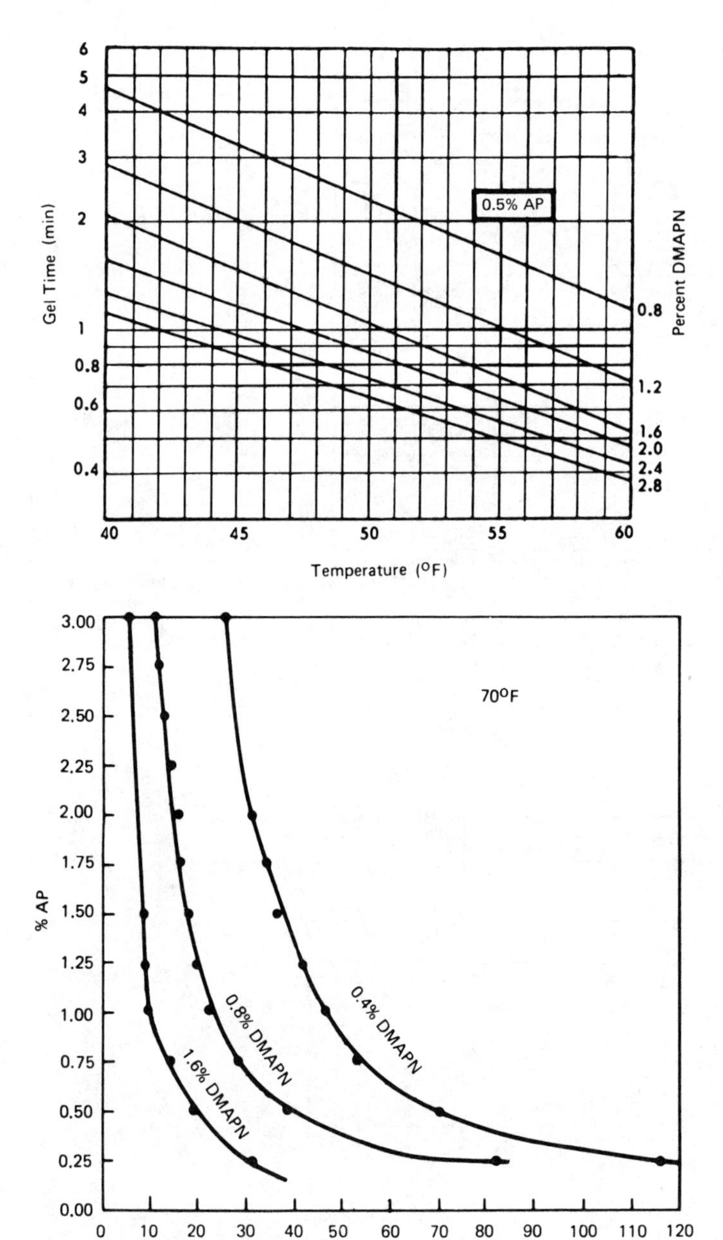

(d)

FIGURE 4.22 (Continued)

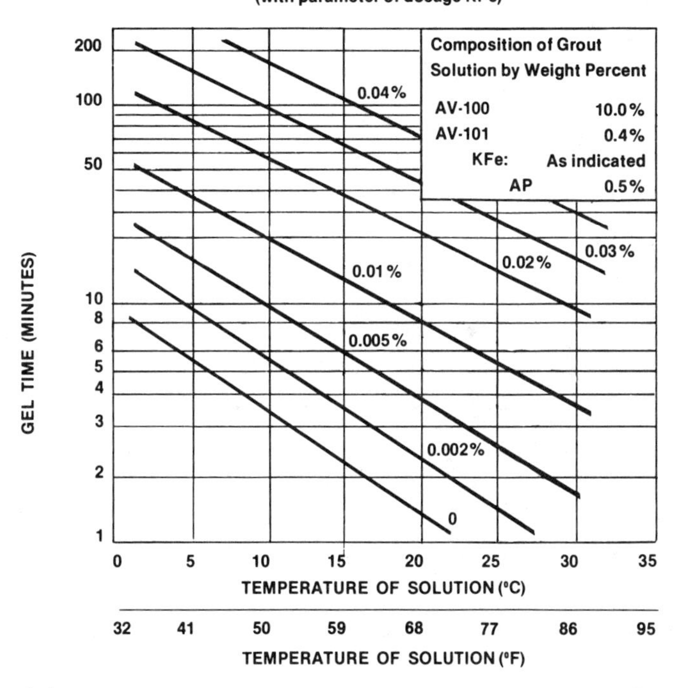

(a)

FIGURE 4.23 Correlation between gel time and temperature (with parameter of dosage KFe). (Courtesy of Avanti International, Webster, Texas.)

do not cause significant pH change in the grout. However, the effects of pH should be considered when grouting in the vicinity of a chemical plant, for example, or in an area previously grouted with cement.

The gel time of a grout solution can also be affected by chemicals dissolved in formation groundwater or by contact with the formation itself. (The effects of groundwater on gel time can be canceled by using groundwater to mix the grout solution. This is discussed in greater detail in following sections.) Sodium and calcium chlorides, in particular, tend to shorten gel times and in fact are sometimes deliberately added for this purpose when working at low temperatures. Figure 4.25 illustrates the effects of adding NaCl to an acrylamide-based grout. Calcium chloride also

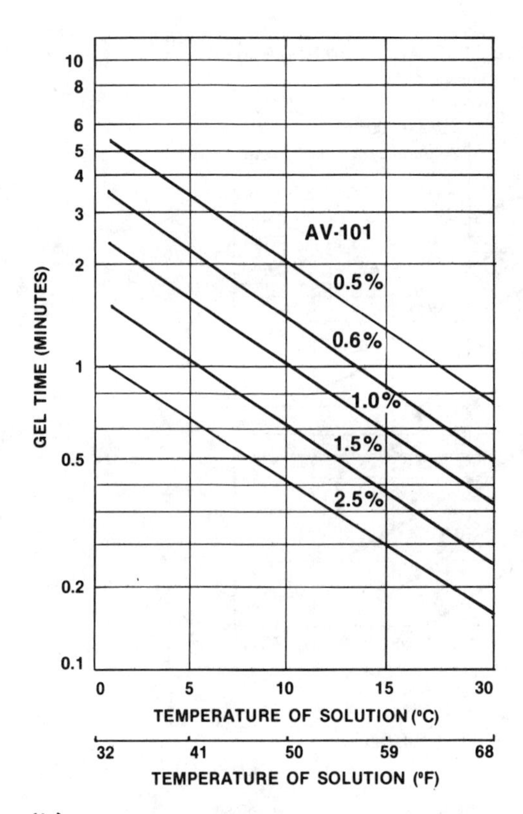

(b)

FIGURE 4.23 (Continued)

has the effect of reducing the rate of water loss from a gel exposed to dehydrating conditions.

Acrylamide gels are considered permanent. The gel is unaffected by exposure to chemicals, except for very strong acids and bases—materials not naturally found in soils. Tests on samples stored under saturated sand for 10 years showed no loss in strength, and many field jobs have shown adequate performance for times reaching well into the second decade. Gels are, however, subject to mechanical deterioration when exposed to alternating drying and/or freezing cycles. They will eventually break the gel into smaller pieces and reduce both the strength and imperviousness of the treated formation. Additives such as antifreeze, glycerine, and calcium chloride may be used to counter the effects of freezing and drying.

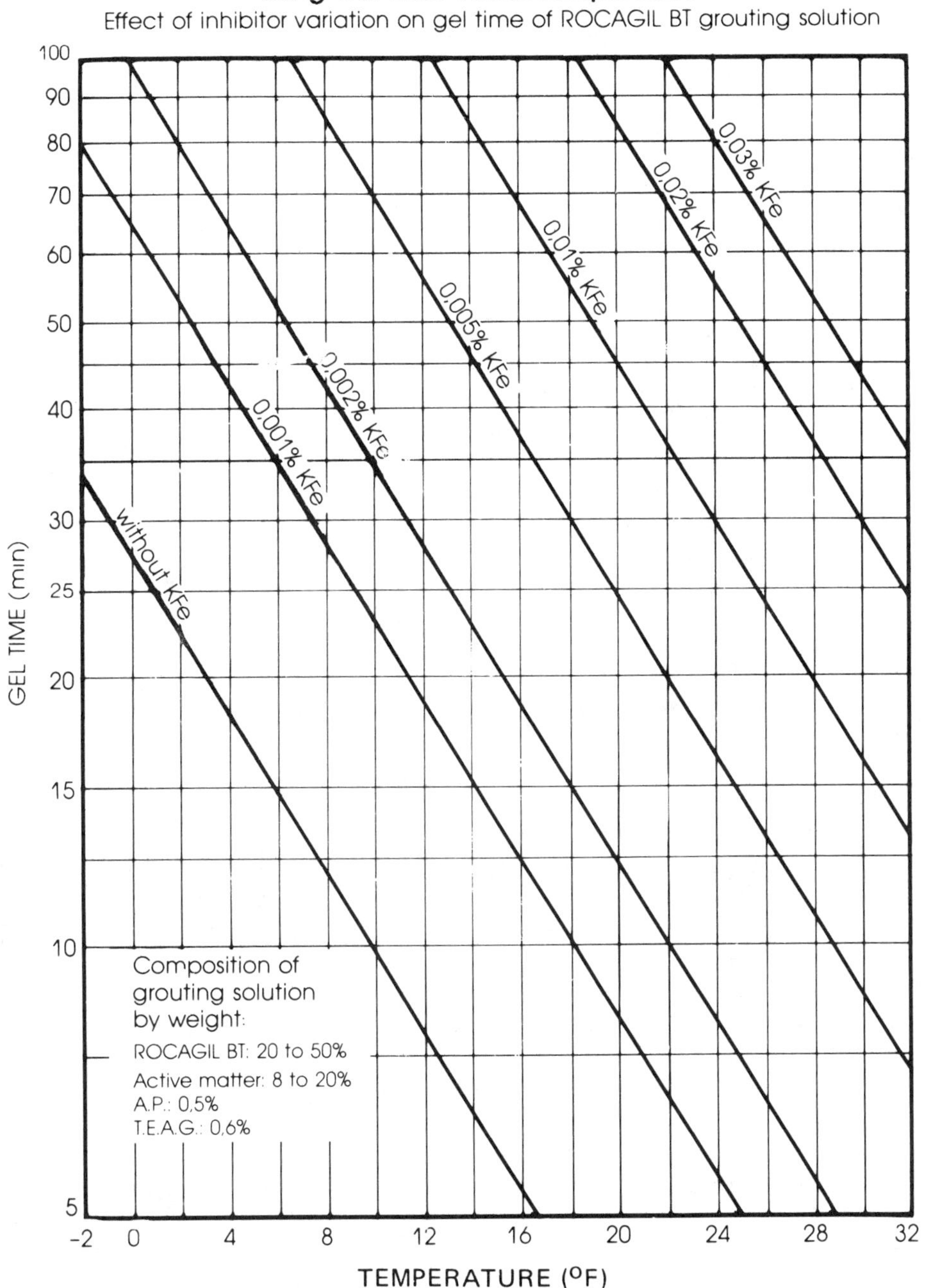

FIGURE 4.24 Gel chart for Rocagil BT. (Courtesy of Rhodia, Inc., Monmouth Junction, New Jersey.)

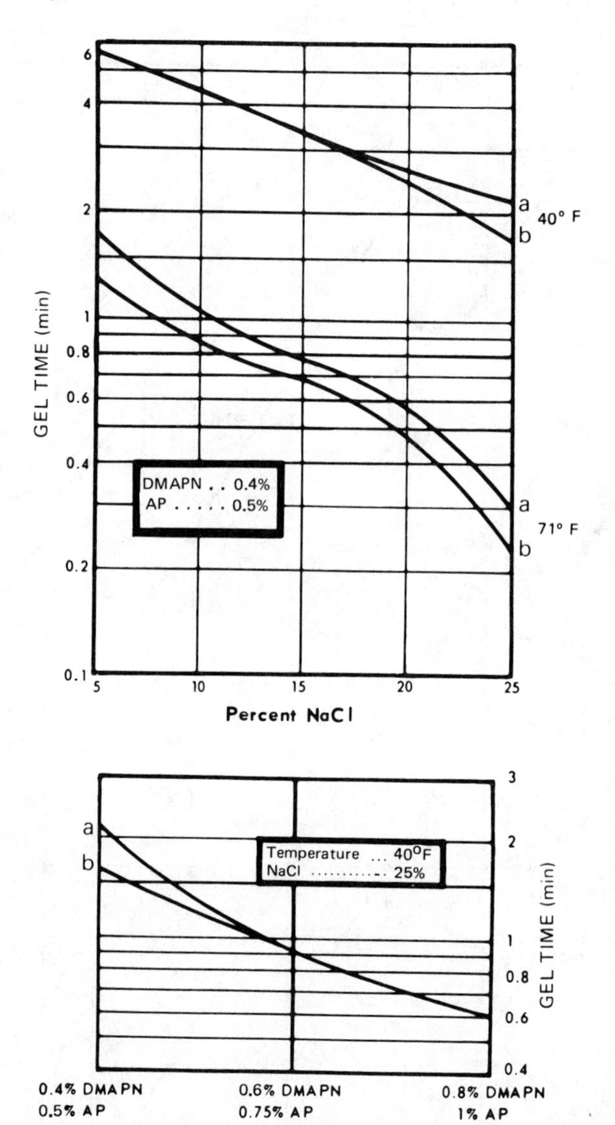

FIGURE 4.25 Effect of NaCl on gel time of acrylamide grout. (a) 6% acrylamide-based grout, (b) 10% acrylamide-based grout.

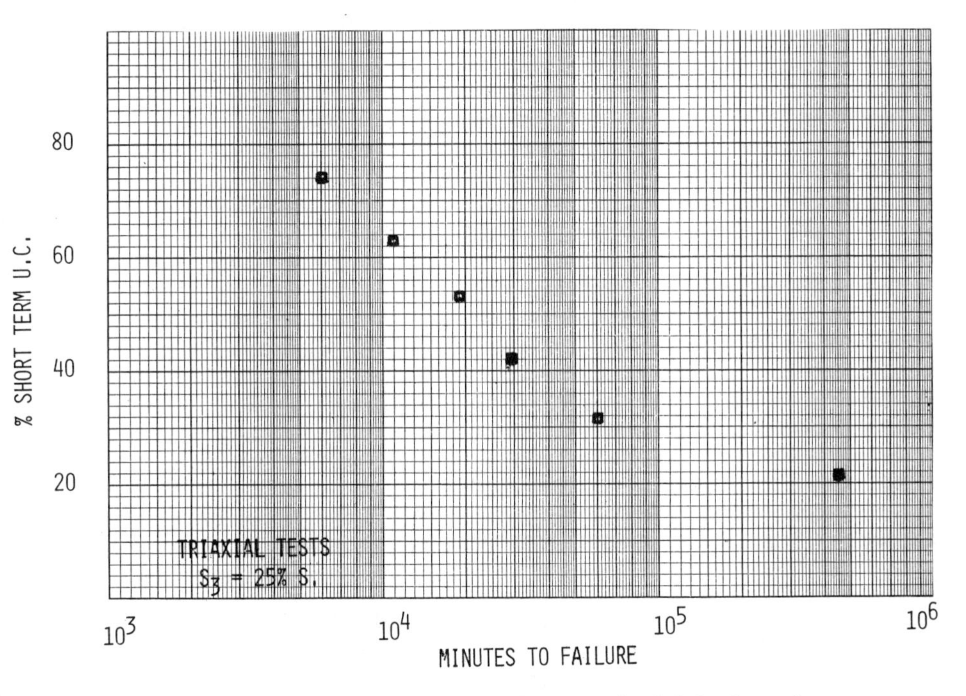

FIGURE 4.26 Acrylamide creep studies. Triaxial tests: S_3 = 25% S.

Wet–dry and freeze–thaw tests performed under laboratory conditions yield data difficult to extrapolate for field use. This is because laboratory tests represent the worst conditions (also those most readily controlled) such as complete drying followed by complete saturation. Under field conditions, it is more probable that both wet–dry and freeze–thaw conditions are partial rather than complete. This consideration must be taken into account when interpreting manufacturers data such as shown in Tables 4.3 and 4.4, which compare acrylamide with acrylate.

Acrylamide gels, like many other grouts, are subject to creep. In the field, this can lead to failure of unsupported grouted soil masses, within hours or days after removal of support (for example, removing breastboards from a grouted tunnel face.)

Creep tests are run by subjecting a grouted soil sample to a sustained load less than the short-term Unconfined Compression strength, until failure occurs. As the sustained load decreases in value, the time to failure increases. Data from such tests plot as shown in Fig. 4.26, and indicate as asymptote at what may be

TABLE 4.3 Freeze–Thaw Cycling Tests Using Manufacturing Data Comparing Acrylamide with Acrylate[a]

	Acrylamide gel weight loss						AC-400 gel weight loss					
	Days						Days					
	7	14	21	28	35	Net (grams)	7	14	21	28	35	Net (grams)
Sample												
1	−0.1	−0.3	−0.3	0	+0.1	−0.6	0	+0.2	−0.2	−0.3	−0.1	−0.4
2		0	−0.1	0	0	−0.1		−0.8	−0.2	−0.1	−0.2	−1.3
3			0	−0.1	−0.2	0.3			−0.7	−0.7	−0.0	−1.4
4				−0.1	−0.4	−0.5				−0.9	−0.1	−1.0
5					−0.2	−0.2					−0.3	−0.3

[a]Five 100 gram samples of 10% acrylamide gel and 10% AC-400 gel were cycled from 20°C to 10°C in 7-day periods through 5 cycles.

TABLE 4.4 Wet–Dry Cycling Tests Using Manufacturing Data Comparing Acrylamide with Acrylate[a]

	60% relative humidity							100% relative humidity						
	Day							Day						
	7	14	21	28	35	42	Net	7	14	21	28	35	42	Net
Cycle	Wet	Dry	Wet	Dry	Wet	Dry		Wet	Dry	Wet	Dry	Wet	Dry	
Acrylamide	+6.2	1.2	+1.1	−0.7	+2.6	−1.3	+6.7	+2.6	−0.4	+0.5	−0.3	+1.4	−0.2	+3.6
AC-400	+4.0	−1.2	+1.1	−0.7	+1.3	−0.7	+3.8	+2.7	−0.4	+1.0	−0.5	+2.4	−0.2	+5.0

[a]Sets of four 100 gram samples of 10 percent acrylamide gel and 10 percent AC-400 gel were cycled at 20°C through wet (water) and 60 percent and 100 percent relative humidity in 7-day cycles for 6 weeks with weight gain (+) or loss (−).

called a "creep endurance limit." For acrylamides these values will range from as low as 20% Unconfined Compression for triaxial tests at low lateral pressure to as high as 40% for triaxial tests at "at-rest" lateral pressures. (For silicates, values may be taken from Fig. 4.15.)

Acrylate grouts appeared on the domestic market in the early 1980s, in response to an industrial need for a less toxic substitute for acrylamide. In virtually all areas of behavior acrylates are similar to acrylamides. They are not quite as strong, a little higher in viscosity, and gel-time control is not as good. However, they are far less toxic (LD_{50} = 1800), not neurotoxic, and with no known carcinogenic problems. This latter characteristic is fostering the use of acrylates to replace acrylamide in many applications.

Acrylates are not new as soil treatment agents. As early as the mid 1950s, calcium acrylate was researched by the United States Army Corps of Engineers [20], and detailed data on acrylate was available in chemical publications [21].

AC-400 was the first acrylate grout marketed, and Tables 4.3 and 4.4 are taken from literature of the manufacturer, Geochemical Corporation. Pertinent gel time properties of AC-400 are shown in Figs. 4.27 and 4.28, also taken from the manufacturer's literature.

This product was evaluated extensively for the distributor at Northwestern University (Technical Report No. HB-13, September 1981). Investigative work was also done at Rutgers University ("Effects of Freeze–Thaw Cycles—student term project, December 1986).

Viscosity of the acrylate grouts as used in the field is for most of the induction period, as shown in Fig. 4.29.

Typical short term strength data for AC-400 is shown in Fig. 4.30. Long-term tests (tests on samples stored for various periods at no load) indicate that grouts gain their full strength immediately upon gelation, and retain that strength (as long as freeze–thaw and wet–dry conditions do not exist). Creep endurance limit is about 30% (see Fig. 4.31). The Northwestern report concludes that AC-400 is "excellently suited for water sealing operations, but its usefulness in structural grouting processes is limited."

The actual composition of the various acrylate grouts is considered proprietary data by the manufacturers (patent applications must, of course, reveal that information). It is believed that all of the grouts are mixtures of acrylate salts, selected for an optimum combination of low cost and solubility and water absorption characteristics. All require a cross-linking agent. For AC-400 this is shown in the manufacturer's literature to be methylene-bis-acrylamide, and is the source of the small toxicity hazard associated with the acrylates. All acrylate grouts use the same redox catalyst system

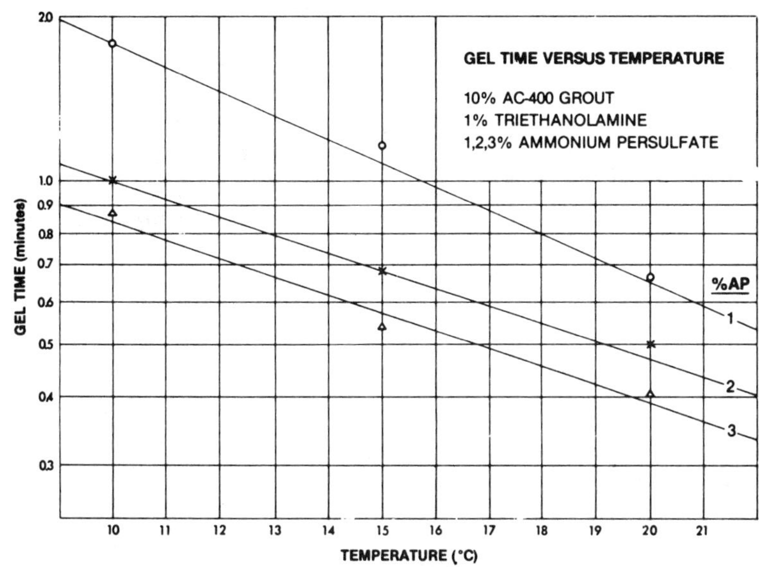

FIG. 2 GEL TIME VERSUS TEMPERATURE FOR 10% AC-400 GROUT

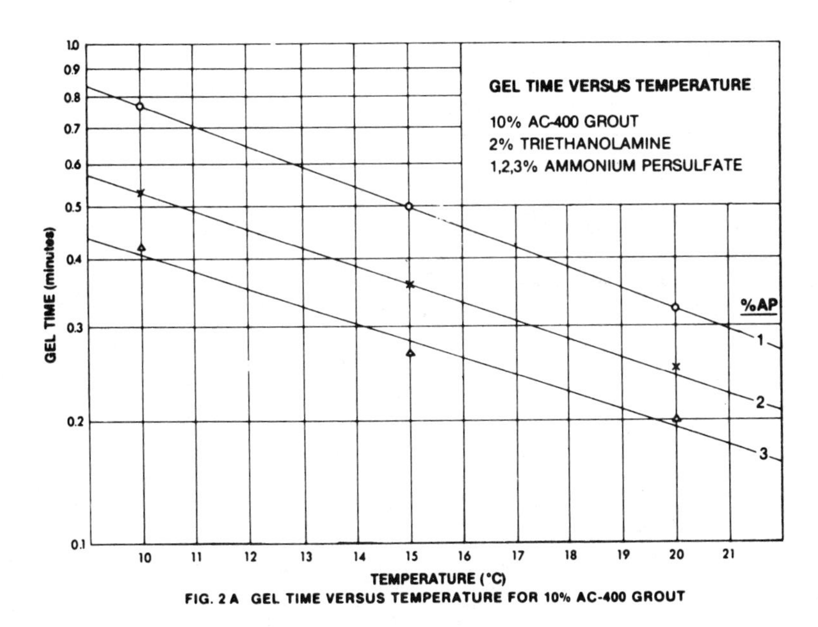

FIG. 2 A GEL TIME VERSUS TEMPERATURE FOR 10% AC-400 GROUT

FIGURE 4.27 Gel time versus temperature. (Courtesy of Geochemical Corp., Ridgewood, New Jersey.)

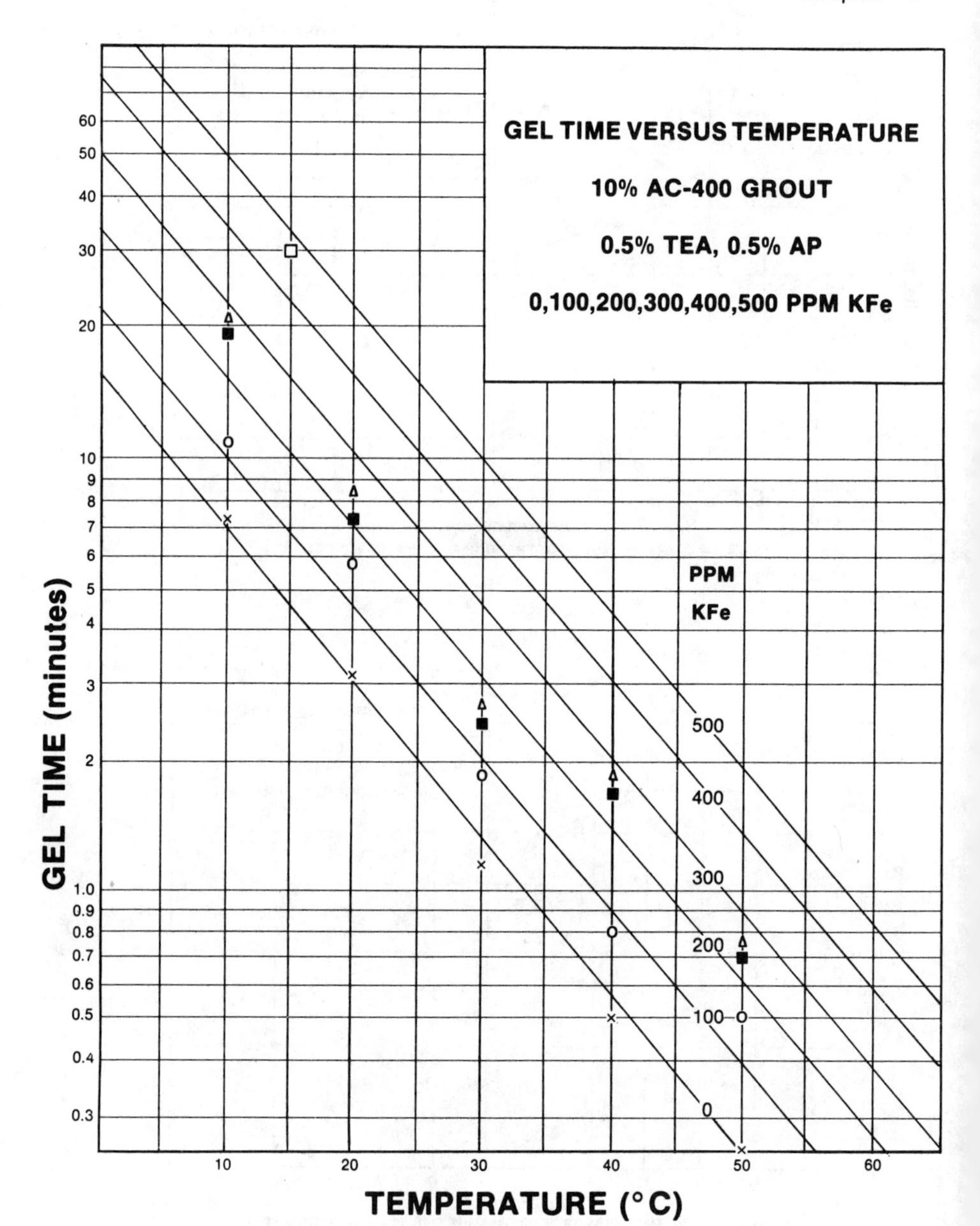

FIGURE 4.28 Gel time versus temperature. (Courtesy of Geochemical Corp., Ridgewood, New Jersey.)

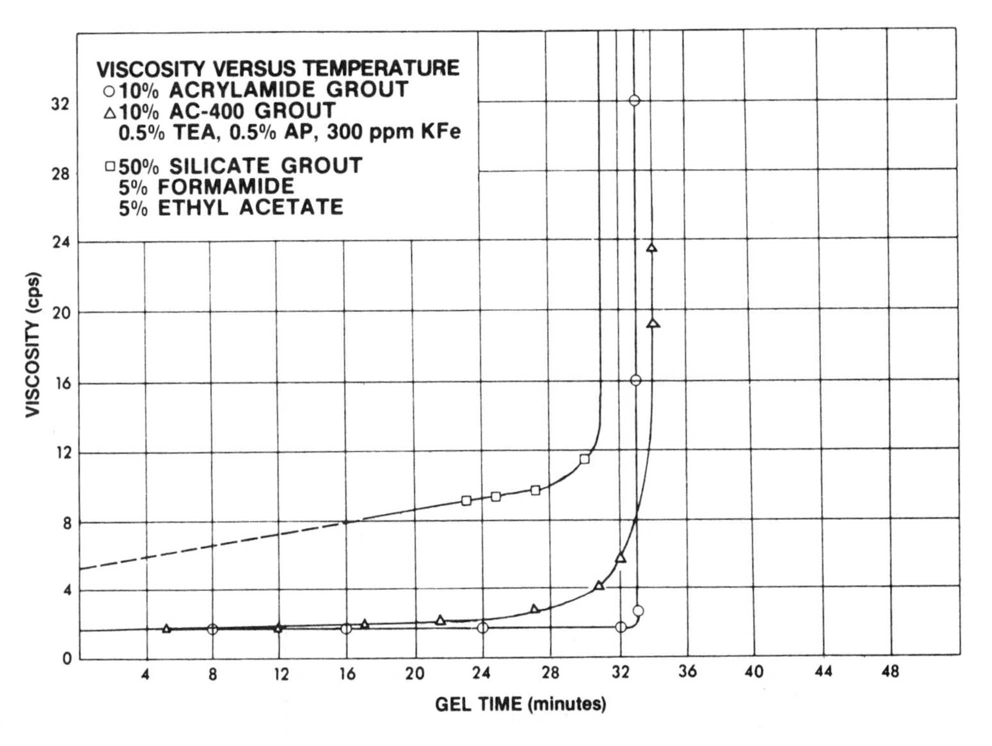

FIGURE 4.29 Gel time versus viscosity for chemical grouts. (Courtesy of Geochemical Corp., Ridgewood, New Jersey.)

(triethanolamine and ammonium persulfate), and the same inhibitor (potassium ferricyanide).

Terragel 55-31 is described by the manufacturer (Celite, Inc., Cleveland, Ohio) as "a blend of liquid acrylate monomers selected for minimum toxicity." Tests at Rutgers University ("Compression Testing for Terragel Grout Cylinders"—student term project, 1985) gave the data shown in Fig. 4.32, indicating that the short-term strength falls off with time to a much lesser value.

AV-110 FlexiGel and AV-120 DuriGel are described by their manufacturer (Avanti International, Webster, Texas) as "water solutions at acrylic resins." These products use sodium persulfate instead of ammonium persulfate. They are mixed with 1 to 3 parts of water for field use and (according to the manufacturer) have viscosities under 2 cp and strengths (stabilized sand) of 100 psi

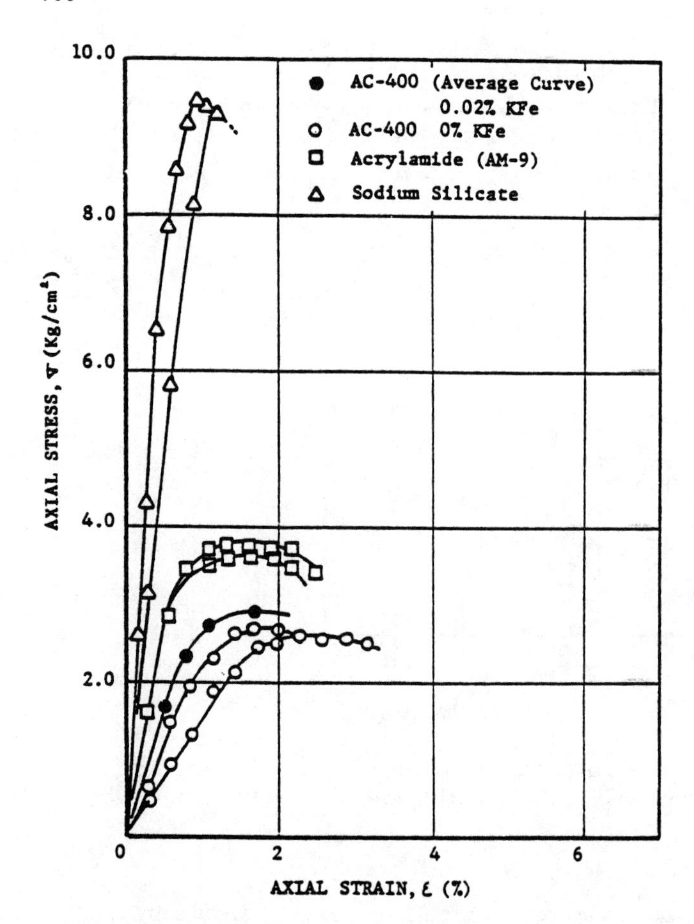

FIGURE 4.30 Stress–strain relationship of AC-400, acrylamide and sodium silicate grouts in unconfined compression. (From Northwestern University, Technical Report HB-13, September 1981.)

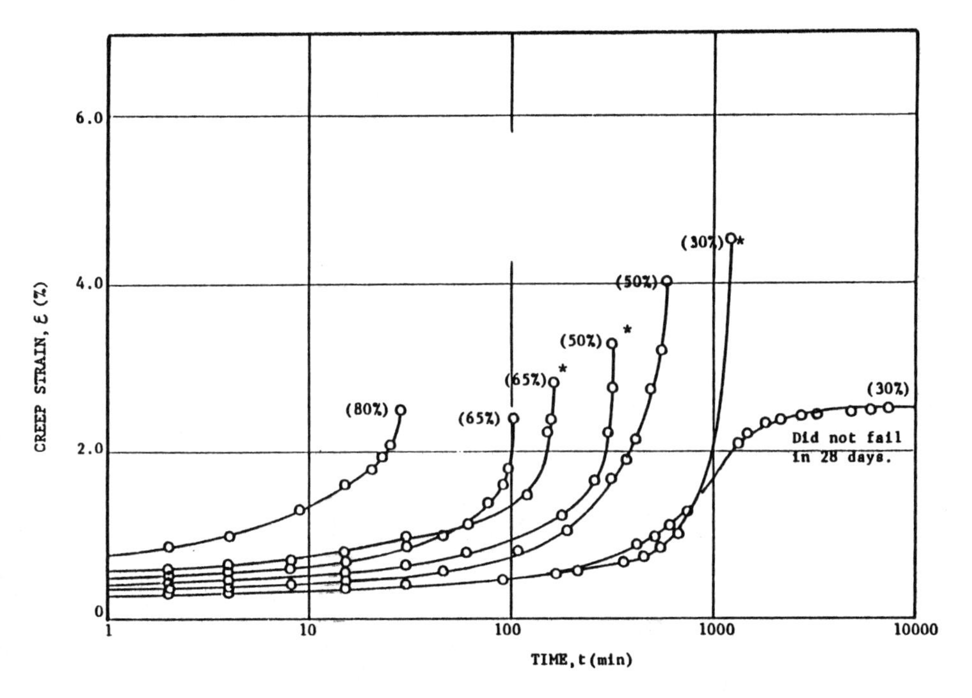

FIGURE 4.31 AC-400 time dependent, stress–strain behavior of 7-day cured grouted sand specimens. Key: * = grease added, (o) = stress level. (From Northwestern University, Technical Report HB-13, September 1981.)

and higher. Some of the formulations may swell as much as 200% in the presence of water, but swelling pressures are low.

Gel time control is shown in Fig. 4.33 (see pages 111–112).

The LD_{50} is 1800.

Lignosulfonate Grouts

Lignosulfonates are made from the waste liquor by-products of the wood processing industries (such as paper mills). Since the grout may not be controlled, it can be widely variable in content, depending on the specific source of trees and the particular wood processing. Not only will the liquor vary from country to country

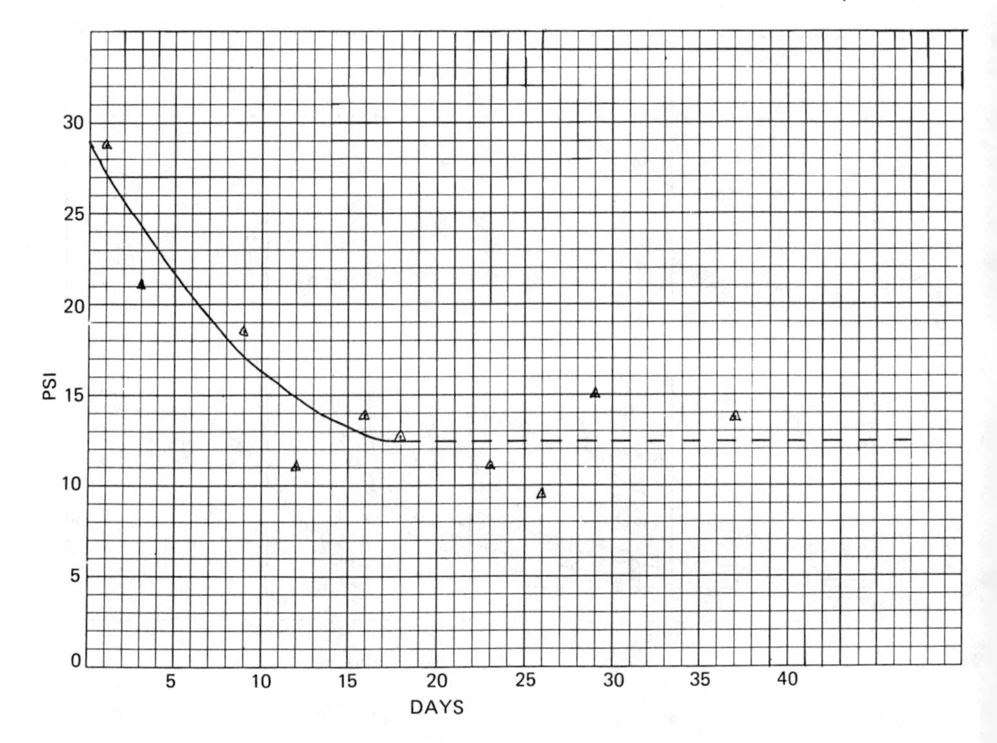

FIGURE 4.32 Aging tests of Terragel 55-31. (Unpublished report, Rutgers University, 1984.)

but also with different mills closely located to each other and even with the same mill at different times of the year. Lignosulfonate solutions used for grouting purposes range from the raw liquor trucked directly from the mill to the job site to dried, precatalyzed powders with other additives to assist in gel quality and quality control.

Reference 5 points out the complexity of the chemistry related to the "chrome lignin" grouts as shown by the excerpt.

> The chemistry lignin is very complex, and there are numerous works on this topic. Many formulas have been developed and suggested over the last twenty years, but all authors seem to agree that lignins are derived from basic units belonging to the (C_6H_5CCC) family, that is to say, they possess a benzine nucleus with a lateral chain of 3 atoms.
>
> According to KLASEN, these units might be linked to coniferylic alcohol:

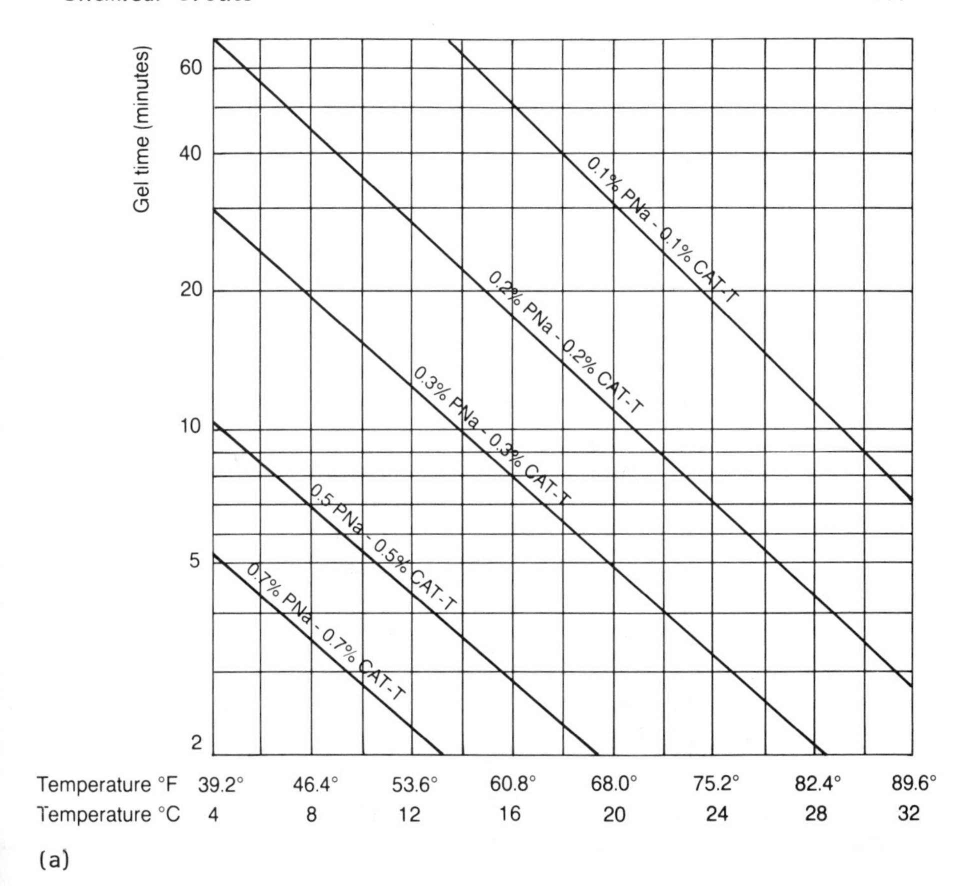

(a)

FIGURE 4.33 Gel time relationship for AV-110. (Courtesy of Avanti International, Webster, Texas.)

OCH_3

HO— [benzene ring] —CH = CH - CH_2OH

According to FRUDENBERG, the basic structure might be one of the following:

OCH_3

HO— [benzene ring] —C(OH)(H) – C(OH)(H) – C(OH)(H) – H

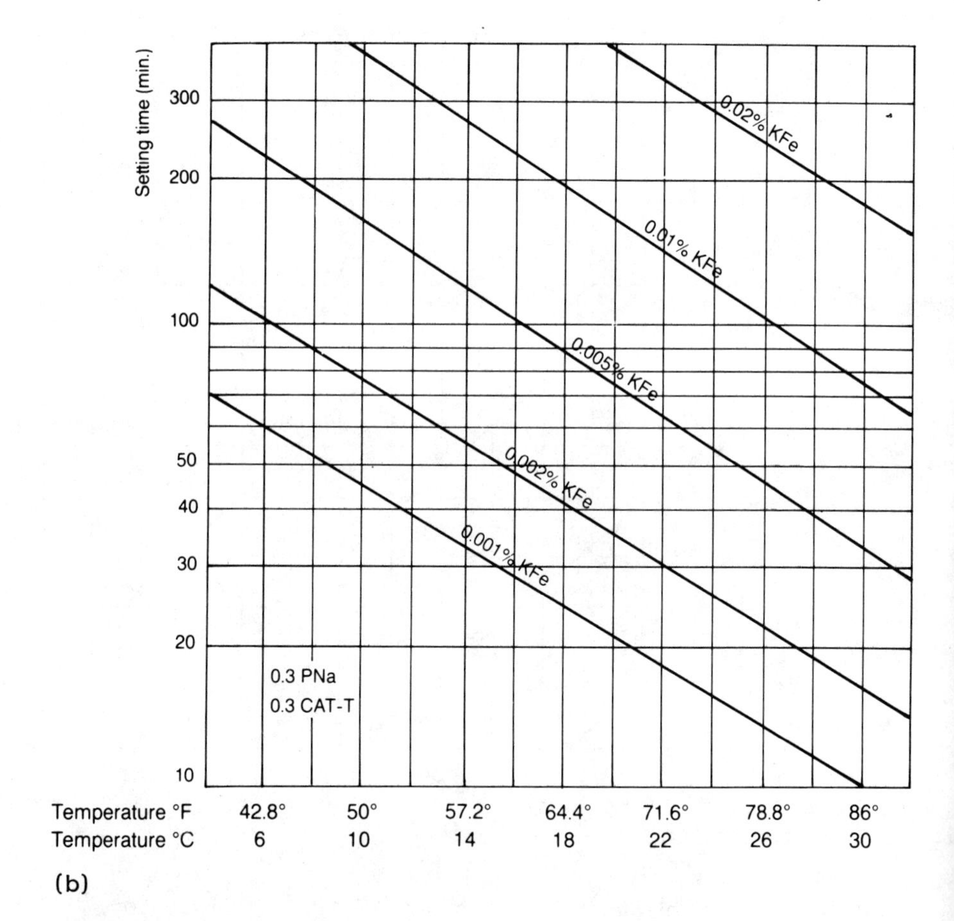

(b)

FIGURE 4.33 (Continued)

$$HO-C_6H_3(OCH_3)-CH(OH)-CH_2-CH{=}O$$

$$HO-C_6H_3(OCH_3)-CH(OH)-C({=}O)-CH_3$$

Thus, the following general formula can be accepted:

OCH_3

HO —(benzene ring)— C — C — C —

The molecular mass would be somewhere between 2000 and 100,000.

The point of fixation for the HSO_3 at the time of sulfonation of the lignin by the bisulfite solution is also poorly defined.

According to K. KRATZ and E. RISNYOVSKY, it is attached to the lateral chain, on one of the 3 carbons, for example:

OCH_3 SO_3H OCH_3 SO_3Me

OH —(ring)— C - C - C - + Me ⟶ HO —(ring)— C - C -

lignosulfonic acid lignosulfonate

Whereas, according to BJORKMAN, it is attached, at least partially, in an alpha position on the benzene nucleus.

OCH_3 OCH_3

OH —(ring)— C - C - C - + Me ⟶ HO-(ring)— C - C - C -

SO_3H SO_3Me

lignosulfonic acid lignosulfonate

Me[1] being Na^+, NH_4^+, $1/2\ Ca^{++}$ or $1/2\ Mg^{++}$.

Lignosulfonate grouts always consist of lignosulfonates and a hexavalent chromium compound—hence the name *chrome-lignins*. Most generally, calcium lignosulfonate is used with sodium dichromate. Sodium, magnesium, and ammonium lignosulfonates are also commercially available. Of these, the sodium compound is considered unstable. In an acid environment, the chromium ion changes its valence from plus 6 to plus 3, oxidizing the lignosulfonate to produce a gel. If the lignosulfonate is of itself sufficiently acid (pH of 6 or below), no other additives are needed. If the pH is above 6, acids or acid salts are generally added to control pH. The effects of pH on setting time, all other factors held constant, is shown in Fig. 4.34 [5].

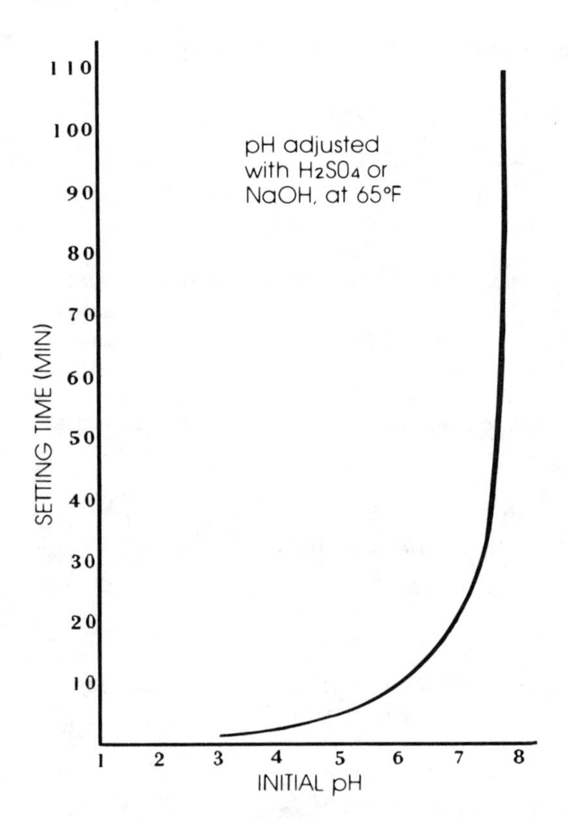

FIGURE 4.34 Effect of pH on setting time of lignosulfonate grouts (pH controlled with H_2SO_4 and NaOH).

As noted before, the composition of the product is variable, but in a unit weight of solids approximately three-quarters will be lignosulfonates, one-fifth reducing sugars, and the rest ash. Concentrations of solids for field use vary between 200 and 600 g/liter. Initial solution viscosities for these concentrations vary as shown in Fig. 4.35, and the relationship between temperature and viscosity is shown in Fig. 4.36.

The viscosity of any specific grout mix increases at an increasing rate from the instant of catalysis to the formation of a gel. Typical of this relationship is the diagram shown in Fig. 4.37 for a Halliburon Co. product.

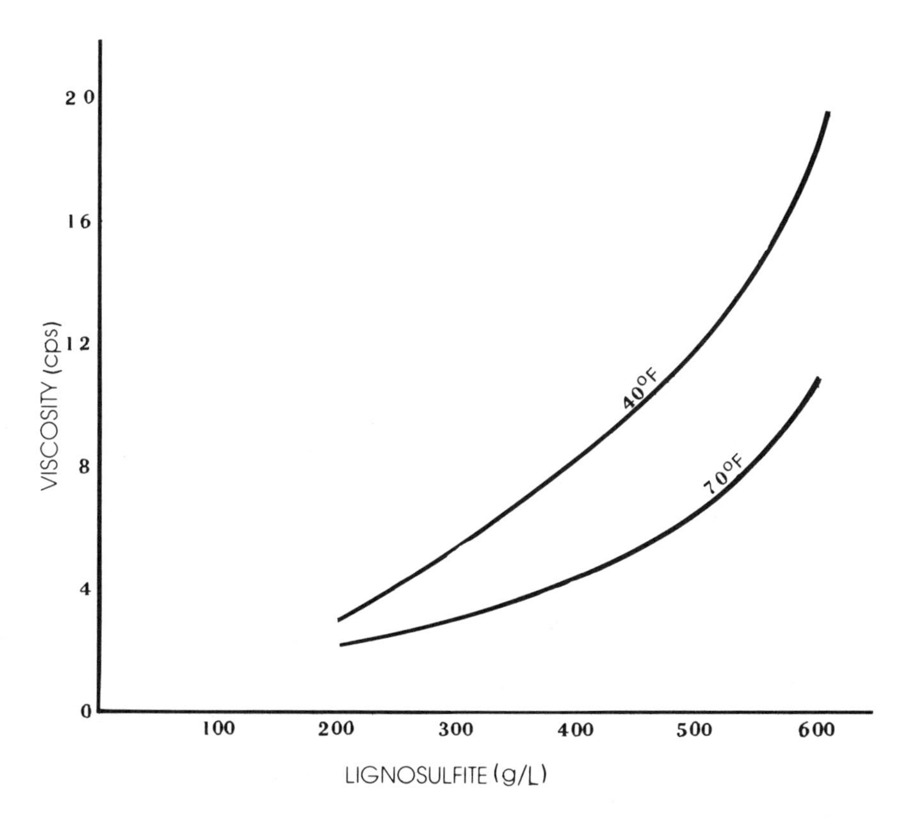

FIGURE 4.35 Viscosity versus solids concentration for lignosulfonate grouts.

The setting time of a catalyzed grout solution varies with the concentration of solids, decreasing as the percent solids increase. The relationship is shown in Fig. 4.38 [10]. Setting times also vary with the dichromate content, decreasing as the dichromate increases. This is illustrated in Fig. 4.39 [5].

The strength of soils stabilized with lignosulfonates is of the same order of magnitude as the acrylate grouts. In common with other grouting materials, the lignosulfonates show higher strengths in finer materials and are subject to creep and consolidation phenomena. The creep endurance limit is of the order of one-fourth to one-half of the UC value, and that value itself is responsive to

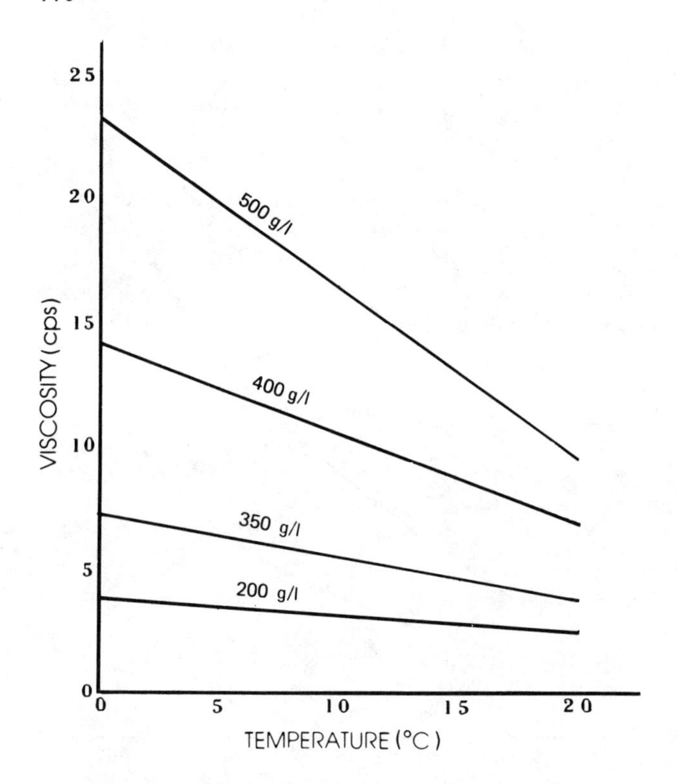

FIGURE 4.36 Relationship between temperature and viscosity for lignosulfonate solutions.

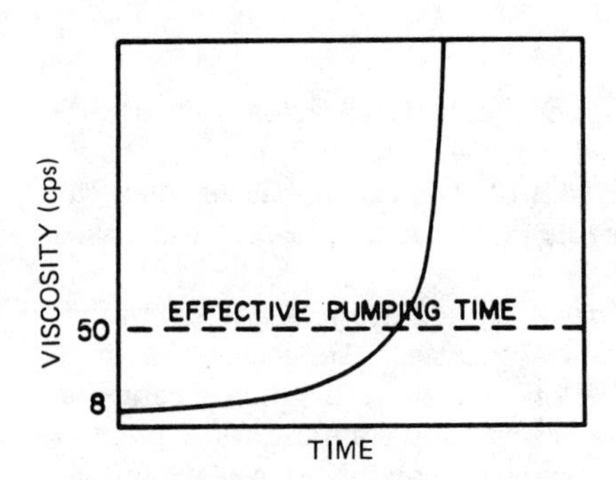

FIGURE 4.37 Viscosity versus setting time for Blox-All (a chrome-lignin grout used by Halliburton). (Courtesy of Halliburton Services, Duncan, Oklahoma.)

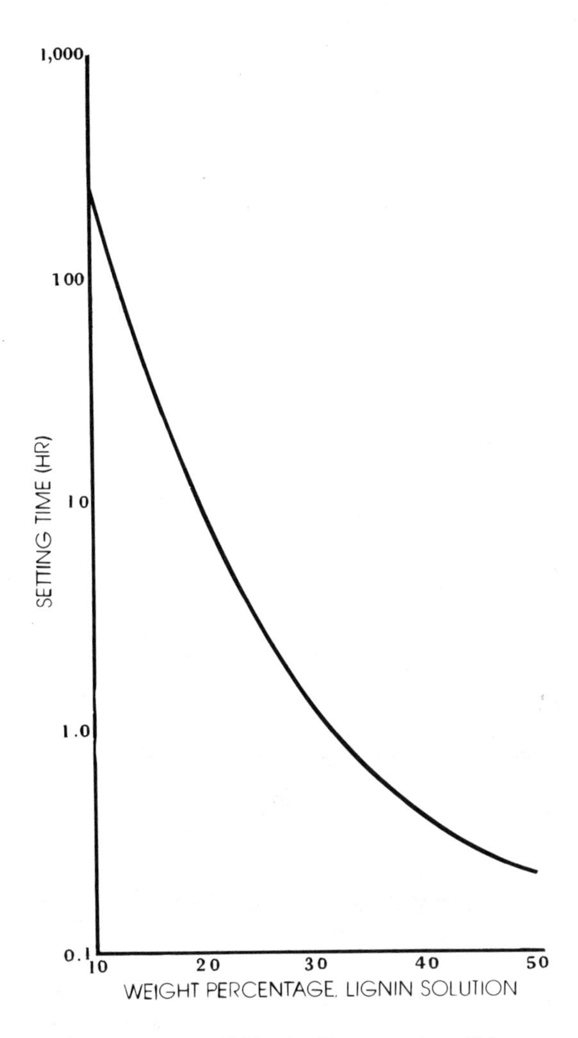

FIGURE 4.38 Effect of percent solids on setting time for lignosulfonate grout.

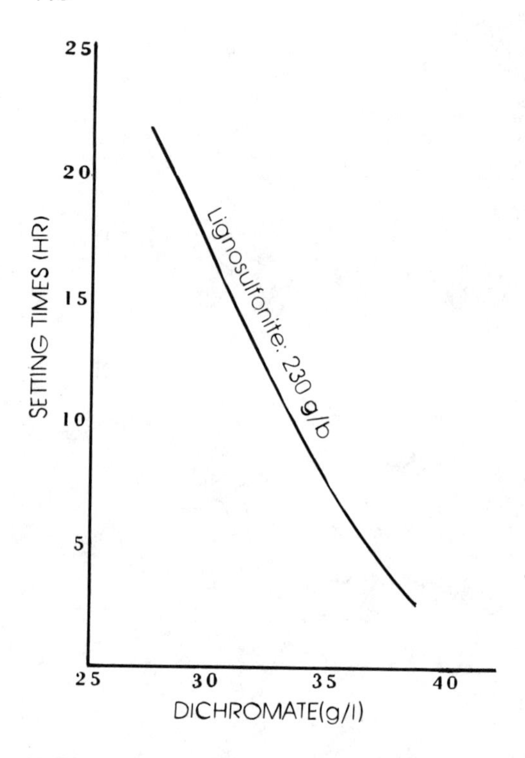

FIGURE 4.39 Effect of dichromate content on lignosulfonite setting time.

strain rate. As with other grouts, UC values should be used for comparison, not design.

The strength of lignosulfonate gels increases dramatically (tripling or quadrupling) as the solids content varies from 300 to 500 g/liter. The strength also increases with increasing dichromate content, although this may be partially due to shorter gel times (which will also generally increase the gel strength). A dramatic increase in gel strength is obtained by lowering the pH, which promotes quicker and more complete chemical reactions. This is shown in Fig. 4.40 [5]. All these effects are carried over in subdued form to stabilize soil samples.

Gelled soils under the water table which are not subject to alternate wet–dry or freeze–thaw cycles have good stability. Under these conditions, the lignosulfonates are considered permanent materials. However, they respond very quickly to wet–dry

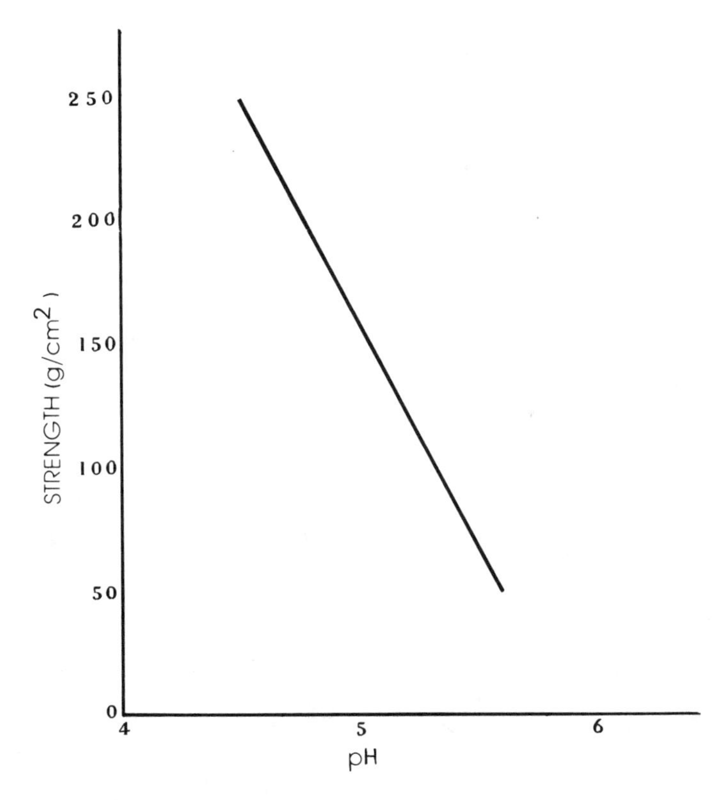

FIGURE 4.40 Relationship between pH and strength for lignosulfonate gels.

and freeze–dry conditions (Fig. 4.41) and should not be used under these conditions for permanent installations.

In the early 1960s a product Terra Firma (Intrusion–Prepakt, Cleveland) was marketed in the United States and abroad. The patent covering this product [11] defines a calcium lignin sulfonate with reduced sugar content, sodium dichromate and aluminum sulfate catalysts, and copper sulfate and calcium chloride accelerators. The product was marketed as a dry, precatalyzed powder. It was mixed with water in varying proportions for field use. Although Terra Firma has disappeared from the marketplace, it is still of interest because the data taken from advertising literature show very concisely the typical properties of lignosulfonate grouts (Fig. 4.42).

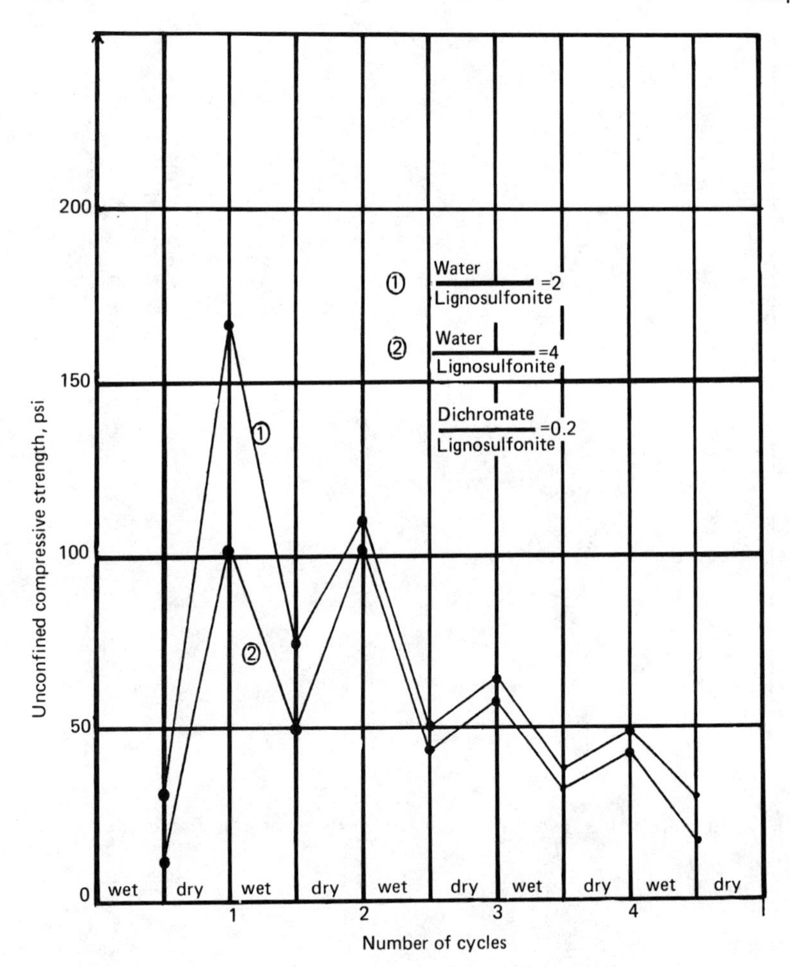

FIGURE 4.41 Effect of drying/wetting cycles on strength of lignochrome gel. (From Ref. 5.)

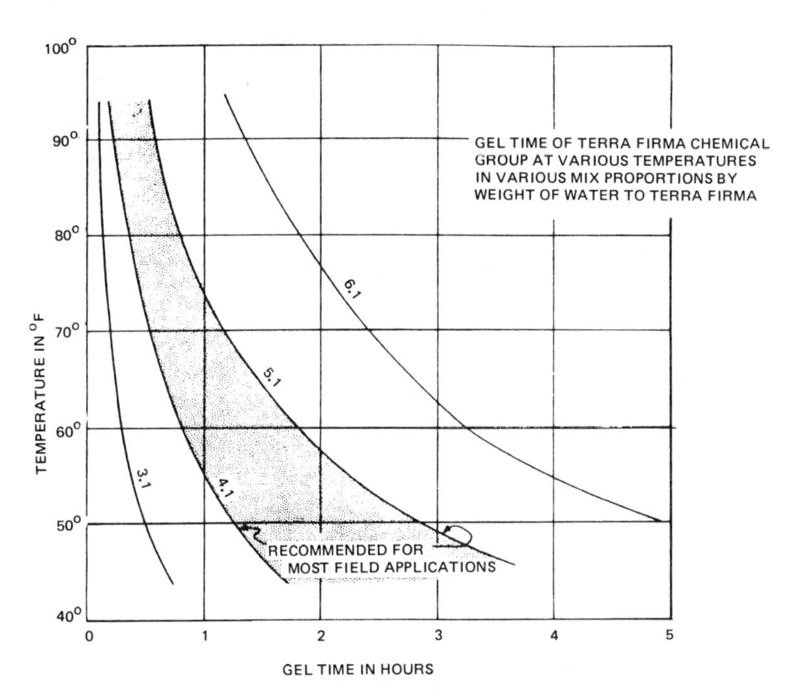

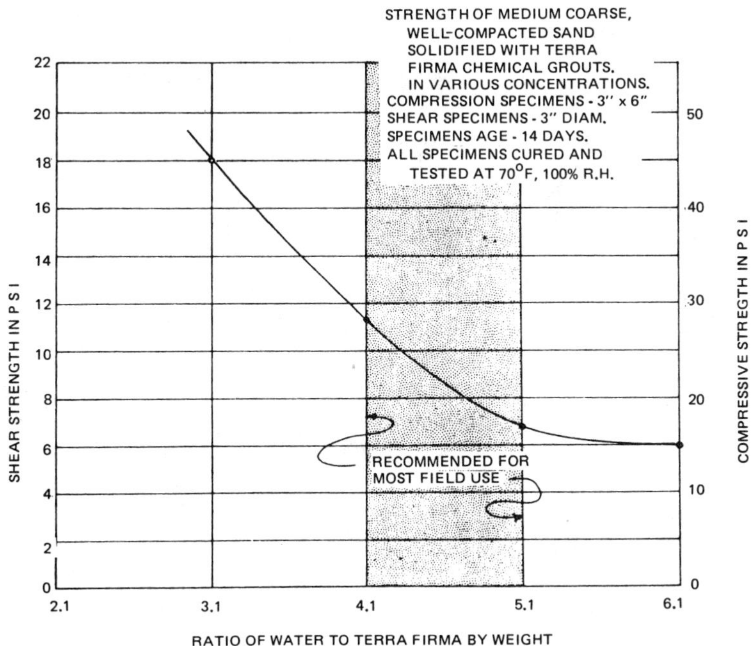

(a)

FIGURE 4.42 Properties of Terra Firma. (Courtesy of Intrusion-Prepakt, Cleveland, Ohio.)

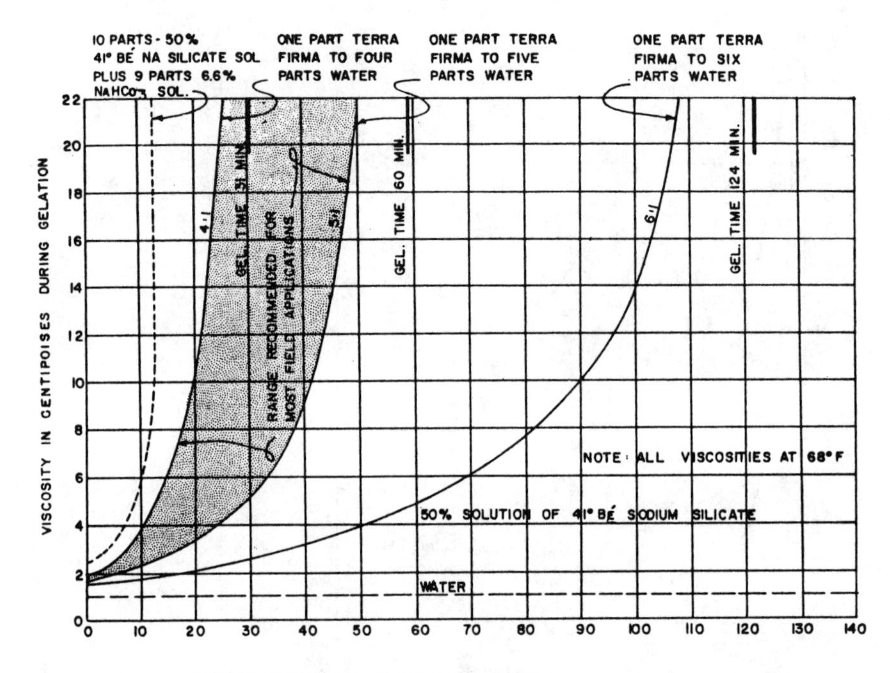
10 PARTS - 50% 41° BÉ NA SILICATE SOL PLUS 9 PARTS 6.6% NaHCO3 SOL.
ONE PART TERRA FIRMA TO FOUR PARTS WATER
ONE PART TERRA FIRMA TO FIVE PARTS WATER
ONE PART TERRA FIRMA TO SIX PARTS WATER
GEL. TIME 31 MIN.
GEL. TIME 60 MIN.
GEL. TIME 124 MIN.
RANGE RECOMMENDED FOR MOST FIELD APPLICATIONS
4:1
5:1
6:1
NOTE: ALL VISCOSITIES AT 68°F
50% SOLUTION OF 41° BÉ SODIUM SILICATE
WATER
VISCOSITY IN CENTIPOISES DURING GELATION
ELAPSED TIME AFTER SOLUTION IN MINUTES

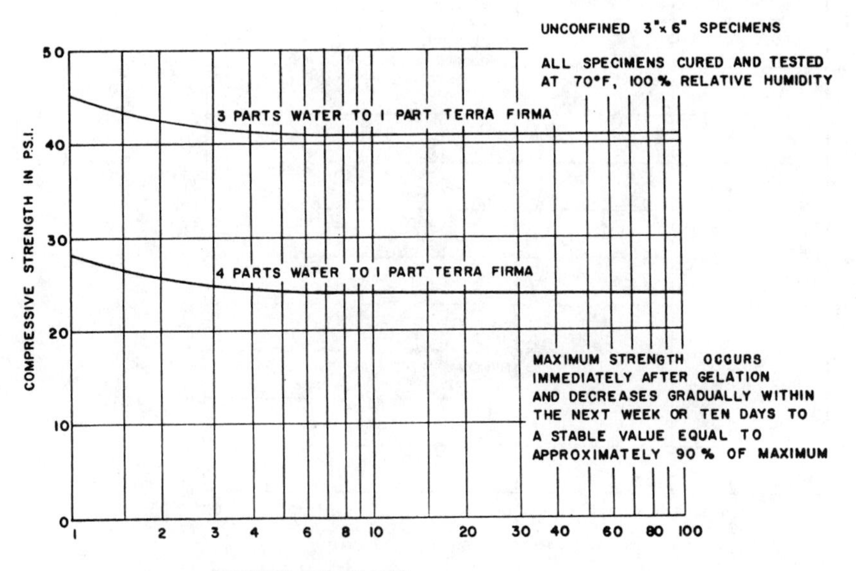
COMPRESSIVE STRENGTH OF WELL COMPACTED MEDIUM COARSE SAND SOLIDIFIED WITH TERRA FIRMA CHEMICAL GROUTS IN VARIOUS CONCENTRATIONS.
UNCONFINED 3" x 6" SPECIMENS
ALL SPECIMENS CURED AND TESTED AT 70°F, 100% RELATIVE HUMIDITY
3 PARTS WATER TO 1 PART TERRA FIRMA
4 PARTS WATER TO 1 PART TERRA FIRMA
MAXIMUM STRENGTH OCCURS IMMEDIATELY AFTER GELATION AND DECREASES GRADUALLY WITHIN THE NEXT WEEK OR TEN DAYS TO A STABLE VALUE EQUAL TO APPROXIMATELY 90% OF MAXIMUM
COMPRESSIVE STRENGTH IN P.S.I.
SPECIMEN AGE IN DAYS

(b)

Since lignosulfonates are by-products of other processes, they are relatively inexpensive and can compete on a cost basis with any other grouting material. The major drawback lies in the toxicity of the dichromate salt, as well as the health hazards of the benzene-type molecule present in all lignosulfonate grouts. Hexavalent chromium is highly toxic. (The U.S. Public Health Service has established 0.05 ppm as the permissible limit in drinking water.) It is reduced to the nontoxic trivalent form during the reaction, but the reduction is not necessarily complete, particularly at the higher range of pH and when using long gel times. Thus, gels which should be innocuous (laboratory research, with careful measurement and proportioning of ingredients, has produced lignosulfonate gels with virtually no free hexavalent chromium) may leach toxic materials into the surrounding environment. This factor is one of those contributing to the phasing out of the product Terra Firma. No lignosulfonate grouts are currently in domestic use.

Phenoplast Grouts

Phenoplast resins are polycondensates resulting from the reaction of a phenol on an aldehyde. They set under heat over a wide pH range at elevated temperatures. Such products have been used for many years in oil well drilling, since the required temperatures are provided by the geothermal gradient in deep holes. At ambient temperature, the reaction for most phenols requires an acid medium. Most soils are neutral or slightly basic, and a grout requiring acid medium. Most soils are neutral or slightly basis, and a grout requiring acid conditions is undesirable.

There are several materials which will react at ambient temperatures, without requiring an acid medium. One of these (the only one generally used for grouting) is resorcinol, and it is most commonly reacted with formaldehyde. A catalyst is required, whose main function is to control pH. Sodium hydroxide is commonly used.

The theoretical proportion for complete reaction is two molecules of resorcinol with three molecules of formaldehyde. Optimum mechanical properties are obtained with that ratio for any given amount of dilution. The only control of setting time is the dilution of the grout components. Typically, the relation between dilution and setting time is shown in Fig. 4.43 [4]. Setting time varies greatly with solution pH, being shortest for any given grout concentration at a pH slightly above 9. Since mechanical properties of the gel are optimum at the shortest setting time for any given grout concentration, this fixes the amount of catalyst. Since the catalyst functions primarily as a pH control, many different soluble materials including hydroxides, carbonates, and phosphates can be used. They will

FIGURE 4.43 Resorcinol–formaldehyde—dilution versus setting time.

give different setting times, equal to or longer than those obtained with sodium hydroxide. Thus, we have a grout whose three ingredients always have fixed proportions to each other, and the only field variable is the dilution water. Setting time is also affected by temperature, approximately doubling for every 10°F drop.

The initial viscosity of resorcinol–formaldehyde grouts ranges from 1.5 to 3 cP for concentrations normally used for field work, and like the acrylate and acrylamide grouts, the viscosity remains constant at those low levels until gelation starts. Again, like the acrylics, the change from liquid to solid is almost instantaneous.

The strength of soils stabilized with resorcinol–formaldehyde grouts is directly proportional to the resin content, as shown in Fig. 4.44. Values are comparable to the high-concentration silicates. The phenoplasts in general appear to be less sensitive to rate of testing strain than other grouts, and their creep endurance

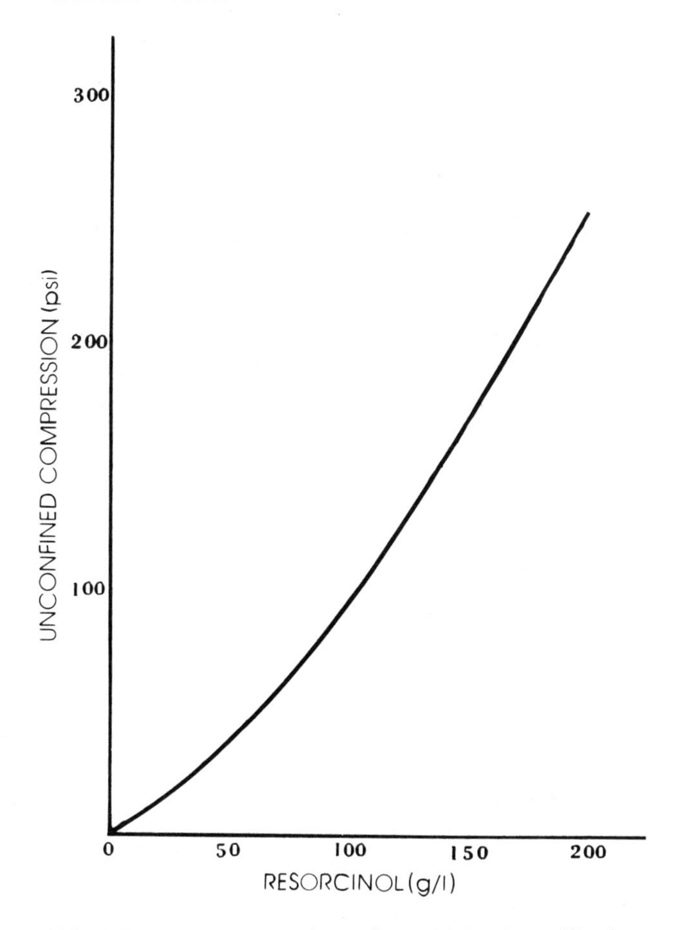

FIGURE 4.44 Resorcinol–formaldehyde—dilution versus strength.

limits appear to be a higher percentage of their UC values. Resistance to wet–dry cycles is poor and can lead to complete disintegration. (Although test data are not available, resistance to freeze–thaw cycles is probably poor also, particularly at low resin concentrations.) Except for grouts placed in a dehydrating environment, the phenoplasts are considered permanent materials.

Phenoplasts always contain a phenol, a formaldehyde, and an alkaline base. All three components are health hazards and potential environmental pollutants. Resorcinol is a phenol, and although not as hazardous as some phenols, it is still toxic and caustic. Formaldehyde is considered a dangerous material and at low atmospheric concentrations can lead to chronic respiratory ailments. Sodium hydroxide is, of course, well known as a caustic material.

Phenoplast gels, if proportioned properly, will not leave an unreacted excess of either of the major components. (Excess of either resorcinol or formaldehyde can free itself into groundwater or air and become a potential hazard.) Thus, only the catalyst could possibly leach out to cause environmental pollution. Gels from properly proportioned constituents are generally inert (i.e., nontoxic and noncaustic).

Because of the fixed relationship of ingredients, strong gels are always associated with short gel times and weak gels with long gel times. This relationship is shown in Fig. 4.45, combining data

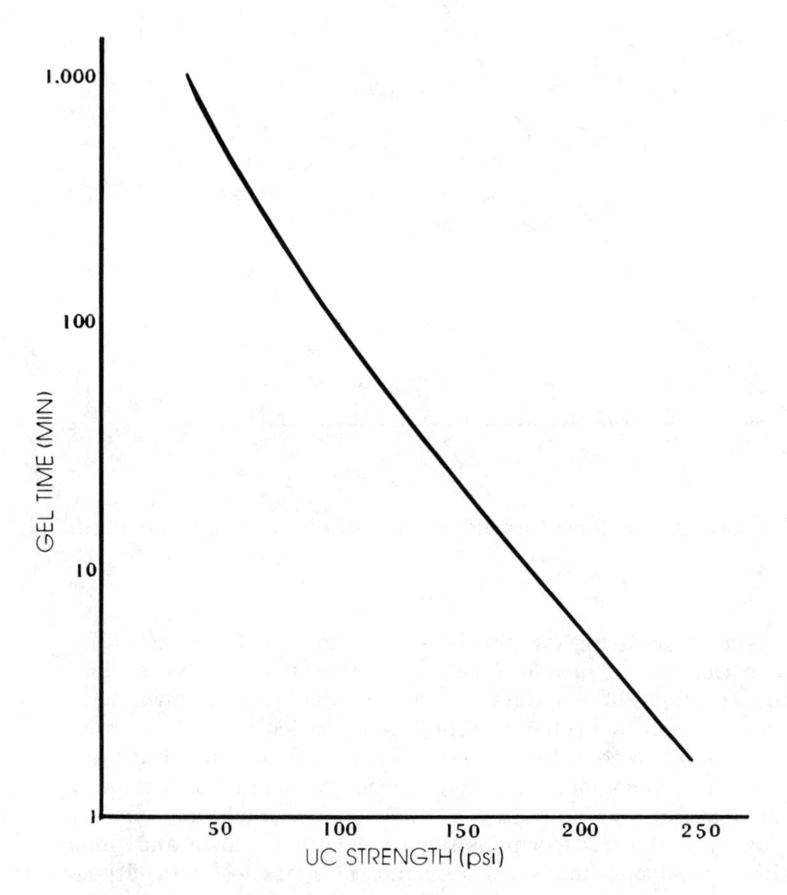

FIGURE 4.45 Resorcinol–formaldehyde—setting time versus strength.

from Figs. 4.43 and 4.44. This property of the grout is undesirable. Waterproofing operations, for example, do not need strong grouts but very often do need short gel times. One method of solving this problem is to combine the phenoplast system with another grout, such as a silicate or a chrome-lignin, and control the gel times with the latter materials. They act as vehicles to move the phenoplast into the formation and then hold it in place over the time it takes for gelation.

There is also another problem, that of obtaining a longer gel time (for example, half an hour) with a high grout concentration to give high strength. This problem cannot be solved by using another grout as a vehicle. However, it could be solved by the use of inhibitors, if they could be developed, and the long gel time problem could be solved by the use of accelerators. Chemical companies recognized the problem and in the mid 1960s began looking for ways to produce a commercial product with all the advantages of the resorcinol–formaldehyde system and without the disadvantages.

Several commercial products are now available. One of these, Rocagil 3555, is described as consisting of partially sulfonated tannin (a natural polyphenol), which is reacted with formaldehyde in an alkaline environment. This product is similar in action and properties to the resorcinol–formaldehydes, except for a higher initial viscosity (5 to 10 cP). Another product, Geoseal (Borden, Inc., Great Britain), is described in the patent [12] as consisting of a vegetable tannin extract which can be from mimosa; formaldehyde or paraformaldehyde; sodium carbonate, metaborate or peroborate; water-soluble inorganic salts of sodium, calcium, magnesium, or aluminum; hydroxyethyl cellulose (a thickening agent); and bentonite.

With all these possible additives the end product is quite different from the original phenol–formaldehyde concept. While Geoseal does succeed in providing gel time control without dilution of the phenol (see Fig. 4.46) and can be used at short gel times (as low as 10 s, according to the manufacturer's literature) the viscosity of the initial solution is dependent on concentration, varying from 2 to 10 cP for field use (resorcinol–formaldehyde solution viscosities are relatively independent of solids content); the viscosity increases continuously from catalyzation to gel; and the final gel is considerably weaker than the resorcinol–formaldehyde formation. Figure 4.47 shows the relationship between strength and concentration.

A third product, Terranier (Rayonier, Inc., Seattle), could be grouped with polyphenols or lignosulfonates, since it contains both a polyphenol base and a dichromate salt. However, Terranier has been phased out of the U.S. market, and its properties (which were very similar to those of the lignochromes) are not discussed separately.

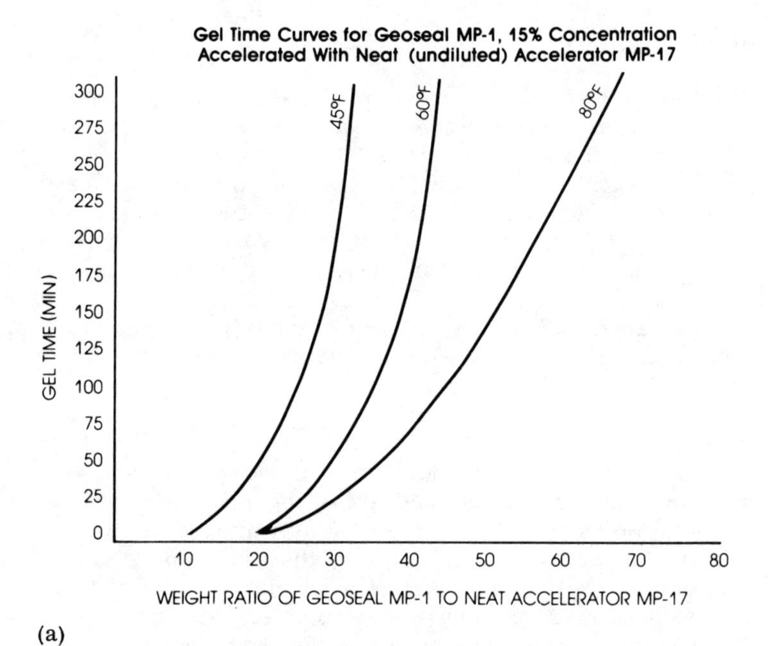

(a)

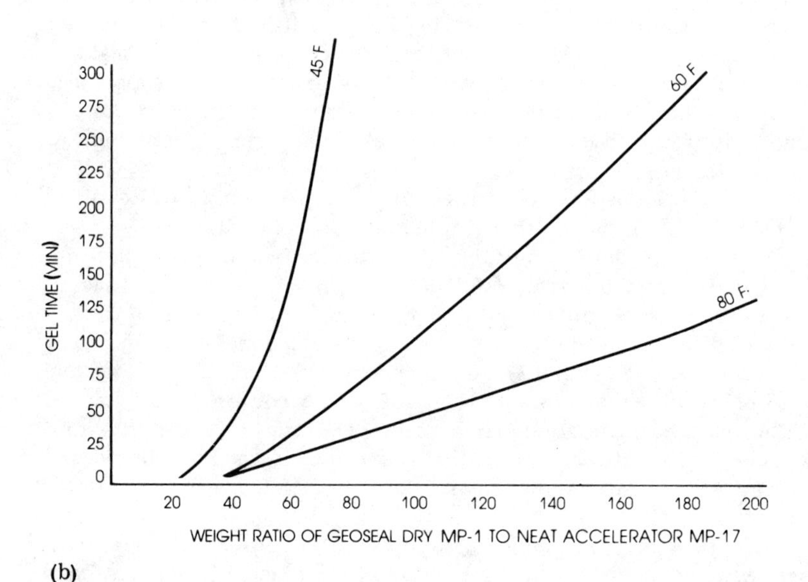

(b)

FIGURE 4.46 Gel time control for Geoseal. (Courtesy of Borden Chemical, Los Angeles, California.)

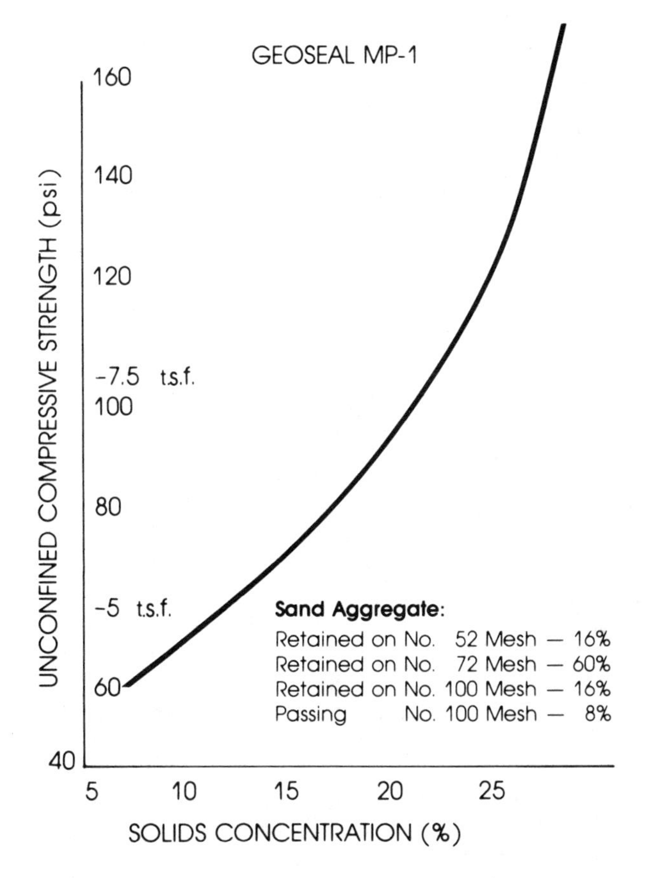

FIGURE 4.47 UC strength versus concentration for Geoseal. (Courtesy of Borden Chemical, Los Angeles, California.)

Aminoplast Grouts

Aminoplasts are grouts in which the major ingredients are urea and formaldehyde. (In addition to urea, melamine, ethylene, and propylene urea, analine and other chemically related materials can be used. Other polymers such as paraformaldehyde, glyoxal, and furfural can replace formaldehyde to make an aminoplast resin. These other combinations, however, have not been used as grouts.) The formation of a resin from these two materials requires heat, and initial suggestions for use (as with the phenoplasts) were in the oil industry. Also paralleling the phenoplasts, the aminoplasts require an acid environment to complete the reaction. However, there are no exceptions to this requirement, as there are for phenoplasts. Thus, all urea–formaldehyde grouts will set up only under acid conditions, a distinct limit to the utility of these grouts. They should be used only when it is known that ground and groundwater pH is well below 7. Such conditions exist in coal mines and may exist elsewhere.

Urea solutions have very low viscosities, similar to the acrylics and phenoplasts. The reaction with formaldehyde, in addition to requiring elevated temperatures, is rapid and difficult to control. However, there are intermediate stages in the reaction when the urea is still soluble in water. Such materials, called precondensates or prepolymers, are readily available from industry, since urea–formaldehydes are used in large quantities as adhesives. Of course, prepolymers are more viscous than the initial urea solution, but products are made which permit the final grout to be used at viscosities in the 10 to 20 cP range. The trade-off in viscosity is made to obtain a product easy to handle, with good gel time control.

The prepolymer is a material in which the reaction has been suspended, either by inhibitors, pH control, or both. The reaction continues once the pH is brought down. Catalysis is therefore done with an acid or an acid salt.

Soils stabilized with urea–formaldehyde have strengths comparable to the phenoplasts and like those materials are less sensitive to testing strain rate than other chemical grouts. (For optimum mechanical properties and to keep free formaldehyde levels low, one molecule of urea should be provided with three molecules of formaldehyde.) Little data are available, but it is probable that aminoplasts break down comparatively quickly under cyclic wet–dry and freeze–thaw conditions. The creep endurance limit is probably a relatively high percentage of the UC. Except as noted above, the resins have good stability and are considered permanent.

Grouts using urea solutions are toxic and corrosive due to the formaldehyde and acid (catalyst) content. Processes using a prepolymer contain much less free formaldehyde than those using a monomer and therefore present much less of a hazard.

The gel, or resin, formed at the conclusion of the reaction is inert, but it generally contains small amounts of unfixed formaldehyde. In an enclosed space such as a mine drift, this can become a nuisance and a hazard.

Several commercial products are available. One of the grouts in the Rocagil (Rhone Poulenc, France) series is an aminoplast advertised specifically for use in coal mines.

Herculox (Halliburton Company, Duncan, Oklahoma) has been used in the United States for several decades. It is a proprietary material and specific engineering data are lacking, but it appears to be a prepolymer.

Diarock (Nitto Chemical Ind., Japan), marketed internationally for about a decade beginning in the mid-1960s, appears (from the manufacturer's literature) to be basic urea–formaldehyde resin (not a prepolymer). Cyanaloc 62 (American Cyanamid Company, Wayne, New Jersey) is a prepolymer marketed as a concentrated liquid which is diluted with water for field use. Sodium bisulfate is the catalyst normally used, and the relationship between catalyst concentration and setting time is shown in Fig. 4.48. For very short gel times or for shorter times at low temperatures, sodium chloride can be added. The gel time is actually the time required for formation of a soft gel (with consistency similar to the acrylics). Over the next few hours to a day or more, depending on the gel time, the gel cures to a much harder and stiffer consistency.

Cyanaloc, as used for applications requiring strength, will have initial viscosities from 13 to 45 cP, depending on temperature, as illustrated in Fig. 4.49.

Since urea prepolymers are readily available domestically, and are relatively inexpensive, these materials are probably the base for proprietary grouts used by specific grouting firms for their own projects. The pH problems must be overcome, probably through the use of either pre-acidification of the formation and/or very acidic catalysts and short setting times. These grouts cannot be used in areas of high pH, such as those already grouted with cement.

Water Reactive Materials

Materials that gel or polymerize upon contact with water offer obvious possibilities for use as grouts. Materials that can be foamed also offer obvious possibilities.

Many foams were investigated in the late 1960s for their potential as soil grouts [13]. The study concluded that on formation methods alone polyethylene, polyvinyl, CCA, urea–formaldehyde,* and syntactic foams can be eliminated from consideration and that of the four other major types of foam (silicone, phenolic, epoxy, and polyurethanes) the polyurethanes excel, having the best mechanical properties and the widest range of conditions of formation.

Polyurethanes are produced by reacting a polyisocyanate with a polyol (or with other chemicals such as polyethers, polyesters, and glycols, which have hydroxyl groups).

---

*Urea–formaldehydes are, of course, used as grouts but not as foamed grouts.

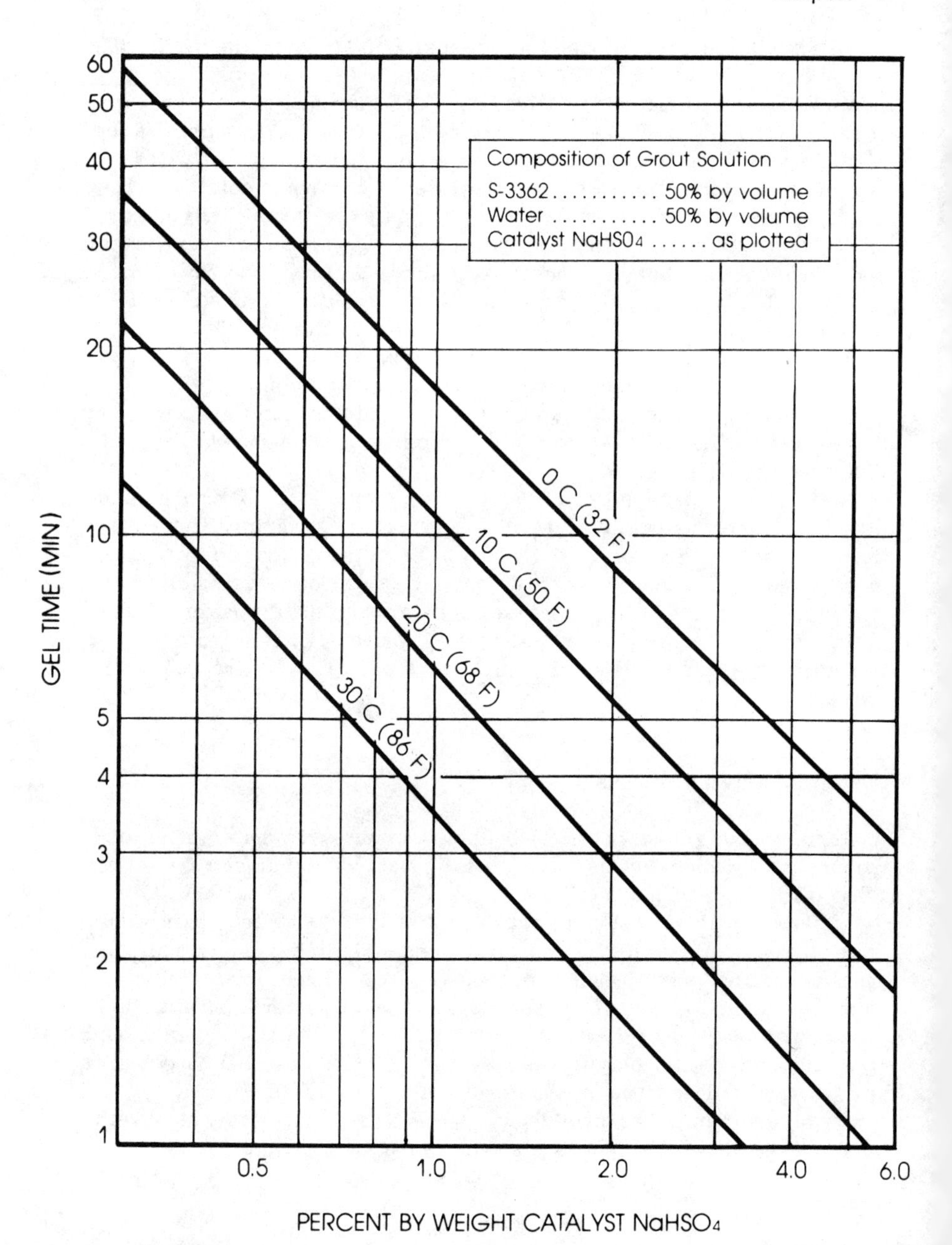

FIGURE 4.48 Catalyst concentration versus gel time for urea-formaldehyde grout.

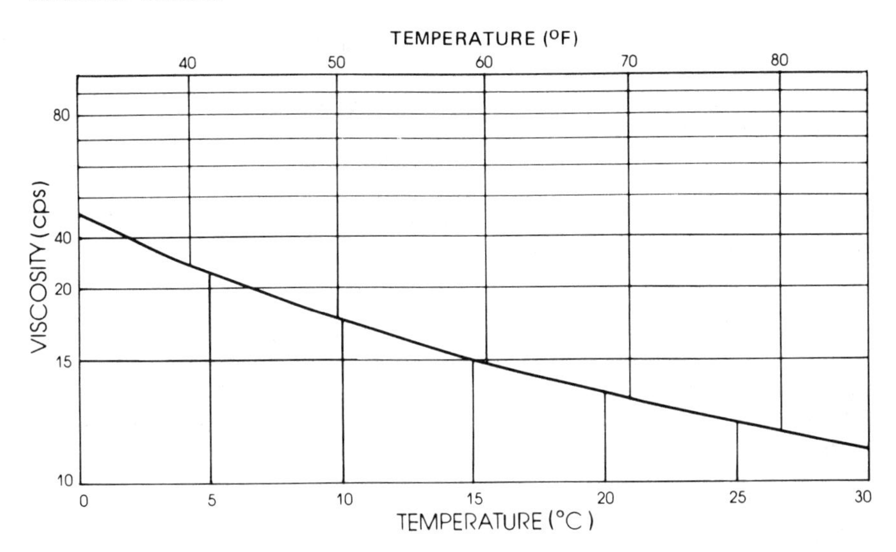

FIGURE 4.49 Temperature-viscosity relationships for a 50% solution of 'Cyanaloc 62, urea-formaldehyde grout.

Catalysts, generally tertiary amines and tin salts, may be used to control the reaction rate.*

Surface-active agents are used to control bubble size. The foam structure itself is produced by a blowing agent, reacting chemically to produce carbon dioxide.

The urethane linkage is not the only one occurring in the polymer. The isocyanate groups also react with carboxylic groups, hydrogen and nitrogen ions, and water. (The physical properties of the polymer depend on the type and amount of linkages which occur, all of which tend to occur simultaneously.) Thus, water plays a role in both the formation of a polymer and the reaction which produces the foam.

Under closely controlled conditions, a manufacturing process, for example, all the chemical elements may be mixed simultaneously. For grouting purposes, use of a prepolymer is advantageous. The prepolymer is formed by adding a hydroxy-containing compound to an excess of polyisocyanate. The prepolymer contains attached

---

*This would serve the purpose of synchronizing the gelation and foaming reactions.

isocyanate groups, which can later be foamed and reacted with water and more hydroxyl groups.

In a paper based on Ref. 13 the authors describe [14] the formulation of six polyurethane grouts. One of these contained 56.5% toluene diisocyanate, 18.9% triethylene glycol, 16.8% 2-ethyl-2-(hydroxymethyl)-1,3-propenediol, 4.5% castor oil, 0.9% L-531 surfactant, and 1.8% diacetone alcohol with 0.6% adipic acid blowing agents. (Percentages shown are by weight.) This liquid has a viscosity of about 13 cP, was mixed for 26 s, and set 45 s later. The results of six unconfined compression tests at 0.01 in./min strain rate are shown in Table 4.5. These tests are made with dry soils and cannot realistically be projected to soils below the water table. However, large-scale tests made in saturated sands yielded cores which averaged over 1100 psi in UC value and between 10^{-6} and 10^{-7} cm/s in permeability.

Three commercial products are available as water-reactive grouts. One of these, TACSS (Takenaka Komuten Company, Japan), was introduced in Japan in 1967, and the research leading to its development probably predates Refs. 13 and 14. TACSS offers a choice of a number of different prepolymers, with viscosities ranging from 22 to 300 cP. (Prepolymers with viscosities as low as 5 cP were originally marketed but have been withdrawn because of flammability.) TACSS formulations are currently marketed in the U.S. by DeNeff America Inc., St. Louis, Missouri.

The catalysers have viscosities of 5 to 15 cP. The mixture of TACSS with its catalyst (the manner of normal injection), has a viscosity well above those of most other grouts. Penetrability is, of course, directly related to viscosity. Secondary penetration occurs when contact with water initiates foaming, building up local pressures as high as 400 psi. The manufacturer's literature claims possible penetration into silts. Typical reaction times are shown in Fig. 4.50.

A second polyurethane grout CR-260 (3M Company, St. Paul, Minnesota) was marketed primarily for sewer sealing applications. It is applied internally through a packer inside the sewer and acts as a gasket in the spigot. (This type of application is covered in greater detail in later chapters.) Newer formulations are intended for use as soil stabilizers.

The grout solutions become less viscous as they mix with water. Additives may be used to prevent an immediate reaction. The resulting product, such as Chemical Grout 5610, has a low enough viscosity to penetrate sands. Typical properties are shown in Fig. 4.51. Strength of the diluted grout is low, comparable to those of the phenoplasts.

Multi-Grout AV-202 (Avanti International) is a similar product, also of relatively high viscosities, described as "water soluble,

TABLE 4.5 Results of Six Unconfined Compression Tests (0.01 in./with Strain Rate)

Chemical system and specimen	Foam density (g/cc)	Water content after soaking (%)	Compressive strength (psi)[a]	E, initial tangent modulus of elasticity (psi)	Curing time (days)	
					Dry	Submerged
(1)	(2)	(3)	(4)	(5)	(6)	(7)
39T1	0.63	0	2070	7.9×10^5	7	7
39T2	0.63	0.7	3500	9.3×10^5		7
39V1	0.63	0	3568	10.8×10^5	14	
39V2	0.63	1.4	3060	3.9×10^5		14
39U1	0.64	0	3993	8.7×10^5	28	
39U2	0.64	1.6	4179	5.4×10^5		28

[a] To convert pounds per square inch to newtons per square meter, multiply by 6894.76.

Source: Reference 14.

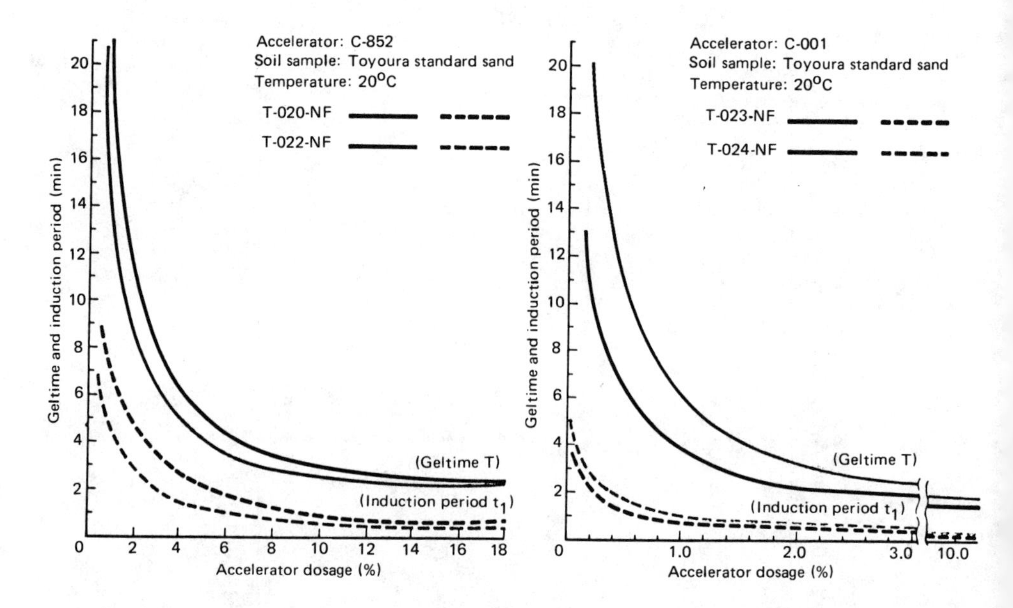

FIGURE 4.50 Typical gel times for the TACSS (polyurethane) system. (Data from manufacturer's literature, courtesy of De Neef America, Inc., St. Louis, Missouri.)

hydrophillic polyurethane prepolymers." Gel time characteristics are shown in Fig. 4.51.

Urethanes pose severe handling problems. The material, when vaporized, is flammable and can cause serious respiratory system effects. In common with the acrylics, the end product (if fully reacted) poses no health problems.

Other Chemical Grouts

Combining two available grouts so as to obtain simultaneously the optimum properties of each is an idea which occurs naturally when trying to select a material for a specific job. To do so, the materials must of course be chemically compatible, particularly in the pH criterion for reaction. Thus, phenoplasts and aminoplasts may be combined, and in fact these combinations have already been mentioned twice. Once the reference was to the use of a chrome-lignin to carry a phenol–formaldehyde, using the gel time control of the former and the high strength of the latter. The second reference was to Terranier, a derivative of natural products which contains a

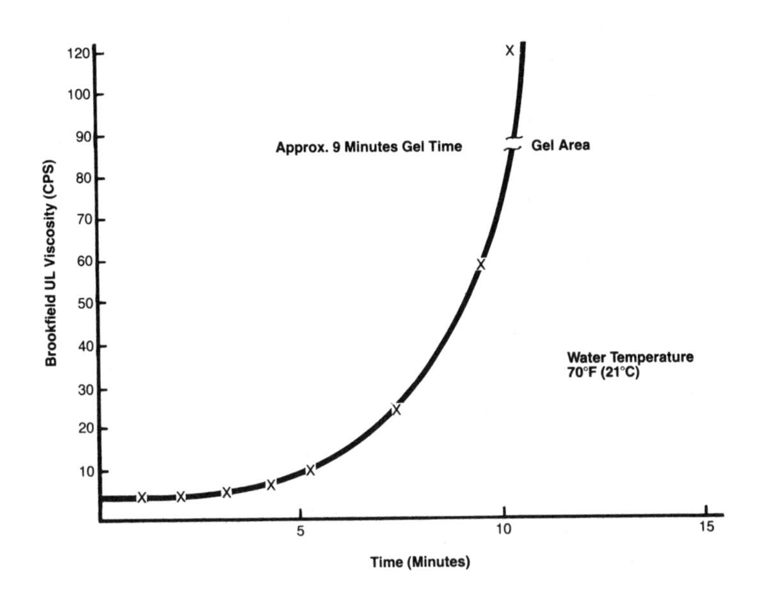

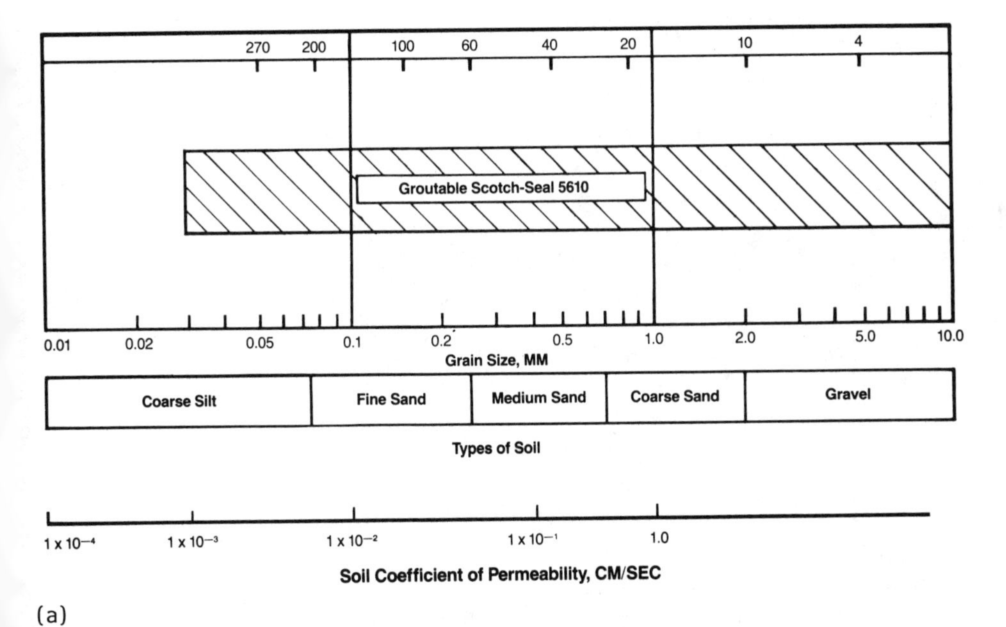

(a)

FIGURE 4.51 Typical properties of Chemical Grout 5610. (From manufacturer's literature courtesy of 3M Company, St. Paul, Minnesota.)

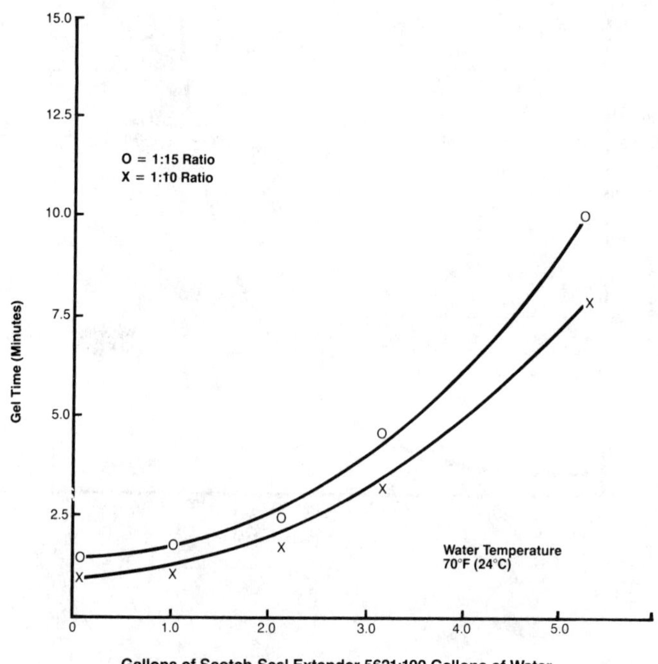

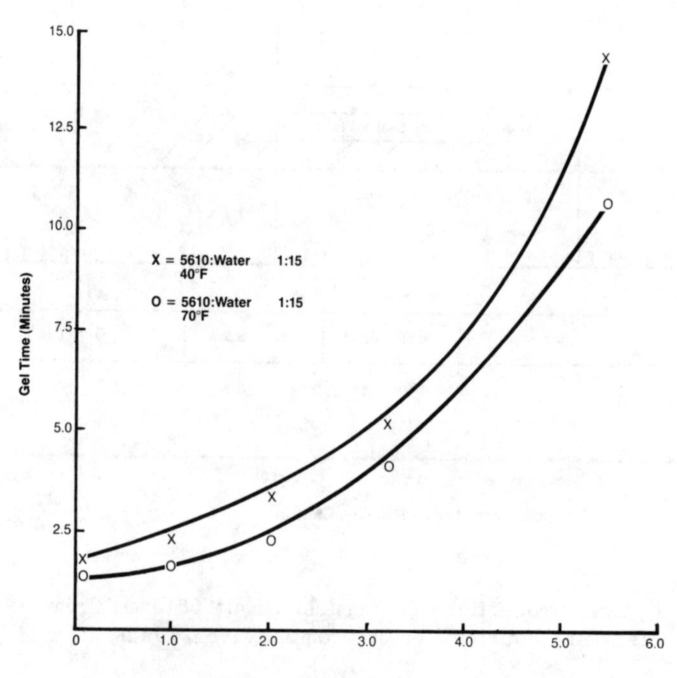

(b)

FIGURE 4.51 (Continued)

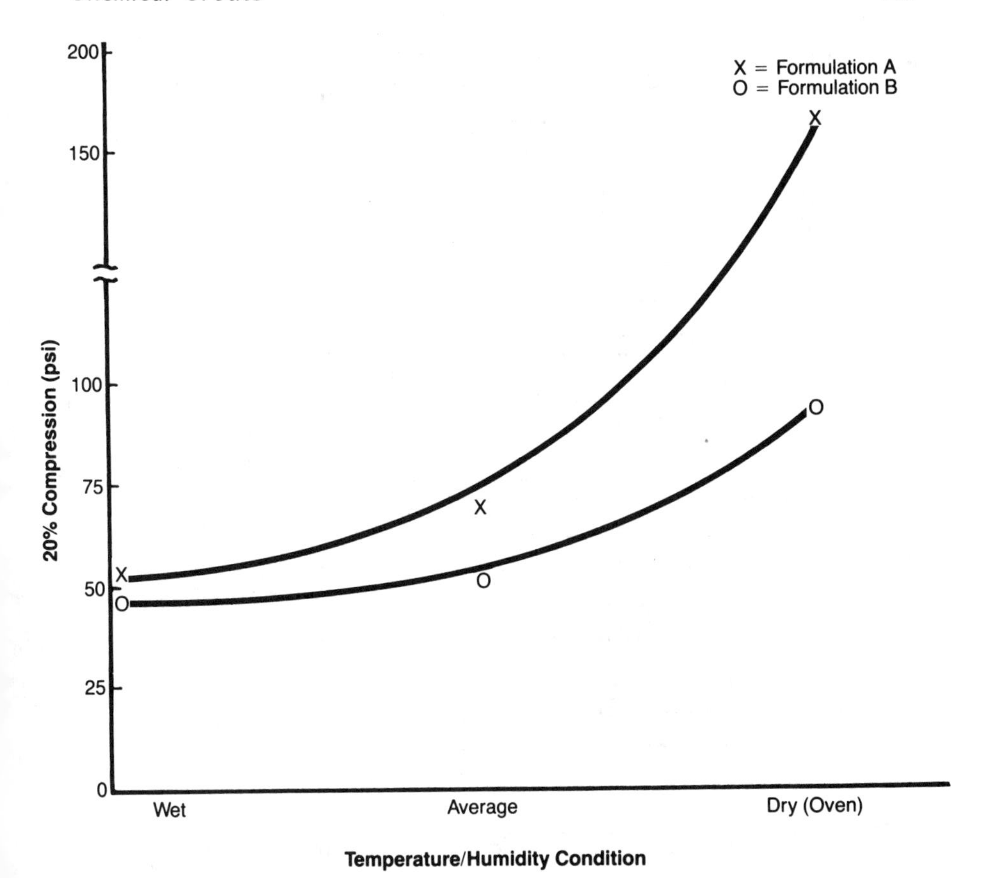

(c)

FIGURE 4.51 (Continued)

phenol and a dichromate and whose gel properties are intermediate compared to the two groups.

The silicates and the acrylamides both require basic conditions for gel formation, but the gelation mechanisms are less related than those of the pheno- and aminoplasts. Silicates and acrylamides can be used together, but the latitude of workable proportions is small. Siprogel (Rhone Poulenc, France) is a commercial product using both materials, available as separate solutions containing

1. Acrylamide, a catalyst for the silicate, and an activator for the acrylamide
2. Sodium silicate and a catalyst for acrylamide

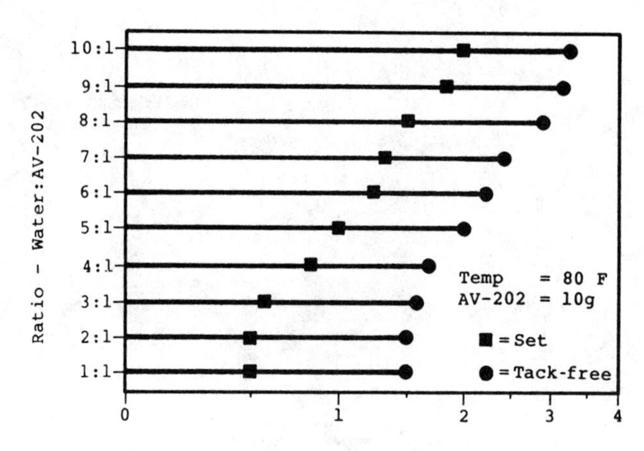
10:1
9:1
8:1
7:1
6:1
5:1
4:1
3:1
2:1
1:1
Ratio - Water:AV-202
Temp = 80 F
AV-202 = 10g
= Set
= Tack-free
0
1
2
3
4
Time - minutes

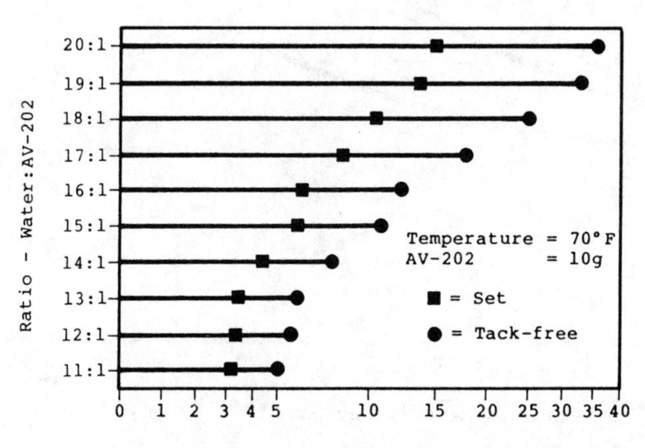
20:1
19:1
18:1
17:1
16:1
15:1
14:1
13:1
12:1
11:1
Ratio - Water:AV-202
Temperature = 70°F
AV-202 = 10g
= Set
= Tack-free
0 1 2 3 4 5 10 15 20 25 30 35 40
Time - minutes

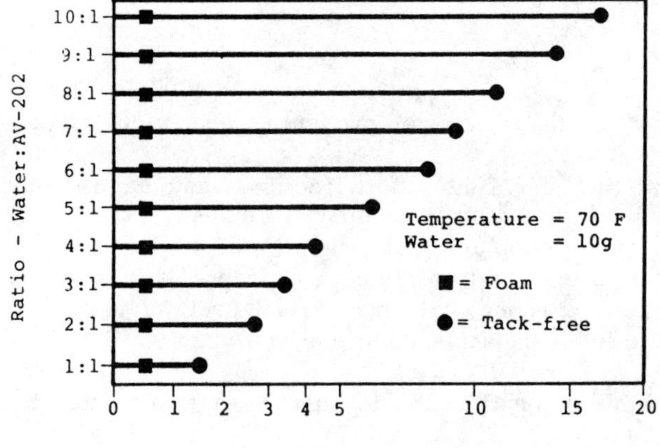
10:1
9:1
8:1
7:1
6:1
5:1
4:1
3:1
2:1
1:1
Ratio - Water:AV-202
Temperature = 70 F
Water = 10g
= Foam
= Tack-free
0 1 2 3 4 5 10 15 20
Time - minutes

The initial viscosity of the solutions when just mixed is about 2 cP. The viscosity increases from that time, as is typical of silicate gels. The setting process must be in sequence:

1. The silica gel starts.
2. The acrylamide gels.
3. The silica finishes the gelation process.

The proportions of catalysts are set to maintain this sequence.

The strength of sands grouted with Siprogel is similar to that of silicate grouted sands. However, the gel is sensitive to wet–dry cycles and absorbs water readily if submerged and unsupported.

A more natural mating of materials occurs if two different monomers can be polymerized by the same catalyst. Nitto SS30R (Nitto Chemical Ind., Japan) is such a grout. It consists of acrylamide and sodium methacrylate monomers with cross-linking agent. The same activator and catalysts are used as for other acrylamide-based grouts. Properties of the liquid grout and the final gel are very similar to those of the other acrylamide systems.

Polyesters and expoxies have been used to seal cracks in rock formations and to anchor rock reinforcement members in drilled holes. The mechanical properties of these materials are much better than those of portland cement. However, the chemicals are (relative to other grouts) very expensive and very viscous. Rendering them less viscous by using diluents so that they will penetrate sands makes them even more expensive. Because of their limited application as grouts, polyesters and epoxies are not detailed further.

Emulsions offer a possible method of getting an otherwise viscous material to penetrate a fine formation. Asphalt emulsions have been used for this purpose [23]. For an emulsion to act as an effective grout, the emulsion must break down at the proper time and place, so that the "grout" phase of the emulsion is deposited. Triggering the breakdown may become involved and expensive and tends to nullify the major advantage of bitumen emulsions, which is low initial cost.

Polyester and epoxy emulsions have also been considered, and while this does solve the problem of penetration into fine materials, the product still remains very expensive.

FIGURE 4.52 Gel time characteristics. (From distributor's literature courtesy of Avanti International, Webster, Texas.)

Soluble Additives to Chemical Grouts

Other components may be added to chemical grout solutions either for dilution or to change the solution or gel properties. Water is the most common additive, being used to dissolve solids or to dilute concentrated solutions.

Water taken from a supply system will not contain dissolved salts that will appreciably alter the action of chemical grouts. Groundwater, on the other hand, may significantly affect setting time and other grout action because of the dissolved salts it may contain. (Because laboratory data on gel times are usually made with distilled water, gel checks in the field must always be made using the actual field source of water.) Groundwater, and especially seawater, will contain sodium chloride. This chemical acta as an accelerator for the acrylic grouts and the urea–formaldehyde grouts. In addition, it will lower the freezing point of the grout solution. Calcium chloride is deliberately used to lower the freezing point of a grout solution. This is particularly valuable to the powder acrylamides. Dissolving the powder in water drops the water temperature 10 to 15°F. If the original water temperature is 40°F or less, the liquid turns to frozen slush. Calcium chloride, when dissolved in water, raises the solution temperature in addition to lowering its freezing point. With the acrylic grouts, calcium chloride also acts as an accelerator and slows the rate of water loss under desiccating conditions. Most soluble salts will tend to lower the freezing point of water solutions. A more effective procedure is to add commercial antifreeze.

Before use, any soluble salt whose effects are not known should be checked for its effect as either an accelerator or an inhibitor.

Solid Additives to Chemical Grouts

By contrast to soluble additives, which are expected to be reactive, solid additives are usually thought of as inert. The two materials that are used most often are the two suspended-solids grouts, cement and bentonite. Bentonite is inert in terms of engaging in chemical activity with the grout with which it may be mixed. Cement, however, is a reactive material with a high pH and may become a partner to the chemical reactions. When used with acrylic and silicate grouts, cement acts as an accelerator. It is held in place by the gel and then sets, drawing its water of hydration from the gel. If used in sufficient quantity, the final set strength can be of the order of 3000 to 5000 psi.

Bentonite, dispersed in water at the proper concentration, becomes a thixotropic fluid. When mixed with a chemical grout,

TABLE 4.6 Relative Ranking of Solution Grouts as to Their Toxicity, Viscosity, and Strength

Grouts		Corrosivity or toxicity	Viscosity	Strength
Silicates	Joosten process	Low	High	High
	Siroc	Medium	Medium	Medium high
	Silicate–bicarbonate	Low	Medium	Low
Lignosulfates	Terra Firma	High	Medium	Low
	Blox-all	High	Medium	Low
Phenoplasts	Terranier	Medium	Medium	Low
	Geoseal	Medium	Medium	Low
Aminoplasts	Herculox	Medium	Medium	High
	Cyanalog	Medium	Medium	High
Acrylamides	AV-100	High	Low	Low
	Rocagel BT	High	Low	Low
	Nitto-SS	High	Low	Low
Polyacrylamide	Inejctite 80	Low	High	Low
Acrylate	AC-400	Low	Low	Low
	Terragel			
	Flexigel			
	DuriGel			
Polyurethane	CR-250	High	High	High
	CR-260			
	TACSS			
	CG5610			
	AV202			

Source: Ref. 16.

bentonite adds this property to those which the grout already has. In contrast to a grout–cement mixture, in a grout–bentonite mixture the bentonite tends to hold the grout in place until a gel forms.

The use of any solid additive will limit the penetrability of the grout to that of the suspension. This disadvantage must be outweighed by some advantages gained through the use of solid additives. These may be factors such as better gel time control or added strength. The use of solids to reduce the unit costs of chemical grouts is seldom justified.

4.5 SUMMARY

The properties of an ideal chemical grout can be readily defined in terms of chemical, mechanical (physical), biological, and economic factors. No single product meeting all of the desirable criteria exists. For field work, the choice of grout must be made by assessing those grout properties most advantageous to the specific project, and matching those properties with available commercial grouts.

Commercial grouts are products that are available on the open market and whose properties are known and documented. All of the materials discussed in detail in previous sections are commercial grouts. Proprietary materials are those whose use is generally limited to one organization (often for bidding advantages) and whose composition is not publicly revealed. Obviously, no such products can be discussed in detail, although they do exist and even crop up in technical literature from time to time.

The economic and physical characteristics of a grouted soil mass can vary over a wide range for any single grout. Thus, numerical comparisons are difficult to make and generally inexact. Comparisons over a broad range, however, can be reliably made and are useful. Such comparisons for commercial grouts are shown in Table 4.6.

REFERENCES

1. *Handbook of Chemistry and Physics*, 43rd Ed.

2. R. H. Karol, *Soils and Soil Engineering*, Prentice-Hall, Englewood Cliffs, N.J., 1960, pp. 67–77.

3. R. H. Karol, *Soils Engineering Research—Creep Tests*, Nov. 30, 1959, American Cyanamid Co., Wayne, N.J.

4. *Chemical Grouting*, Department of the Army, Office of the Chief of Engineers, EM 1110-2-3504, May 31, 1973, Washington, D.C.

5. *Chemical Grouts for Soils, Vol. I and Vol. II, Available Materials*, Report No. FHWA-RD-77-50, June 1977, Federal Highway Administration, Washington, D.C.

6. C. Caron, French Patent 1,154,835, Jan. 18, 1957, Soletanche.

7. P. Peeler, U.S. Patent 2,209,415, Oct. 20, 1959, Diamond Alkali.

8. C. Wayne Clough, W. H. Baker, and F. Mensah-Dwumah, *Development of Design Procedures for Stabilized Soil Support Systems for Soft Ground Tunnelling*, Final Report, Oct. 1978, Stanford University, Stanford, California.

9. C. Caron, Physico-chemical study of silica gels, *Ann. ITBTP-Essais Mesures, 81*:447–484 (March, April 1965).

10. R. W. Burrows, III, J. Gibbs, and H. M. Hunter, General Report No. 15, Nov. 20, 1952, U.S. Bureau of Reclamation, Design and Construction Division, Denver.

11. S. J. Rehmar and N. L. Liver, Process of and Material for Treating Loose Soil, U.S. Patent 3,053,675, Sept. 11, 1962, Intrusion Prepakt, Inc.

12. W. A. Lees, Improvements in or Relating to a Composition and Process for the Stabilization of Soil, British Patent No. 1,183,838, Borden Chemical Co., U.K., Ltd.

13. T. S. Vinson, "The Application of Polyurethane Foamed Plastic in Soil Grouting," Doctoral thesis, Unviersity of California at Berkeley, 1970.

14. T. S. Vinson and J. K. Mitchell, Polyurethane foamed plastic in soil grouting, *J. Soil Mech. Foundation Div.*, ASCE paper 8947, SM6, June 1972.

15. G. S. Littlejohn, Text material for Grouting Course sponsored by South African Institution of Civil Engineers at the University of Witwatersrand, Johannesburg (July 4–6, 1983).

16. R. H. Karol, Grout Penetrability, *Issues in Dam Grouting*, Geotechnical Engineering Division, ASCE, New York, April 30, 1985.

17. W. H. Baker, "Planning and performing structural chemical grouting," Proceedings of the ASCE Specialty Conference Grouting in Geotechnical Engineering, Feb. 1982, ASCE, New York.

18. R. J. Krizek and M. Madden, Permanence of chemically grouted sands, *Issues in Dam Grouting*, Geotechnical Engineering Division, ASCE, New York, April 30, 1985.

19. B. Al-Assadi, "Syneresis in Sodium Silicate-Based Grouts," Graduate M.S. thesis, Rutgers University, New Brunswick, New Jersey.

20. *Field Evaluation of Calcium Acrylate*, U.S. Army Waterways Experiment Station, Corps of Engineers, Vicksburg, Miss., Technical Report No. 3-455, June 1957.

21. R. P. Hopkins, "Acrylate Salts of Divalent Metals, Ind. Eng. Chem., American Chemical Society, *47*(11):2258–2265, Nov. 1955.

22. R. H. Borden, "Time-Dependent Strength and Stress–Strain Behavior of Silicate-Grouted Sand," Ph.D Dissertation, Department of Civil Engineering, Northwestern University, Evanston, Illinois.

23. L. R. Gebhart, "Experimental Cationic Asphalt Emulsion Grouting," *J. Soil Mech. Foundation Div., ASCE*, *98*(SM9):859–868, Sept. 1972.

5

Grouting Technology

5.1 INTRODUCTION

The successful use of chemical grouting to solve a field problem hinges on many factors. They include proper analysis of the problem, selection of the most effective grouting materials and method of placement, and finally having a gel form in the proper location and of the proper extent. This series of actions is not haphazard. Data and experience exist from which, by logic or other deductive processes, intelligent choices can be made which are related to optimum chances of success. The summation of this data may, collectively, be termed grouting technology. That portion of the technology which deals with having a gel form in the proper (or desired) location is the subject of this chapter.

5.2 POINT INJECTIONS

Starting in the late 1950s, serious and sophisticated attempts were made with acrylamide and other chemical grouts to define the parameters that control the size, shape, and location of the solid gel resulting from the placement in the ground of liquid grout. Much of this work can be found only in copies of company reports. Some has been published by technical societies and other journals [1–4]. A complete list of papers devoted to grouting research appears in Sec. IX of Ref. 5.

Under absolutely uniform conditions (such as can be obtained in a laboratory but hardly ever in the field), it is possible to obtain accurate verification of the theory of fluid flow in a uniform,

FIGURE 5.1 Radial flow of grout under uniform conditions.

saturated, isotropic sand. Injection of grout from a point within such a sand mass would be expected to give uniform radial flow. Thus, the shape of a stabilized mass resulting from the injection of grout under these conditions would theoretically be a sphere. (This is substantially true if the grout injection pressure is significantly greater than the static head and if the stabilized volume is small enough so that hydrostatic pressures at the top and bottom of the mass are not significantly different.) Excellent verification has been obtained many times in experiments on a laboratory scale and also in experiments on a field scale.

Figure 5.1 shows a photograph of a vertical section through a grouted mass made in the laboratory under ideal conditions. For this experiment, dyes were used to trace the flow of grout. A total of 6000 cc of grout was injected without interruption at a rate of 500 cc/min into saturated, dense, medium sand. The gel time was 20 min, and each successive 1000 cc was dyed a different color. The photograph shows very clearly the concentric rings; they are seen to be very close to true circles, verifying the theoretical three-dimensional uniform radial flow.

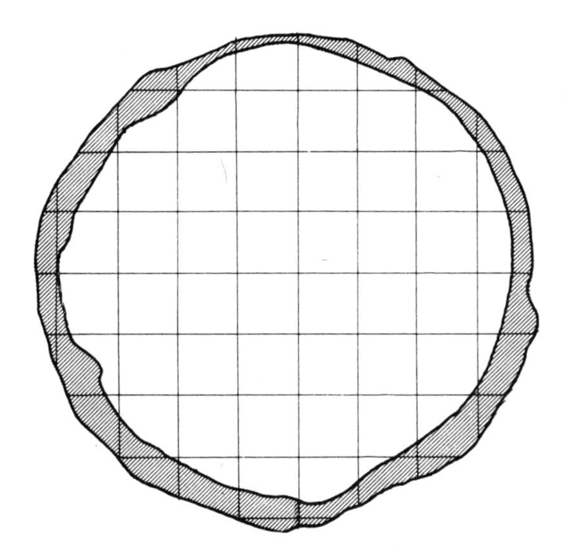

FIGURE 5.2 Extent of grout dilution with groundwater.

The injection shown in Fig. 5.1 took 12 min to complete, and the gel time was 20 min. Thus, the liquid grout stayed in place as a liquid for 8 min after the completion of pumping. Nothing is gained by permitting the grout to stay in place after pumping is finished. However, detrimental effects can occur. One of these is the possible loss of grout by dilution with the surrounding static water. Figure 5.2 is a vertical section (drawn on a 1 in. square grid) of an injection similar to that shown in Fig. 5.1 except that a gel time of 1 h was used, and the groundwater contained a dye. The shaded area is where dye appeared in the stabilized mass. Experiments of this kind verify that the process of placing chemical grouts is a displacement process; that is, grout displaces the groundwater and does not mix with it except at the grout–groundwater interface, and even there the mixing is of low magnitude (this conclusion does not apply in turbulent flow conditions).

There are other generally understood limitations to the discussion above—and to much of the discussion that follows. These include the assumptions that (1) the formation is pervious to the grout that is pumped into it, (2) the pumping rate is slow enough so that the formation does not fracture under the accompanying pumping pressure, and (3) the pumping rate is slow enough so that turbulence does not occur more than a short distance beyond the grout pipe openings.

Another effect detrimental to leaving ungelled grout in the ground occurs because groundwater is not really static. The upper surface of subterranean water, which we call the water table, generally is a subdued replica of the terrain. The elevation of the water table or phreatic surface varies locally with surface precipitation and weather conditions. The mass of groundwater itself also flows in a horizontal direction toward the exposed and subsurface drainage channels.

Under large expanses of level terrain, the rate of groundwater flow is relatively slow and generally of inconsequential magnitude insofar as chemical grouting operations are concerned. In rolling or mountainous country, however, groundwater flow may be rapid enough to affect a grouting operation unfavorably. This is particularly true if the major portion of the flow is occurring through a limited number of flow channels or in formations that do not have overall porosity. Groundwater flow is generally rapid enough to cause problems in the vicinity of an excavation that enters the water table. Even in areas where groundwater flow is normally insignificant, local conditions of high flow rates may be caused by the initial injections of a grouting program.

The effects of flowing water are illustrated by the cross section of a stabilized mass shown in Fig. 5.3. This injection was made in exactly the same fashion as that shown in Fig. 5.1 except that groundwater was flowing slowly. In the photograph, it is interesting to note that the inner concentric rings are not much affected by the groundwater flow. This will always be true as long as the rate of injection of grout is substantially greater than the volume of groundwater moving past the injection plane. If the volume of liquid grout stays in a place for a considerable length of time prior to gelation, even a slow rate of groundwater flow can cause grout displacement and the attendant dilution along the grout–water interface. The loss of grout from the outer concentric rings is clearly visible in Fig. 5.3.

The stabilized mass is still roughly spherical, although it is evident that there has been grout loss from the *upstream* side. It is also evident (from the location of the grout injection point) that the stabilized mass in total is displaced in the direction of groundwater flow.

When groundwater is flowing at a relatively rapid rate, in addition to displacement of the grouted mass in the direction of flow, the shape of the stabilized volume will be modified to conform with the flow net caused by the introduction of a point of high potential at the end of the injection pipe. (This same effect is also caused in slowly moving groundwater when the time lapse between grouting and setting of the grout is very long.) Figure 5.4 is an illustration of a laboratory scale injection showing conformation of a grouted

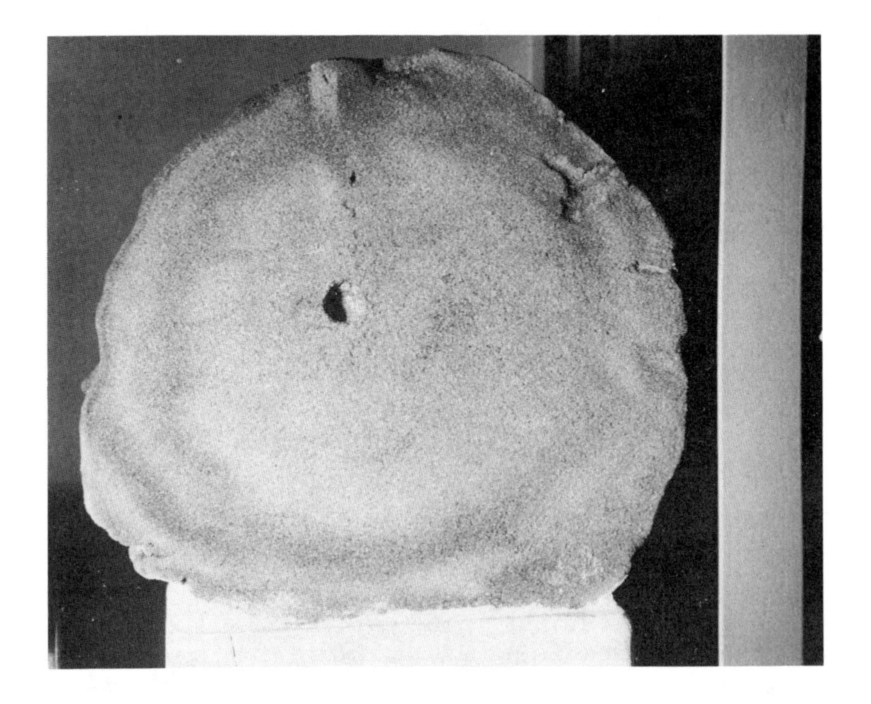

FIGURE 5.3 Effects of flowing groundwater.

FIGURE 5.4 Effects of flowing groundwater.

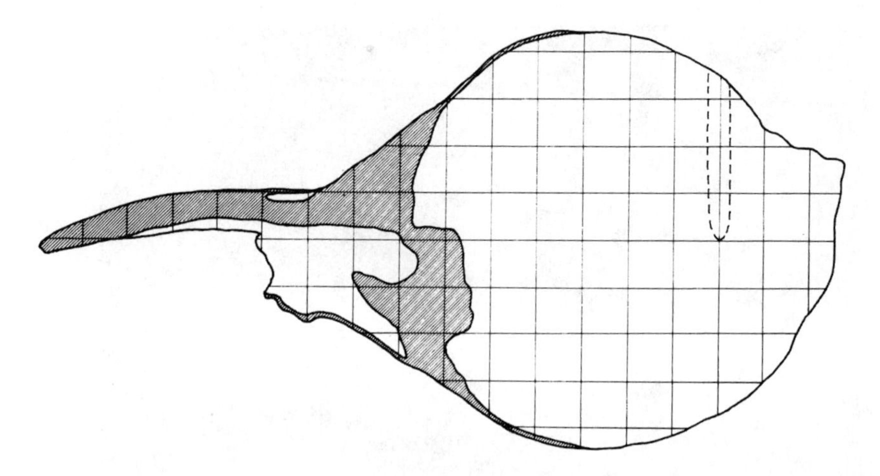

FIGURE 5.5 Dilution zones due to flowing groundwater.

mass to the flow lines. In this particular case, approximately 3000 cc of grout were injected during 6 min. Groundwater flow was about 4 cm/min. Even though the gel time was 9 min (only 3 min longer than the total pumping time), the grouted mass is displaced almost 7 in. (about three-quarters of its vertical dimension) from the location of the grout pipe. It is interesting to note that displacement is about half the groundwater flow distance during the grout induction period.

In a similar experiment in which the groundwater contained dye, the zone stabilized by diluted grout could be seen, by the shaded areas, as shown in Fig. 5.5. The long thin tail is most probably formed from grout washed from the upstream side by flowing groundwater but not yet diluted to the extent where a gel will not form. By inference, some of the grout did dilute beyond the point where it would gel.

When injections are made into stratified deposits, the degree of displacement is related to the formation permeability, and the grouted mass can take odd shapes, such as illustrated in Fig. 5.6. In this experiment, the zones where diluted grout gelled are shown by the shading on the upper and lower sand strata of Fig. 5.7

It is apparent, then, that the major effect of flowing groundwater is to displace the grouted mass from the location where it enters the formation and that a minor effect is to modify the first shape of the grouted mass. The degree of displacement and the shape modification are functions of the relationship among the rate of groundwater flow, the rate at which grout is placed, and the

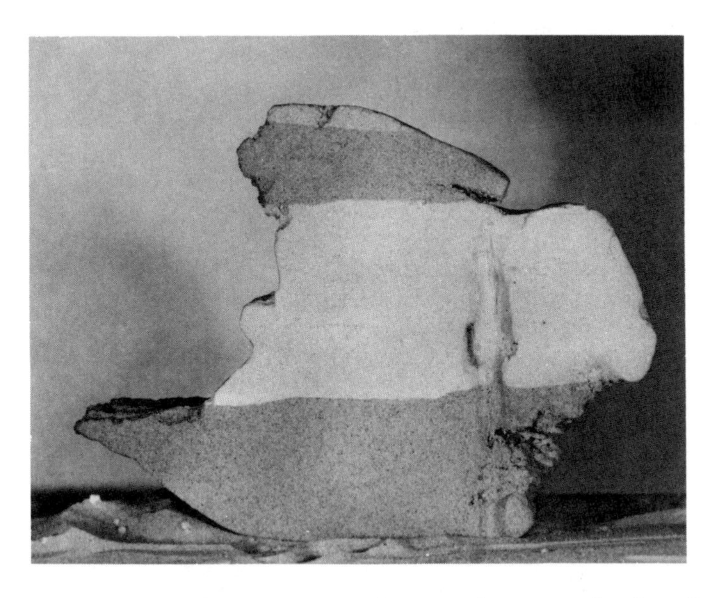

FIGURE 5.6 Displacement of grout by groundwater flowing in stratified soils.

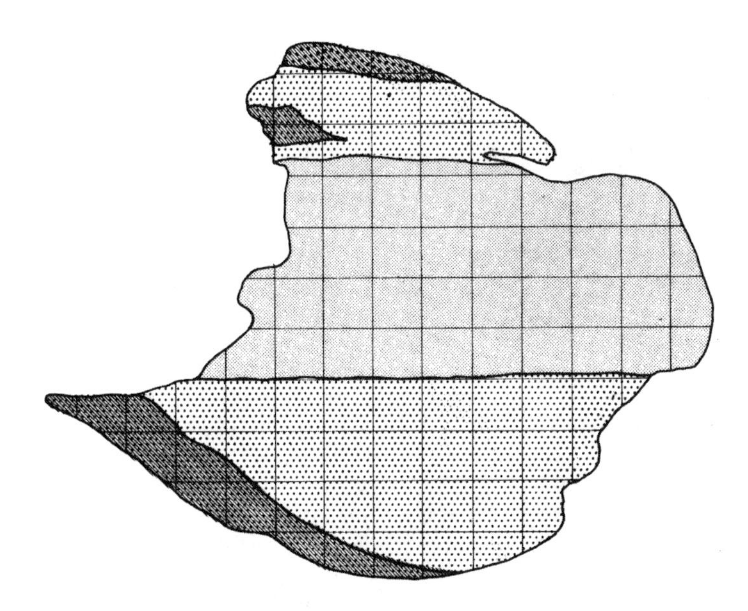

FIGURE 5.7 Areas where dilution with groundwater occurred in Fig. 5.6.

length of time for the grout to gel. By extension of these data, it becomes apparent that grout displacement to the point of ineffectiveness can occur in the field when long gel times are used and when groundwater is moving (relatively) rapidly. Since the groundwater flow rate is generally beyond the control of the grouter, it is the gel time which must be controlled and generally kept to a minimum. Local control of groundwater flow *is* possible, for example, by caulking exposed cracks in excavated rock formations or by bulkheading a tunnel face. Such control measures are often used to facilitate grouting near exposed seepage channels (generally with the objective of reducing turbulent flow to laminar).

In actual practice, uniform soil conditions rarely exist. Even deposits which appear very uniform will generally have greater permeability in the horizontal direction. Thus, the almost perfect sphere shown in Fig. 5.1 would in the field have flattened into an oval shape. In stratified deposits, as shown in Fig. 5.6, the effects are exaggerated and compounded by differences in flow rate between the coarser and finer strata. However, single injections within a large volume of soil generally serve no purpose. It is the relationship among a number of such injections which is important.

5.3 INJECTIONS ALONG A GROUT HOLE

Grout is generally placed in the ground through holes or pipes drilled, jetted, or driven to the desired location. Although there are applications in which grout is injected only at the bottom or open end of the pipe, more often there are several locations along the length of pipe or hole where grout is placed. When this is done, the intent is to form a stabilized cylinder of a desired specific diameter along the length of pipe. The diameter is selected so that stabilized masses from adjacent grout holes will be in contact with each other.

In practice, it is difficult to synchronize the pumping rate and grout pipe pulling (or driving) rate to obtain a uniform grout placement rate along the pipe length. It is common practice to pull (or drive) the pipe in increments and hold it in place for whatever length of time is required to place the desired volume of grout. If small volumes of grout are placed at considerable distances apart, the obvious result is isolated stabilized spheres (or flattened spheres). As the distance between placement points decreases, the stabilized masses approach each other. The stabilized masses will also approach each other, if the distance between placement points remains constant but grout volume increases. Eventually a point will be reached where the stabilized masses become tangent, as shown in Fig. 5.8. With further decrease in placement point

FIGURE 5.8 Tangent stabilized spheres due to excessive distance between injections.

distance, a uniform cylinder is eventually realized, as shown in Fig. 5.9. Experiment and experience have shown that the chances of achieving a relatively uniform cylindrical shape are best when the distance the pipe is pulled between grout injections does not exceed the grout flow distance normal to the pipe. For example, if a stabilized cylinder 5 ft in diameter is wanted, in a soil with 30% voids, 45 gal of grout is needed per foot of grout hole. The pipe should not be pulled more than 30 in. At 30 in. pulling distance, 112 gal should be placed. (The grouting could also be done by injecting 77 gal at 18 in. intervals, etc.)

The masses shown in Figs. 5.8 and 5.9 were made in uniform sands. Even when the proper relationship between volumes and pulling distance is observed, nonuniform penetration can still occur in natural deposits when these are stratified. Resulting shapes can be as shown in Fig. 5.10. Under extreme conditions, degrees of

FIGURE 5.9 Uniform stabilization cylinder formed by pulling the pipe shorter distances than in Fig. 5.8.

permeation can vary as much as natural permeability differences, as shown in Fig. 5.11. Such nonuniformity has obvious adverse effects on the ability to carry out a field grouting operation in accordance with the engineering design.

It would obviously be of major value to be able to obtain uniform penetration regardless of permeability differences in the soil profile. In assessing the cause for penetration differences, it becomes apparent that the grout that is injected first will seek the easiest flow paths (through the most pervious materials) and will flow preferentially through those paths. To modify this condition, other factors must be introduced. The most effective factor that can be modified by the grouter is control of setting time. Thus, if

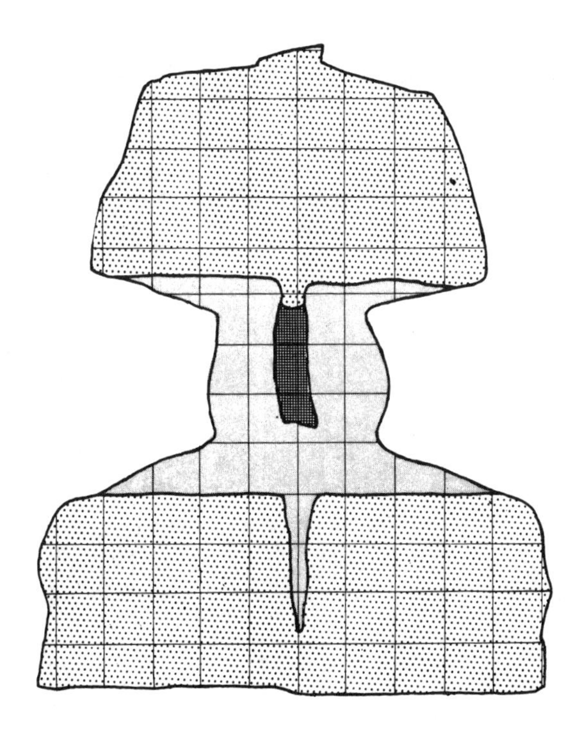

FIGURE 5.10 Difference in grout penetration due to stratification.

the grout were made to set prior to the completion of the grouting operation, it would set in the more open channels where it had gone first and force the remaining grout to flow into the finer ones. Accurate control of gel time thus becomes an important factor in obtaining more uniform penetration in stratified deposits. Just as in controlling the detrimental effects of groundwater flow, more uniform penetration in stratified deposits also requires keeping gel times to a minimum.

FIGURE 5.11 Extreme differences in grout penetration due to stratification ranging from coarse sand to fine sand and silt.

5.4 SHORT GEL TIMES

Many of the early practices and procedures adapted to chemical grouts, when these materials first became commercially available, were based on previously developed cement grouting technology.

Cement grouting is basically a batch procedure. Dry cement is reasonably stable. When mixed with water, it hydrates, and the cement particles (if in contact) join together to form a solid mortar. Since the cement must be mixed with water in order to pump it into the ground, the hydration reaction begins in the mixing tank. Whenever the time for the initial set, the full volume in the tank will set at that time. Thus, every last bit of cement suspension must be cleared from the tank, the pump, the discharge hose, and (hopefully) the grout hole prior to the time the grout begins setting up. Since it is very costly to permit grout to set up in the equipment, the size of batch mixed is correlated with the anticipated pumping rate to allow a good safety factor (generally around 2 but often more). Thus, in cement grouting, if all goes according to plan, the cement suspension is in the ground and at the whim of gravity and groundwater long before it starts to set.

Early work in the 1950s with the new chemical grouts also was done by batching (except, of course, with the Joosten process, which by its nature requires two separate injections). All the ingredients were mixed in one tank, with the catalyst added last, just prior to the start of pumping. From that point on, the necessity to empty the tank before gelation became of paramount importance, often overriding considerations of an engineering nature. Chemical grouts, however, were so much more costly than cement that wastage had to be kept to a minimum. Not only must the chemicals not be wasted in the tank, they must not be wasted in the ground. Movement and possible dilution by gravity and groundwater must be held to a minimum. Thus, while the batch system of placement called for long gel times to provide a safety factor against gelation in the tank, engineering and economic considerations called for gel times approaching the pumping time. Thus, both in the laboratory and in the field it became apparent that better control of grout placement can be obtained when the grout is made to set up at the instant when the desired volume has been placed. Early laboratory experiments were based on determining what happened when the attempts to control these times accurately were unsuccessful.

To be able to work with pumping times which exceeded the gel time, the catalyst was separated from the grout in its own tank. Two pumps were used with separated discharge lines, meeting where the grout pipe entered the soil. Catalysis took place at that point. For gel times shorter than the pumping time, it had been thought that the grouting operation would be halted by excessive pressures at the instant of gelation. Surprisingly, however, it was found that, with a low viscosity, polymer grout pumping of grout into a formation could continue for periods of time substantially longer than the gel time.

Much of the basic research done with short gel times was done with acrylamide. For some time it was thought that the observed phenomena were peculiar to acrylamide-based grouts, but subsequent verification was made that all the chemical grouting materials in current use can be pumped for periods longer than the gel time.

Stabilized shapes resulting from such a procedure are illustrated by a typical grouted mass shown in Fig. 5.12. The lumpy surface is typical of grouted masses in which the gel time is a small fraction of the total pumping time. The mechanism at work during this process is not obvious, and much laboratory and field experimentation was performed to explain the process. One of the most lucid

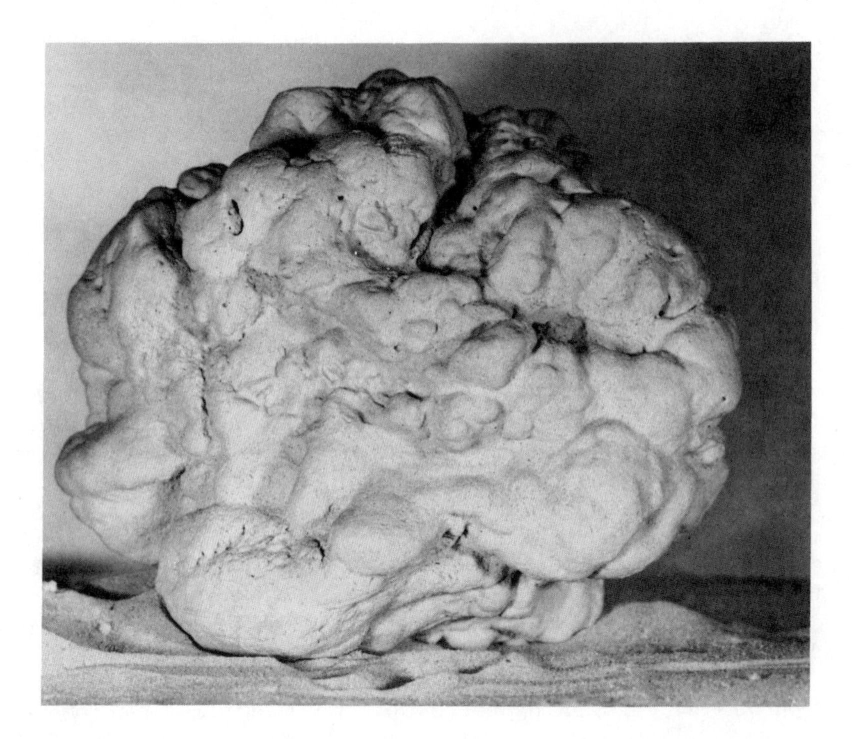

FIGURE 5.12 Knobby surface typical of short gel times in fine sands.

illustrations of the principles involved resulted from a group of experiments in which dye was used to detect the sequential location of the various portions of the total grout volume. One of these experiments will be described in detail.

5.5 THEORY OF SHORT GEL TIMES

The photograph in Fig. 5.13 is a cross section of a stabilized sand mass resulting from the injection of 6000 cc of a 10% acrylamide-based chemical grout into a dense, medium sand. The injection was made through an open-ended pipe under static groundwater conditions at a rate of 500 cc/min per pump. Six different-colored grout solutions were used, 1000 cc of each, so that flow sequence could be traced. An equal volume (see Chap. 6) system was used to

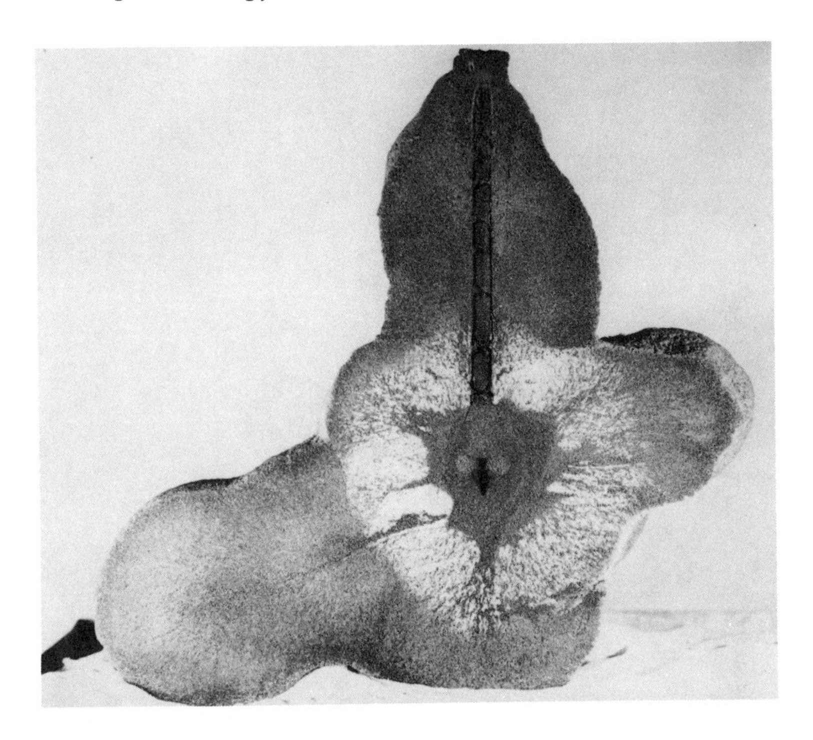

FIGURE 5.13 Stabilized medium sand mass made with dyed grout using short gel times.

place the grout, with catalyst concentrations adjusted to give a gel time of 60 s.

This experiment, like many others of a similar nature, verifies that until gelation occurs, the flow of fluid through the soil mass is entirely in keeping with theoretical concepts. As pumping begins, the grout is displaced radially in three dimensions from the opening in the pipe. The leading surface of the grout mass forms a sphere. The rate of motion of this leading edge is decreasing if the pumping rate remains constant. When gelation occurs, it starts at an infinitely thin shell which is the boundary between the grout and the groundwater. As pumping continues, a finite number of channels are ruptured in this very thin, incipient gel; through these channels, grout continues to flow as pumping continues. (The process cannot take place with a batch system, since it depends on infinitely small volumes of grout reaching the gel state in time succession.)

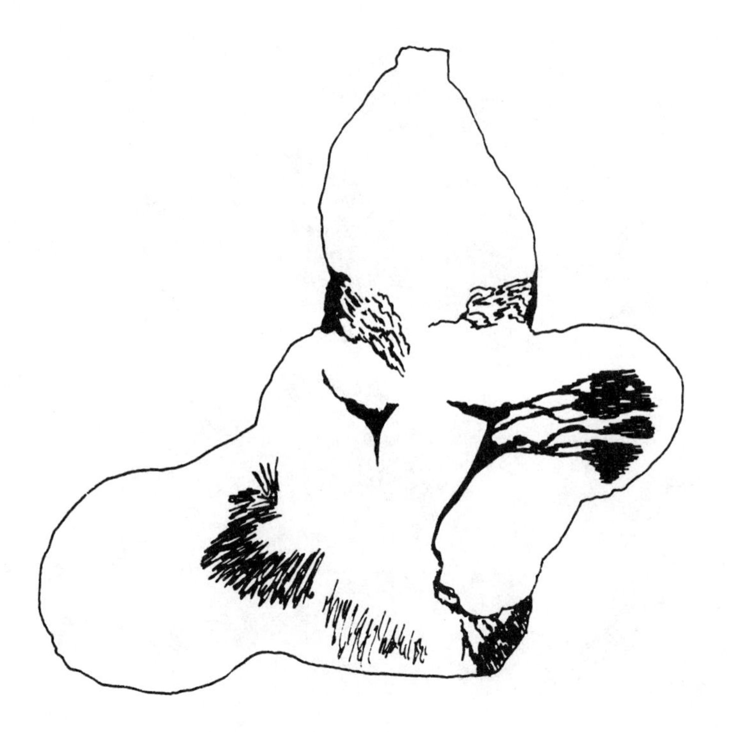

FIGURE 5.14 Third color location from Fig. 5.11.

Much of the grout trapped without the initial thin shell also gels to form a fairly thick shell containing a finite number of open channels. The location of this shell is clearly shown in the photograph. The remainder of the fluid grout within the shell is forced out through the open channels as pumping continues. At the point where each of these channels comes out of the initial shell into unstabilized soil, three-dimensional radial flow again begins. In this fashion, hemispheres begin to grow on the surface of the initial sphere. Eventually, the leading surface of these new partial spheres of grout will gel, and the entire process will be repeated.

If the final location in the grout mass of one color is plotted, the results are difficult to interpret. This is illustrated in Fig. 5.14, which shows the final location on the cross section of the third color. There is difficulty in interpretation because the location plotted also shows the channels used by the succeeding volumes of grout. However, if each color is used as a guide to plot the spread of grout at 0.5 min intervals, the results become meaningful and informative. This is shown by Figs. 5.15 a through 5.15g.

At the end of 0.5 min of pumping, all the grout that had been placed was still liquid, and the flow was radial, so the shape is a sphere, as indicated by the vertical cross section in Fig. 5.15a. At the end of 1 min, the shape is still spherical, but the grout injected first (which is now farthest from the injection point and is the interface between grout and groundwater) has reached the gel time, and a thin shell of gel begins to form at the grout–groundwater boundary. This is shown by the solid black lines in Fig. 5.15b. Photographs of this process in color are included in Ref. 2.

As grout flow continues, a finite number of small channels are opened in the thin shell of gel. These are channels in the soil voids between grains. There is no displacement of soil particles, and the channels cannot be identified visually. Through these openings, fresh grout continues to flow, moving radially from each channel and thus building up hemispheres on the surface of the original sphere. At the same time, much of the original liquid grout trapped inside the thin shell of gel also gels, forming a thick shell, with open channels feeding the zone beyond the initial thin shell. This thick shell is clearly identified in Fig. 5.13. At the end of 1.5 min, the condition is as shown in Fig. 5.15c. Again the solid black area represents gel, and the lighter area represents liquid grout.

At the end of 2 min, the fresh grout flowing through the initial shell reaches its gel time, and thin incipient gel hemispherical shells begin to form at the new grout–groundwater interface. Figure 5.15d shows the condition at 2 min.

As grout flow continues, one or more flow channels open in most or all the second-phase hemispheres, and from these openings fresh grout again flows radially into the formation, beginning to form new hemispherical grout masses. At the same time, the grout trapped within the shells on the second-phase hemispheres also gels, leaving the flow channel open and connected to the center of the grouted mass, which remains fluid (since it is continuously composed of fresh grout). Figure 5.15e shows the condition at 2.5 min.

At 3 min, once again incipient gel shells begin to form, and flow channels are opened in several of them. This is shown in Fig. 5.15f. (Due to minor differences in soil structure and gel formation, flow does not usually occur uniformly, and the final shape may be quite different from a sphere. In this case, the figure shows that at this time flow stopped through the bottom and right-side of the stabilized mass.)

The growth of the stabilized mass continues in the upward and left-hand directions, with the sequence of formation of gel shells

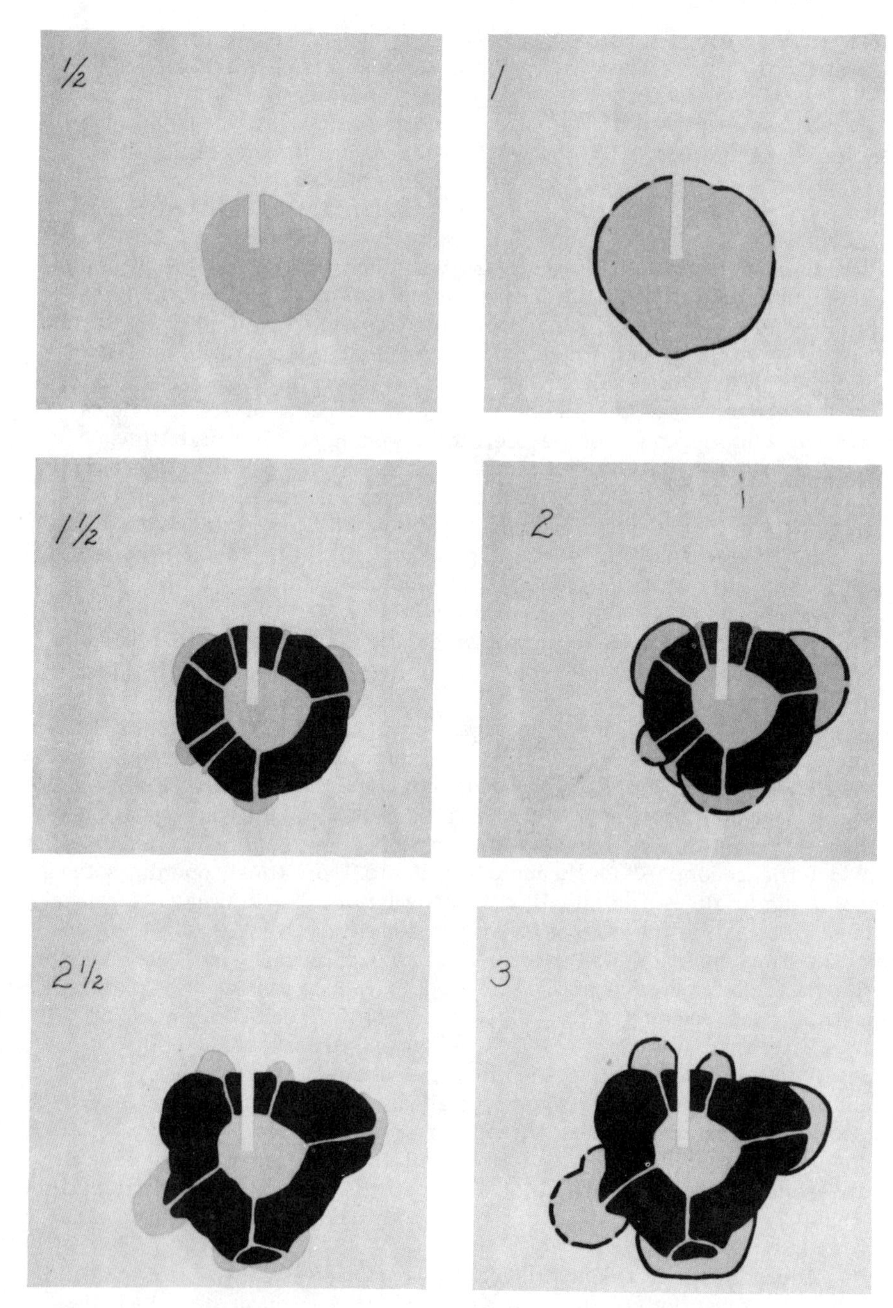

FIGURE 5.15 Sequence of grout spread.

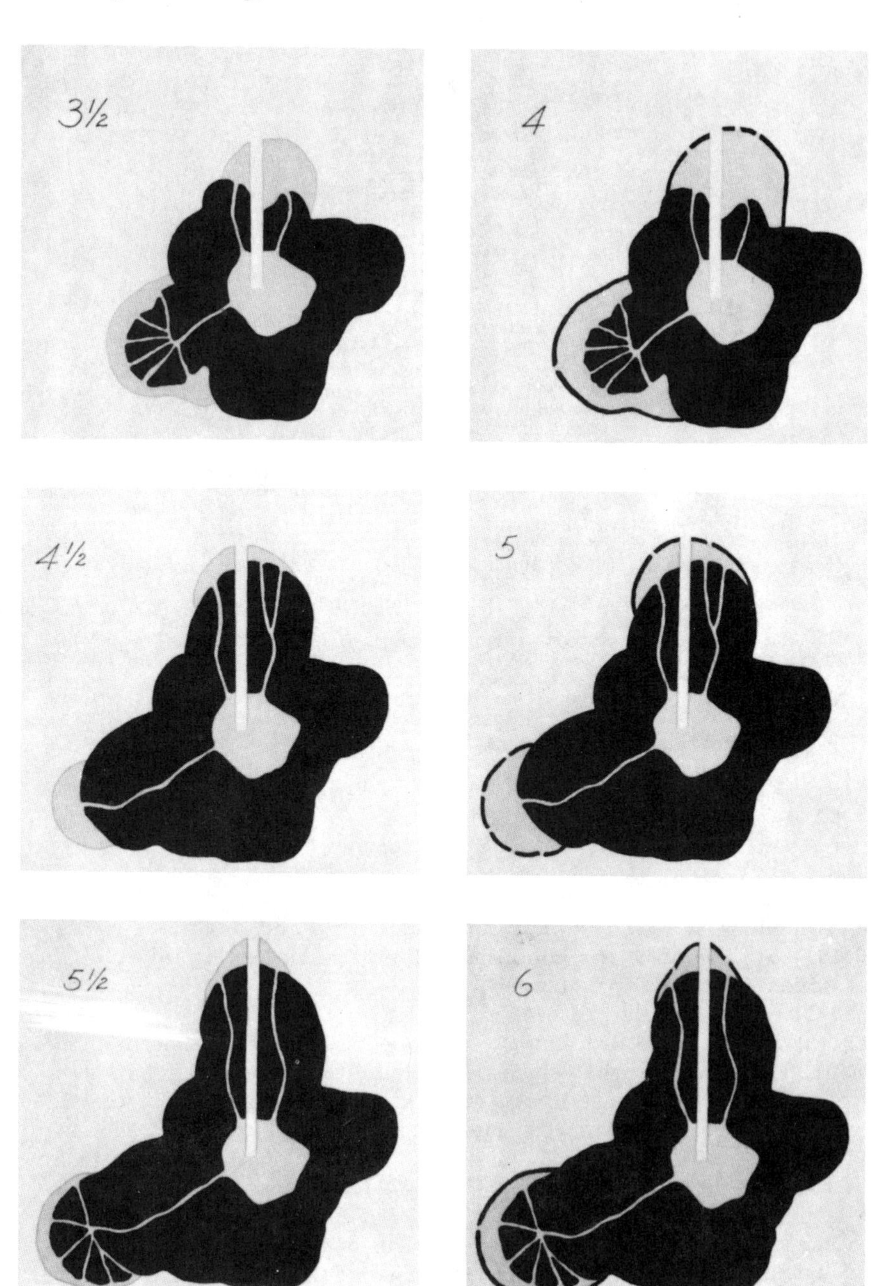

FIGURE 5.15 (Continued)

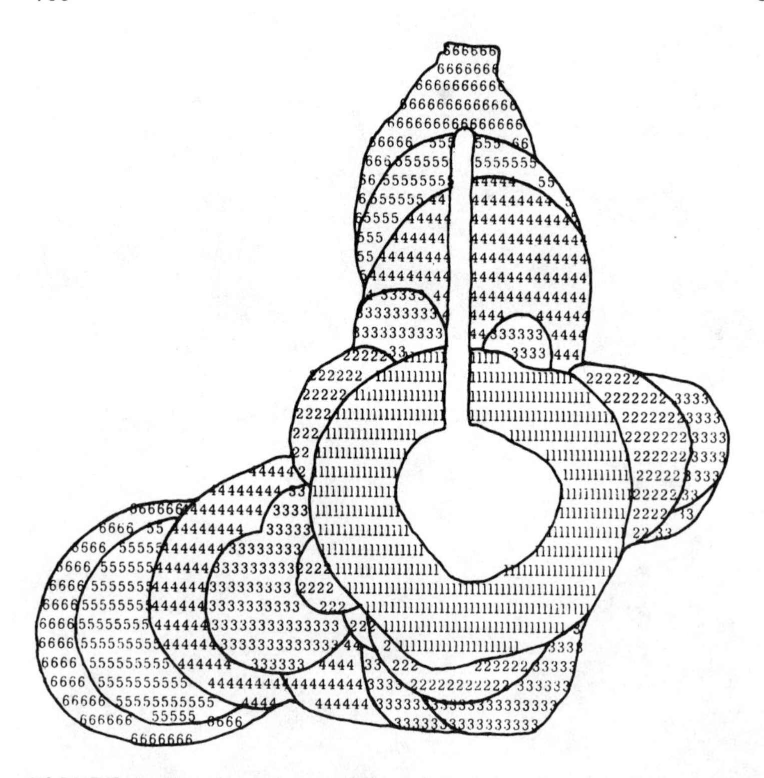

FIGURE 5.16 Grout location related to placement sequence when short gel times are used.

and opening of flow channels repeating again and again. Figures 5.15g to 5.15l show the conditions at 0.5 min intervals up to 6 min, at which time pumping stopped.

The final location of each dye color in the stabilized mass is shown in Fig. 5.16, with numbers indicating the sequence of placement. Opposed to what occurs in a batch system, the grout injected first gels closest to the grout pipe, but more important is the fact that the entire grouted mass surrounds the point of injection. Since the location of that point is known, the location of the grout in the ground is also known. (This statement assumes the grout was properly handled and did in fact gel in the formation.) Thus, it becomes possible to design a grouting operation with much more assurance than if the grout is left to travel at the whim of gravity and groundwater flow.

5.6 FACTORS RELATED TO THE USE OF SHORT GEL TIMES

Experiments of the kind described have indicated that the ungelled zone at the end of the grout pipe (within the first thick shell of the grout formed) tends to grow smaller as pumping continues, leading to the conclusion that eventually it would close and pumping could not continue. In small-scale experiments, however, pumping was possible up to 30 times the gel time, and field experience has shown that closing of the fluid center does not seem to be a problem.

It had been anticipated that after the grout had begun to set up in the ground and the paths through which flow occurred had been drastically reduced that the event would be marked by an increase in pumping pressure. Actually, there were no significant changes in pumping pressure during the laboratory tests. By contrast, similar procedures in the field generally lead to gradually increasing pumping pressure, which for low pumping rates may quickly exceed allowable values and which appear to be roughly related to the inverse of the gel time.

Once the grout has begun to gel in the ground, pumping cannot be interrupted, even momentarily. When gel starts forming in the individual channels running through the grouted mass, it becomes impossible to pump additional grout.

By contrast to laboratory experiments at long gel times, work at short gel times is difficult to duplicate in terms of the shape of the grouted mass. This is due to the fact that the number and location of channels which open in the incipient gel shell is totally haphazard. Many experiments were run in which the second half of the grout contained a dye which could be identified after gelation. At gel times longer than the total pumping time, a typical vertical cross section through a stabilized mass looked like that shown in Fig. 5.17, regardless of the grain size of the soil involved. (Grid lines, wherever they appear on stabilized soil cross sections, are always 1 in. apart.) The white area is the location of the first half of the grout volume; the shaded area shows that half of the grout volume injected last. At gel times much shorter than the pumping time, however, grain size does have an effect. Figure 5.18 shows a vertical cross section through a stabilized coarse sand. Six flow channels can be identified through the initial grout shell. Figure 5.19 shows a vertical section through a stabilized fine sand, and two flow channels or directions can be identified. Figure 5.20 shows a vertical section through a stabilized fine sand and silt, and only one flow channel through the original grout shell can be found.

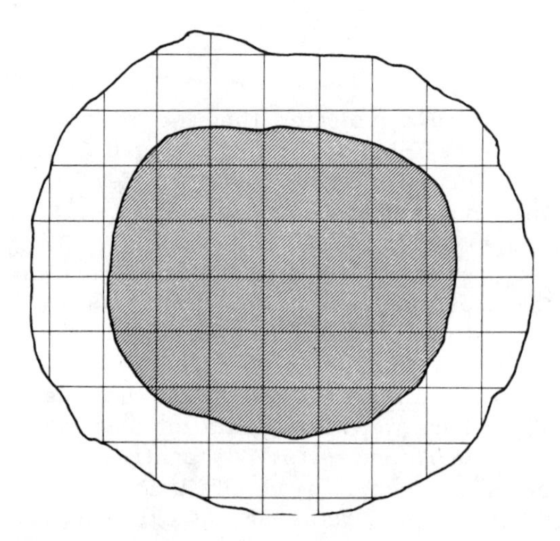

FIGURE 5.17 Vertical section through stabilized coarse sand with gel time longer than the pumping time.

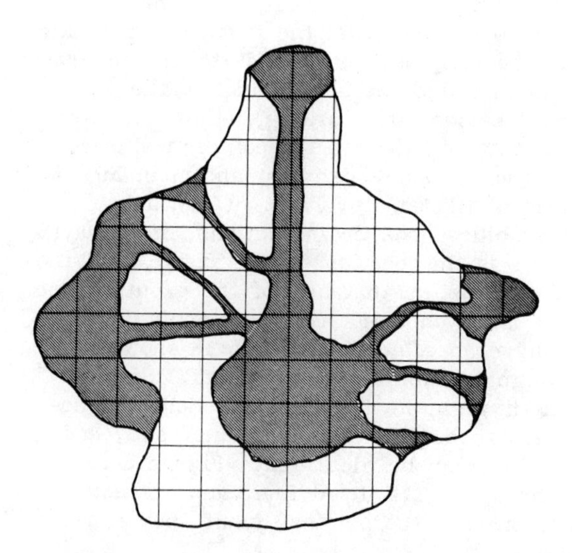

FIGURE 5.18 Vertical section through stabilized coarse sand with gel time shorter than the pumping time.

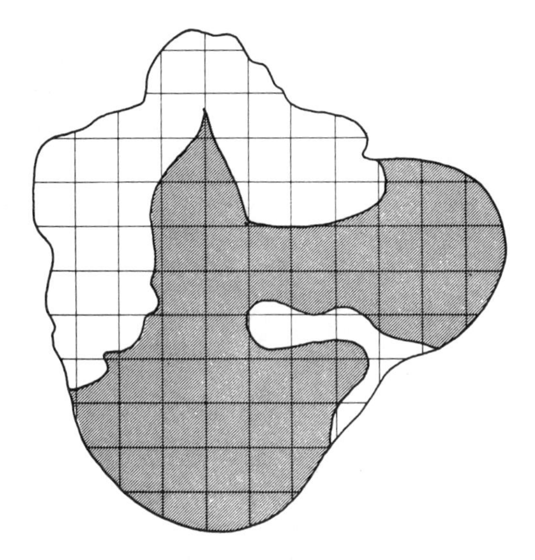

FIGURE 5.19 Vertical section through stabilized fine sand with gel time shorter than the pumping time.

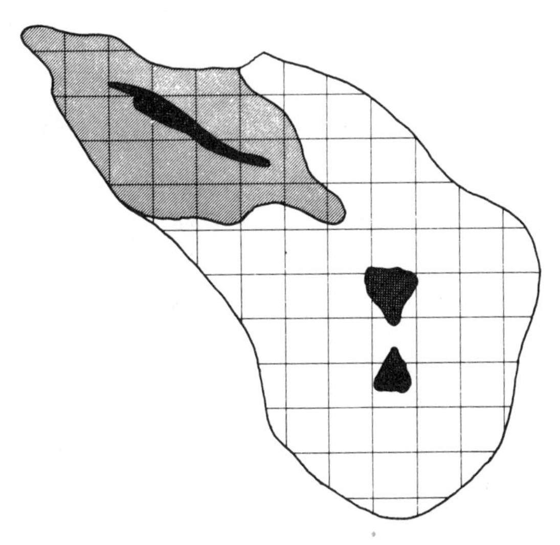

FIGURE 5.20 Vertical section through stabilized fine sand and silt with gel time shorter than the pumping time.

Laboratory evidence is conclusive that there is a trend toward fewer breakthrough points as the grain size decreases. This means that the stabilized mass will be more irregular in shape in fine soils than in coarse ones. If only one breakthrough point occurs and grout feeds through a single flow channel, one possible result could be a long thin sheet or knobby tube of stabilized soil, rather than the anticipated rough sphere surrounding the point of injection. Thus, using very short gel times for no other reason than that the product can be readily pumped at short gel times is not necessarily the best engineering approach.

5.7 UNIFORM PENETRATION IN STRATIFIED DEPOSITS

The design of any specific grouting project is based on the necessity to grout a volume of soil or rock generally well defined spatially. The success of that grouting project depends on the accuracy with which grout can be placed in the desired locations. Thus, the prediction of where liquid grout pumped from the ground surface will gel within the ground mass is an important part of the design and operation processes. Several operating principles can be deduced from the previous sections, acting in the direction of permitting better prediction of grout location:

1. The pipe pulling distance must be related to volume placed at one point.
2. The dispersion effects of gravity and groundwater should be kept to a minimum.
3. Excess penetration in coarse strata must be controlled to permit grouting of adjacent finer strata.

The first criterion requires arithmetic and a knowledge of the soil voids. It is obvious that isolated stabilized spheres will result if the distance the pipe is pulled between injections is greater than the diameter of the spheres formed by the volumes pumped. Graphic trials at decreasing pipe pulling distances readily show that the stabilized shape begins to approach a cylinder as the distance the pipe is pulled approaches the radial spread of the grout. This has been verified both in laboratory studies and by field data. In use, the criterion is simple to follow. For example, if a uniform cylinder 4 ft in diameter is desired, the pipe pulling distance should not exceed 2 ft. To determine the volume of grout needed, a chart such as that in Fig. 5.21 is helpful.

The second criterion requires that grout be placed at a substantially greater rate than the flow of groundwater past the

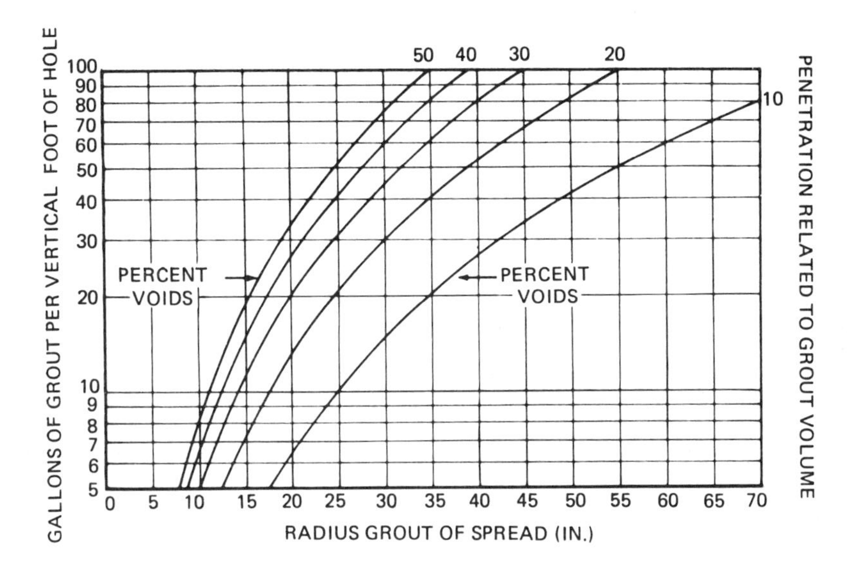

FIGURE 5.21 Relationship between soil voids and grout volumes for various radial spreads of grout.

placement point and that the gel time does not exceed the pumping time. In the formations where chemical grouts would be considered—those too fine to be treated by cement—pumping rates more than 1 gpm are adequate to prevent dispersion under laminar flow conditions. (Turbulent flow does not occur in such soils other than at surfaces exposed by excavation.) The control of gel times not to exceed the pumping time is readily done with dual pumping systems but is difficult and frustrating with batch systems. (See Chap. 6 for a full discussion of pumping systems.)

The third criterion requires that the gel time be shorter than the pumping time. (The alternative is to make additional injections in the same zone after the first injection has gelled. This will probably require additional drilling and will certainly be more costly.) It has been shown that this process is feasible with chemical grouts but obviously cannot work with a batch system. Dual pumps and continuous catalysis are required.

Figure 5.22 shows grain size analyses for four granular materials used in a series of experiments related to grout penetration. Figures 5.23 and 5.24 are injections made into strata of soils A, B, and C. In both cases a pipe was driven to depth and then retracted in steps while grouting continued. In Fig. 5.23, the gel time was considerably longer than the time the pipe remained at each vertical

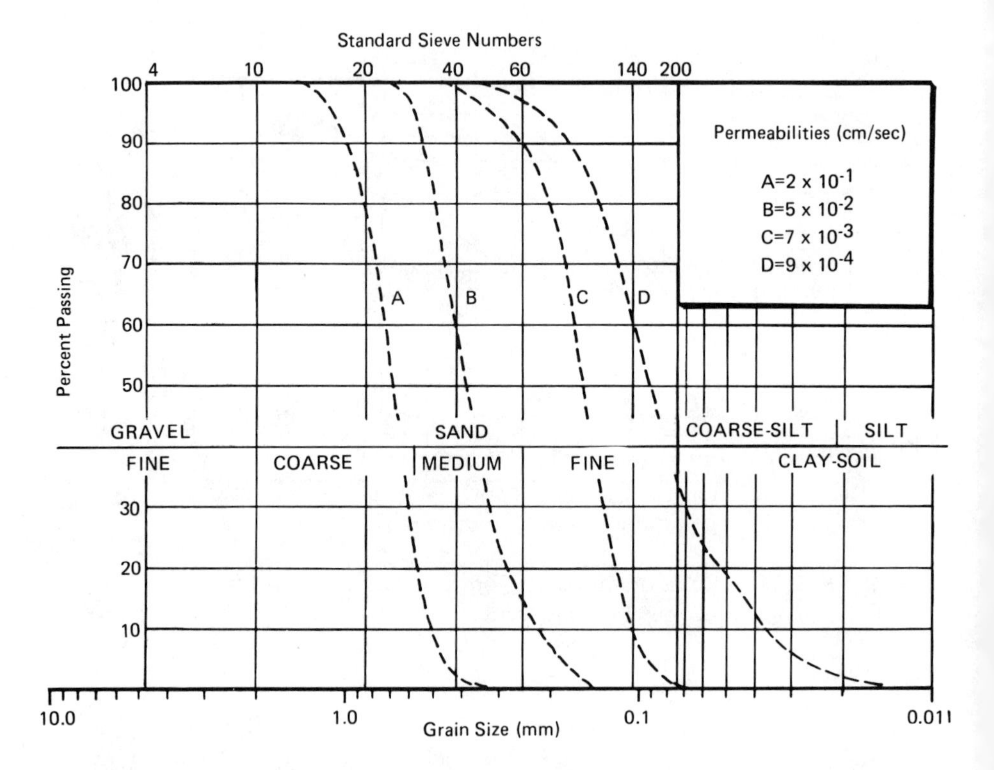

FIGURE 5.22 Grain size analysis.

location but less than the total pumping time. In Fig. 5.24, the gel time was less than the time at each vertical location. The improvement in uniformity of penetration is significant.

When the gel time decreases toward half the time the pipe stays at one location, the uniformity of penetration can be remarkable (at least in these lab experiments). Figure 5.25 is an injection into stratified A, B, and C soils. The difference in penetration is less than 2, even though the permeability differences are about 30. Of course, in the field it is probable that improvements in uniformity of penetration do not match those obtainable under controlled laboratory conditions. Actually, uniform penetration is unattainable in most field projects. It should be thought of as a goal, and the criteria presented here shall be interpreted as directions toward that goal.

FIGURE 5.23 Differential penetration in stratified deposits.

When grout is injected into a coarse stratum confined between two finer strata, the primary flow is horizontal through the coarse stratum. However, the coarse stratum acts as a source for secondary flow to occur vertically into the fine strata when long gel times are used. This phenomenon is shown in Fig. 5.26, where soil B acted as a source for grouting soil C, giving typical triangulated stabilized zones in the finer material. By trial and error, it is possible to adjust the gel time, the pumping rate, and the pumping volume so that most of the finer strata are grouted from the coarse. While such trial-and-error procedures make for interesting lab work, they cannot be implemented in the field because one cannot dig up and examine the grouted zones. Guessing at the proper relationship on a field project is far less likely to lead to success than the use of short gel times.

FIGURE 5.24 Improvement in uniformity of penetration through short gel times.

Even in varved deposits, the use of short gel times significantly improves the uniformity of penetration, as illustrated in Fig. 5.27, made in soils A and C. On the left-hand side, virtually all the grout flowed into the coarse strata, even though the gel time was considerably less than the total pumping time. On the right-hand side, a gel time was used equal to half the time the pipe stayed at each vertical location. Although penetration throughout the total stabilized mass is far from uniform, penetration into adjacent coarse and fine strata is very uniform.

In previous discussion, the advantages of short gel times in flowing groundwater were related to single-point injections. The conclusions reached should apply equally well to stratified deposits through which a grout pipe is retracted in stages.

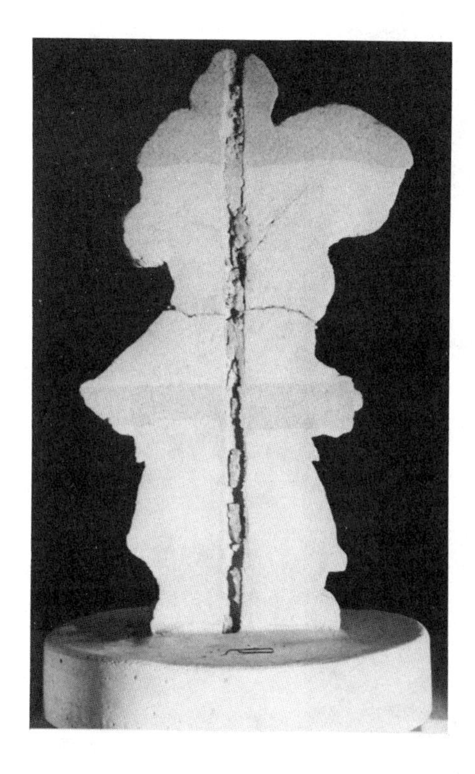

FIGURE 5.25 Very uniform penetration in stratified deposits.

Figure 5.28 shows an artificial profile created with sands A, B, C, and D. Numbers along the sides of the profile show the depth in inches. A horizontal flow of water at an average rate of 0.4 in./min was induced through the soils. Four separate injections were made by driving a pipe to the 17 in. depth and then retraciing in 1 in. stages every 30 s to the 5 in. depth. Gel times for the four injections were 0.25 min, 1 min, 4 min, and 10 min, respectively.

The stabilized masses resulting from the four injections are shown in their relative depth locations in Fig. 5.29. They are vertical cross sections taken in the direction of groundwater flow. The location within the stabilized mass of the injection pipe can be delineated in the photographs but is seen more clearly in the drawings of Fig. 5.30. These drawings are placed on a grid of 1 in. squares for scale purposes.

FIGURE 5.26 Secondary stabilization from coarse strata into finer strata.

The injection shown in Fig. 5.30a is the only one that was made with a gel time less than the pipe pulling time. This was also the only injection in which pure gel (indicated by the solid black areas) was found along the trace of the injection pipe. The gel very clearly outlines the successive positions of the pipe point (the grout pipe had a tapered driving point at the bottom and small holes in the sides just above the point).

It would normally be anticipated that flowing water would displace the entire stabilized mass in the direction of flow. In Fig. 5.30a, the two coarser sands *are* displaced in the direction of flow, but the two finer materials show more stabilized volume upstream. This can be rationalized as follows: In each stratum, the material initially pumped moved downstream in response to the groundwater

FIGURE 5.27 Improvement in uniformity of penetration through short gel time.

flow. In all strata, initial formation of gel thus occurred on the downstream side of the point. Subsequent gel formation had to occur behind this initial gel. In the coarser sands, where groundwater flow was rapid, initial gel formation occurred far enough downstream so that the total mass was displaced in that direction. (A somewhat different but equally plausible rationalization would be that at very short gel times the effects of flowing water are negligible, and the stabilized shapes are those which would occur under static groundwater.) In the finer materials, groundwater flow was slow, initial gel formation occurred closer to the pipe, and the injected volume in these strata were forced to flow upstream because of the gel barrier. Thus, longer gel times should result in downstream displacement for all materials. This is verified by Figs. 5.30b, c, and d.

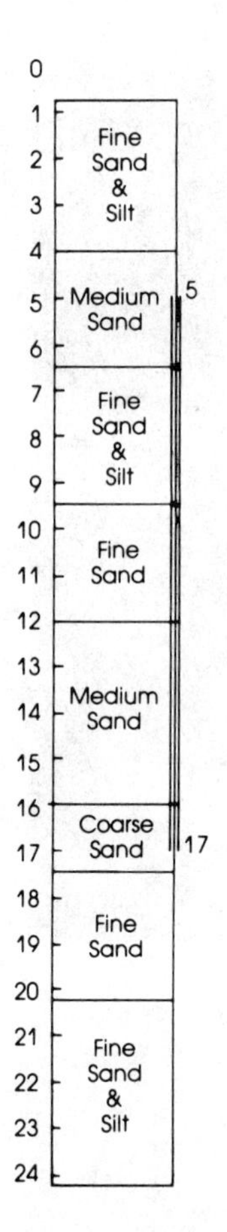

FIGURE 5.28 Profile of soils for experiments whose results are shown in Figs. 5.29 and 5.30.

Figures 5.30a and b show the results of an 0.25 min and 1 min gel time, respectively. Both these gel times are relatively short, but the difference in results is striking. Although dilution losses are almost negligible based on the measured relationship between the grout volume injected and the total volume stabilized, the displacement that occurred with slower gelation would make it extremely difficult to erect an effective cutoff curtain. Of importance also is the fact that at the somewhat longer gel time, all the material placed

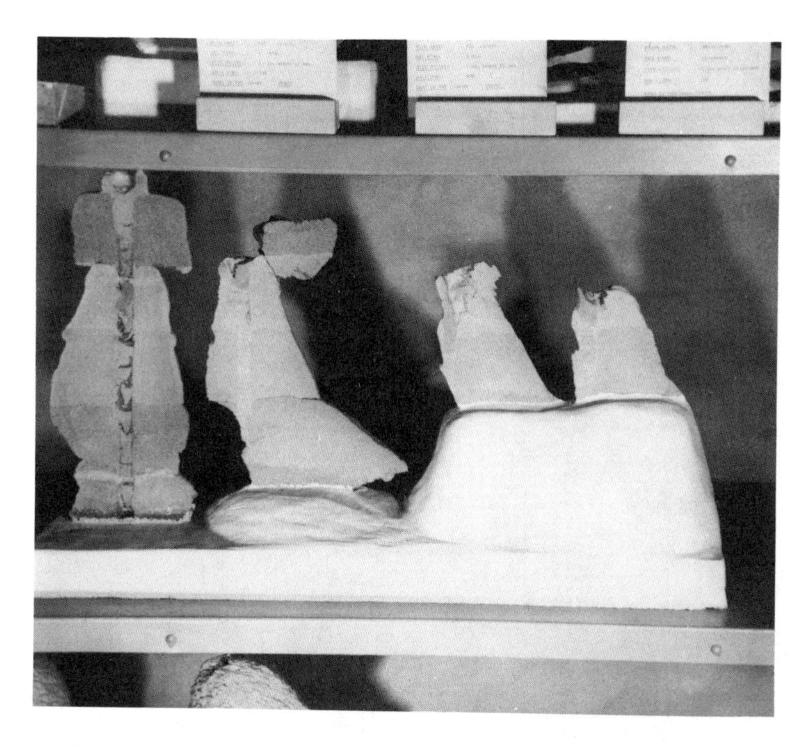

FIGURE 5.29 Photograph of stabilized masses resulting from varying gel times under flowing groundwater.

in the coarsest sand was completely washed away by groundwater flow.

At gel times of 4 min and 10 min (Figs. 5.30c and d), effects were very severe. For both these injections virtually all the grout was washed out of the two coarser sands. The stabilized masses that remain have volumes of half or less of those obtained with the shorter gel times and represent a very inefficient use of material. Further, this would also be an extremely ineffective attempt at erecting a grout curtain.

Laboratory experiments such as the one described indicate clearly that as the gel time decreases, when grouting in stratified deposits through which groundwater is flowing, more uniform penetration, gelation closer to the injection pipe, and smaller dilution losses are achieved. Although there may be some question about

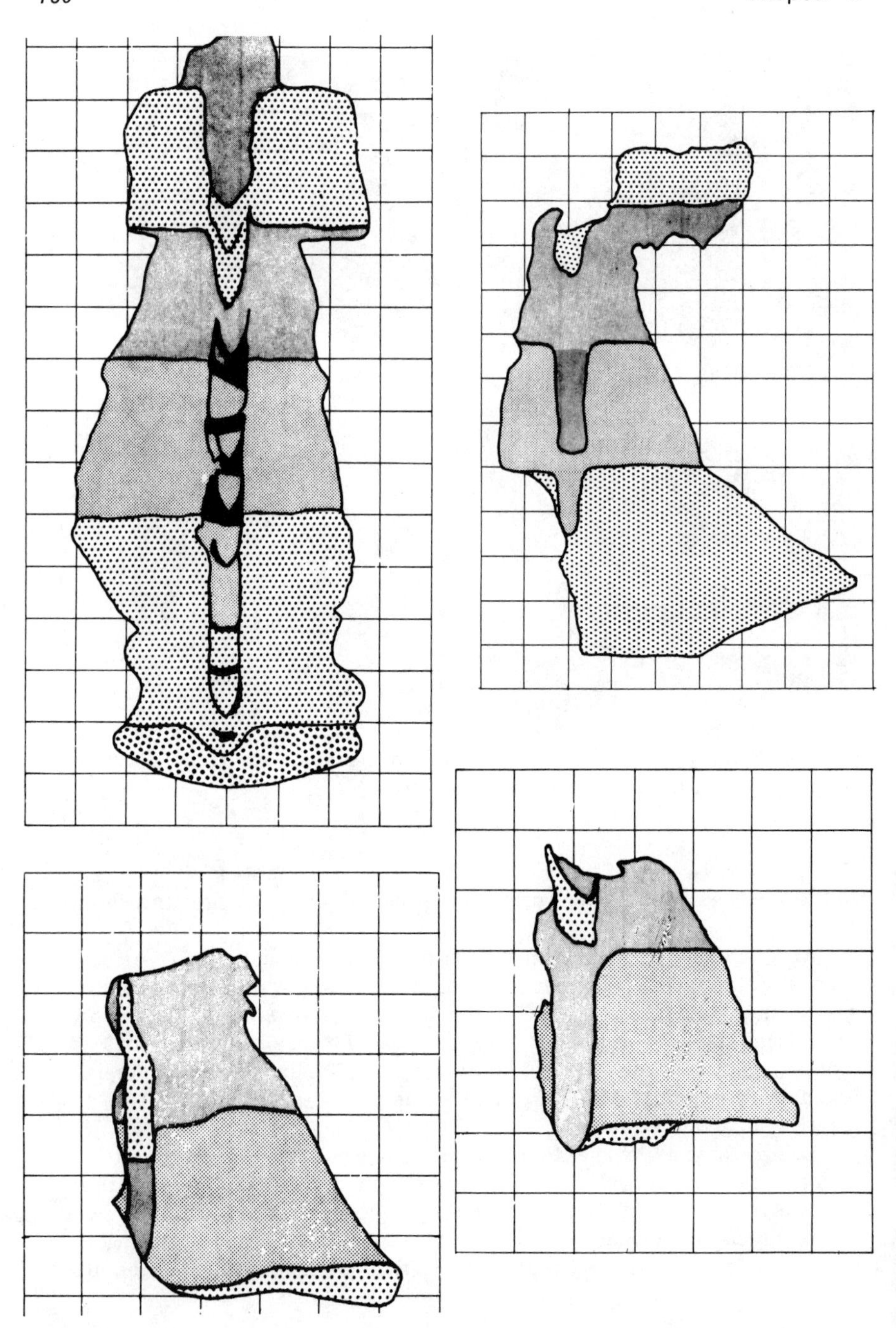

FIGURE 5.30 Drawing of stabilized masses resulting from various gel times under flowing groundwater.

linear extrapolation of small-scale experiments to field scale injections, field experience tends to corroborate using gel times of half or less of the time the pipe remains at each injection stage.

5.8 GROUT CURTAINS

Many field grouting operations have an ultimate purpose of placing a relatively impermeable barrier of considerable horizontal extent (and sometimes also a considerable depth) in a predetermined location. Such barriers are generally called grout curtains or cutoff walls. They consist of one or more interlocking rows of grouted soil or rock cylinders. Each of the individual cylinders is formed by the injection of grout through a pipe or drilled hole which has been placed in the formation. Hopefully these cylinders have relatively uniform cross sections throughout their depth.

To obtain as great a degree of uniformity as possible, it is desirable to grout short sections or stages of any individual hole so as to minimize the opportunity for grout to flow preferentially throughout the hole depth. The actual depth or stage is related to the grout take and to the pumping capacity. If take is very low per unit length of hole, each stage may have to be of considerable depth in order to be able to operate the grout pumping equipment at adequate capacity. When takes are high, stages can be short, usually of the order of 1 to 5 ft. In such cases stage length is often determined by economics since the cost of grouting is directly related to the number of stages per hole.

When grouting in open formations, the empirical relationships previously discussed for obtaining optimum uniformity of penetration should be used. In determining actual gel times it will be found that these are directly related to pumping rates and to stage depth, and for most occasions, all these variables cannot be predetermined with great accuracy. The initial injections of the actual grout curtains are generally used to arrive at values or ranges of values for stage length, pumping pressure, and gel time. These values, too, are subject to modification during the actual grouting operation, since the placement of grout in a portion of the curtain often affects the acceptance of grout in other parts of the curtain.

When grouting in very fine formations, whether these are soil or rock, gel times are often selected so as to coincide with the pumping time for each stage, rather than following the empirical relationships previously established. This is primarily due to the fact that in such formations pumping pressure relationships are generally

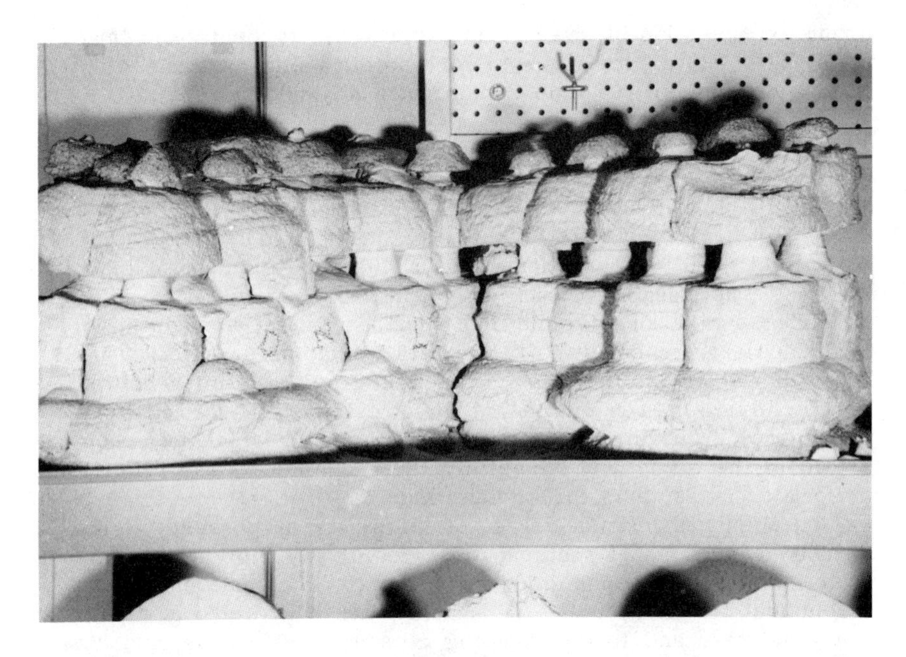

FIGURE 5.31 Grout curtain in stratified deposits using two- and three-row patterns.

severe and the selection of longer gel times often alleviates problems which may otherwise result due to high pumping pressures.

Figure 5.31 shows the results of a laboratory experiment in soils B, C, and D performed essentially to evaluate the results of two-row versus three-row grouting. Experience in the field, verified by laboratory work indicates very clearly that in linear patterns a minimum of three rows of holes is necessary in order to approach complete cutoff. This applies to the first treatment of formations with, essentially, overall permeability. In formations where the pervious zones are few in number, either through natural occurrence or by virtue of previous grouting, complete cut-off is often closely attained with one row of holes using split spacing (a method in which the distance between grout holes is halved by new grout holes until sufficient cut-off is attained). In Fig. 5.31 the right-hand portion of the pattern was made using two rows of holes. In the stratum of fine sand and silt, the average penetration was not sufficient to cause overlapping of the gel masses, resulting in fairly large open passages in this stratum. The left-hand portion of this curtain was grouted in the same fashion as the right, except for the addition of a third (central) row of holes. The photograph shows clearly that the injections in the third row, having no place else to go, were forced into the openings previously left in the least pervious stratum.

Even under the most ideal conditions that could be established in the laboratory, actual uniformity of penetration in stratified deposits may still leave something to be desired. Figure 5.32 is a photograph of an attempt to make a grout curtain with a single row of holes. All the empirical relationships previously developed for optimum uniformity were employed for these injections. In one stratum, however, average penetration is quite small. In this one zone, considerable flow could still occur. Even though uniformity of penetration in all the other strata is quite satisfactory, this curtain as a whole does not approach complete cutoff. This particular experiment clearly indicates the necessity for multiple rows of holes in curtain grouting.

Figure 5.33 shows another grout curtain made under laboratory conditions. This view is the downstream face of a three-row curtain, erected in stratified deposits under conditions of flowing groundwater. Uniformity of penetration in the various strata can be seen to be good. The photograph shows two areas in which overlapping did not quite occur. Records of this particular grouting operation indicated that the pertinent holes in the center row failed to take an adequate amount of grout. In addition to verifying the utility of a three-row pattern, this particular experiment also pointed out

FIGURE 5.32 Grout curtain in stratified deposits made with single row or holes.

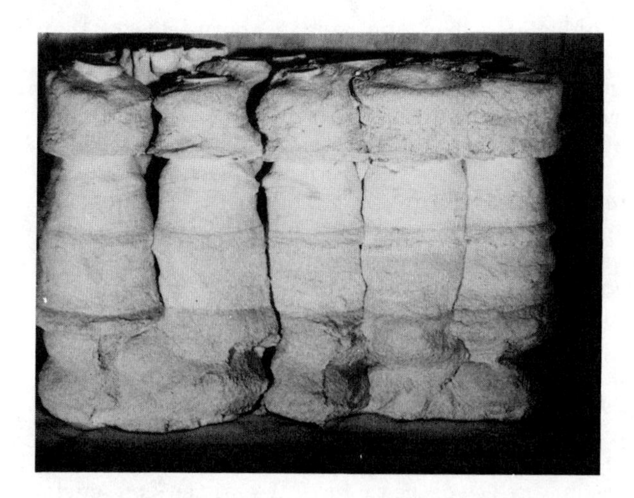

FIGURE 5.33 Outer row of three-row grout curtain pattern.

that the keeping of complete accurate records can often enable the prediction of areas in which a grout curtain did not close.

Field and laboratory experience continually emphasize the need for three-row grout patterns in linear cutoff grouting. In terms of resistance to extrusion and reduction of permeability, very thin cutoff walls would suffice. (A 6- to 12-in.-thick grouted curtain in sands and silts will support unbalanced hydrostatic heads of several hundred pounds per square inch. This is reflected in the fact that slurry trenches are generally very thin also.) In terms of constructing a solid, windowless grouted mass, much more thickness is required in practice, and chemically grouted cutoffs generally have 2 to 5 ft spacing of outer rows and holes.

The spacing between the outer rows of holes and spacing between holes in each row should be selected so as to form a pattern of squares. The grouting sequence is from one outer row to the other by diagonals. Thus, two passes completes the outer rows. The center (third) row of holes is placed in the center of each previously formed square.

The zones where fluid grout contacts previously gelled soil cannot usually be visually distinguished. In laboratory studies, the use of dyes can show very clearly the interlocking of the various grouted columns. Figure 5.34 is a drawing of a horizontal section through an experimental three-row pattern, showing the interlocking at that particular elevation. The small black circles indicate grout pipe location, and the numbers near the circles show the

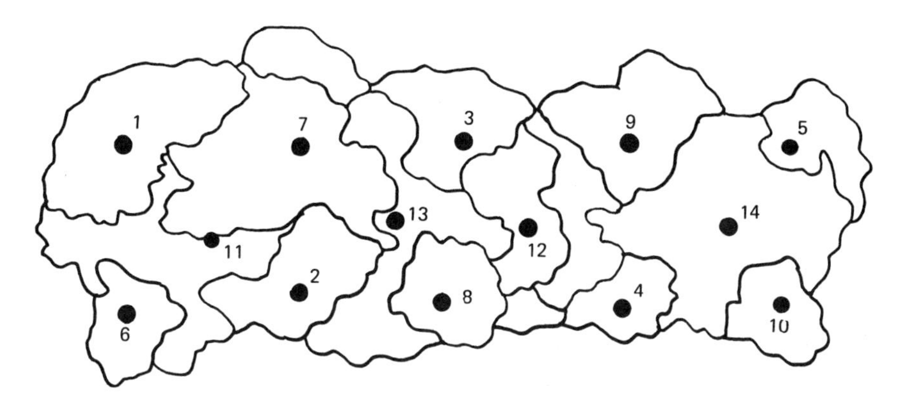

FIGURE 5.34 Interlocking of separate grout injections.

grouting sequence. The need for the third (central) row of holes is obvious.

In closed patterns such as a circle or square, the geometry of the pattern has the effect of providing confinement. Adequate results can generally be obtained with two rows of holes. For these cases the outer circle of holes is grouted first, working alternate holes, and then filling in the spaces. The inner circle of holes in the inner row are placed between hole locations in the outer row. The squence of grouting is to stagger holes and then fill in the gaps.

5.9 SUMMARY

The flow of grout through a natural soil or rock formation is governed by the porosity and permeability of the formation, the rate of which groundwater is moving, and the rate and pressure of the grout flow. These latter factors are related to the grout viscosity, and to its setting time.

The grouter generally has no control over the formation properties other than possible temporary changes in the groundwater level and velocity. Once the specific grout has been selected, ceilings are automatically established for grout flow rate and related pressure. Within those limits, the only controllable factor remaining is the setting time of the grout.

Haphazard (uncontrolled) flow of grout into a natural formation results in one or more of these negative results: (1) excess dilution with groundwater, which prevents gelation altogether, or significantly lengthens the gel time, (2) travel of the grout away from the zones where it is needed, and (3) selective travel into the most pervious zones, leaving open "windows" and channels in the formation.

Experience in the laboratory, verified in the field, indicates that the use of gel times shorter than the pumping time has beneficial effects in minimizing dilution with groundwater and excess grout travel, and in maximizing uniformity of penetration in stratified deposits.

Grout curtains or cut-off walls are constructed to create an impervious barrier in permeable formations. Generally, such curtains must be virtually complete in order to be effective. The use of short gel times is further enhanced by multiple rows of grout holes and the sequence in which each hole is grouted.

REFERENCES

1. R. H. Karol and A. M. Swift, Grouting in flowing water and stratified deposits, *J. Soil Mech. Foundation Div. ASCE*, 125–145 (April 1961).
2. R. H. Karol, Short gel times with chemical grouts, *Min. Mag.*, 148–152 (Sept. 1967).
3. R. H. Karol, Chemical grouting technology, *J. Soil Mech. Foundation Div. ASCE*, *94*(SM1):175–204 (Jan. 1968).
4. *Bibliography on Grouting*, U.S. Army Engineer Waterways Experiment Station, Miscellaneous Paper C-78-8.

6

Field Equipment

6.1 INTRODUCTION

The mechanical items used by the grouter to place chemicals into a formation are many and varied. They include (in order of decreasing size) drilling rigs, casing, drill rod and pipe, pumping systems, jackhammers, tank agitators, pressure hose, valves and fittings, measuring devices for pressure, volume, weight, etc. Much of the total equipment needed can be obtained readily on the open market and used with little or no modifications. In a similar fashion much of the technology related to the use of that equipment is also totally or mainly relevant to chemical grouting. Drilling rigs and drilling methods are excellent examples of tools and technology developed for other purposes but filling neatly the grouter's needs as well. Equipment, tools, and methodologies such as drilling, used by the grouter without significant modification, are not discussed here since details are readily available from other sources.

There are two main areas of equipment design and use that are specific to the chemical grouting industry: Pumping units and grout pipes and the equipment needed will be discussed in detail. In addition, the area of instrumentation deserves attention, since instrumentation requirements are vastly different from those used by cement grouters.

6.2 THE BATCH PUMPING SYSTEM

In its early stage of development in the United States, the chemical grouting industry was burdened by aspects of cement grouting technology forced upon it by grouters with cement grouting experience.

The heaviest burden was the batch system. By common usage definition, a batch is a mixture of all the materials which constitute the grout. For cement grouts, this would include the two ingredients which must always be present, cement and water. Also included might be fillers such as clay and sand and accelerators such as chloride salts. All the components are added to water to make the batch. After mixing, and generally with continuous agitation, the batch is pumped into the formation. It is obviously of importance to finish pumping the batch before it begins to set. Human nature being what it is, the first priority of the grouting crew becomes emptying the batch or dumping it before it sets. Engineering considerations take second place, at best.

There is excellent reason why more sophisticated pumping systems were never developed for cement grouting. The actual catalyst for the cement is the water in which it is suspended. It is not possible to separate the catalyst from the cement and pump each separately. Therefore, cement (and any other solid whose catalyst is water) will be forever limited to placement by batching (see discussion of small, successive batches in the following paragraphs). Chemical grouts, on the other hand, can generally be separated from their catalysts (even when the catalyst is water, as, for example the polyurethanes), and each component may be mixed and placed separately. This single factor is the one which opened new vistas in equipment design and use.

In a batch system, the total batch is catalyzed at the same time, and the entire batch gels or sets at the same time. (There may be minor differences due to dilution and temperature changes which various portions of the grout may undergo after placement in the formation.) Gel times can only be as short as the anticipated time of pumping the batch into the formation. When the grout sets, it all sets simultaneously and may block all the major flow channels and prevent pumping of additional grout. Therefore, gel times are generally very long (hours) in order to permit placement of a significant total volume. In contrast, if the grout and its catalyst are mixed in separate tanks, moved by separate pumps, and the two solutions meet first at the grout pipe, (1) there is never a danger of grout setting up in the tanks, pumps, or discharge hoses, and (2) each infinitesimal grout volume is catalyzed separately in time sequence, and at any given instant only a very small volume of grout reaches its gel time. Separation of the solutions makes feasible the use of gel times shorter than the time required to empty the tanks. Sequential time catalysis is the prerequisite for continuous pumping at times much longer than the gel time.

In general, the batch system is a relatively ineffective tool for placing chemical grouts. In two special sets of conditions, batching of chemicals may be acceptable. One of these is the requirement for very high pumping pressure, coupled with the need to place only a limited volume of grout. Such conditions might be met in deep shaft work when grouting fine sandstone or siltstone to form a cutoff around the shaft. Under such conditions, the volume required for each hole may be placed in one batch with a gel time equal to the pumping time.

The other condition where batching may be acceptable is actually an approximation of the two-pump process. If many very small batches (such as 5 gal) are mixed and placed sequentially, without stopping the pump, relatively small, but finite, volumes of grout reach the gel stage in time sequence. If this is done with one of the weaker grouts such as the acrylates or the chrome-lignins, it may be quite feasible to pump continuously at a gel time less than the pumping time for two successive batches. Obviously, as the batches decrease in size, the system approaches the more desirable multipump systems.

6.3 TWO-TANK SINGLE-PUMP SYSTEMS

The simplest way to ensure that the grout does not gel in the mixing tank is to use two tanks, one of which holds all the grout components except the catalyst, while the other tank holds only the catalyst. Fluid from each tank is fed to the suction of a pump. In the pump catalysis takes place, and the catalyzed grout is discharged through as short a hose as practical to the grout pipe. If some unforeseen occurrence interrupts the ability to pump, grout can set up in the pump and discharge hose, the same as it might in a batch system. However, the two-tank system is more effective than the batch because it does permit the use of short gel times and permits the operator to work at pumping times which exceed the gel time.

When using two-tank systems, it is convenient to work with twice the desired final grout concentration and to dilute the catalyst concentration in a volume of water equal to the volume of the grout solution. This simplifies the control of flow to the pump suction, since hand valves may be used to keep the levels in the two tanks equal.

A variation of this system is the interposition of a third tank between the pump and the other two tanks. Solutions are metered from the first two into the third, where catalysis takes place. This makes it easier to control fluid volume ratios when it is not practical to use equal volumes in the first two tanks.

The major advantage of two-tank single-pump systems is that they are far less expensive to construct than the more sophisticated multipump chemical grout plants normally recommended and used.

The major disadvantages are the inability to work at very short gel times (no shorter than the length of time to move grout from the pump suction through the pump and discharge hose and into the formation) and the inability to vary the gel time quickly and accurately, as might be dictated by changing field conditions.

6.4 EQUAL VOLUME SYSTEMS

If each of the tanks described in Sec. 6.3 is provided with its own pump and if the two pumps are exactly alike and operated by a common drive, the grout plant is commonly identified as an equal volume system. As in any system using two tanks, one tank contains all the grout components except the catalyst, and the other tank contains only the catalyst. In each tank, the components are mixed at twice the desired final concentration, and total volumes are kept equal for the two tanks. Thus, when the pumps blend the two liquids on a one-to-one volume basis, the concentration of each component is automatically reduced to its final desired value.

Equal volume systems eliminate the problems associated with passing catalyzed grout through pumps and discharge hose. They also permit the use of very short gel times and can be used to pump for periods which greatly exceed the gel time. The major disadvantage is the difficulty in accurately changing the gel times while pumping because the volume of liquid in each tank is constantly decreasing, making it difficult to add catalyst or inhibitor at a specific concentration. For applications where gel time changes are not needed during pumping, equal volume systems are economical and perform very satisfactorily. Such applications include many seepage problems in fissured rock and porous concrete, where grout volume requirements are low.

Figure 6.1 is a low-volume high-pressure equal volume system designed specifically for work in porous and fractured concrete. The pumps are operated by an air cylinder, which (by control of the input air pressure) permits the pumps to slow down and then still as the grouting pressure approaches and then reaches the predetermined allowable maximum. (As an index to size, the pressure gage on the air supply regulator is 2 in.) Figure 6.2 is a unit very similar in design to that shown in Fig. 6.1 but with lower pressure capacity and higher volume output. It, too, is operated by a reciprocating air cylinder, which is readily adapted to linear-motion piston pumps.

The combinations of components which will result in adequate equal volume systems are almost limitless. Except for centrifugals, which have too much slippage at low volume, any positive displacement pumps whose materials are compatible with the grout components

FIGURE 6.1 Low-volume equal-volume system.

FIGURE 6.2 Equal-volume system.

may be used. Other forms of motive power may of course also be used but require separate controls to avoid overpressuring of the formation.

Despite its name, the two pumps in the system need not be of equal volume capacity. It is perfectly possible to build satisfactory systems using pumps with 2:1, 3:1, or other ratios. This may be advantageous when the specific grout can be more easily handled at other than 1:1 ratios.

6.5 METERING SYSTEMS

To overcome the major deficiency of the equal volume system, it is necessary to add a mechanism that permits rapid and accurate control of gel time changes during pumping. Since gel times are generally more responsive to catalyst concentration than to other variables, a mechanism for concentration is indicated. Further, changes in catalyst concentration must be made in a fashion that does not change the gel properties or affect other variables which might influence gel time or gel strength (for grouts whose properties are closely related to catalyst concentration and/or gel time, effects cannot be avoided). Thus, while changes in the catalyst concentration can be made simply by changing the speed of the catalyst pump, this cannot be done with typical equal volume solutions, since it would cause significant (inverse) changes in the concentration of other grout components. The only way in which this problem can be overcome is to use a very concentrated catalyst solution. In this fashion, very small volumes of catalyst, not enough to significantly dilute other grout components, will still provide a wide range of gel times. While there may be other equally effective methods of designing a chemical grout metering pump set, the most obvious direction is to use very concentrated catalyst metered into the grout discharge line by a small, accurate, variable-volume pump.

Figure 6.3 is a line diagram showing the major components of a chemical grout metering system. Not shown on the diagram are two additional necessary elements: (1) instruments to indicate actual discharge flows from each pump and (2) a pressure-actuated shutoff system to prevent inadvertent overpressuring of the formation.

Figure 6.4 shows a field metering system rated at 10 gpm and 225 psi, built in accordance with the diagram of Fig. 6.3. Glass tube flow meters are used on the discharge end of each pump to give constant readings of discharge volumes. This instrumentation is important whenever pumps with slippage and wear characteristics that affect output are used. Each pump has a valved line leading

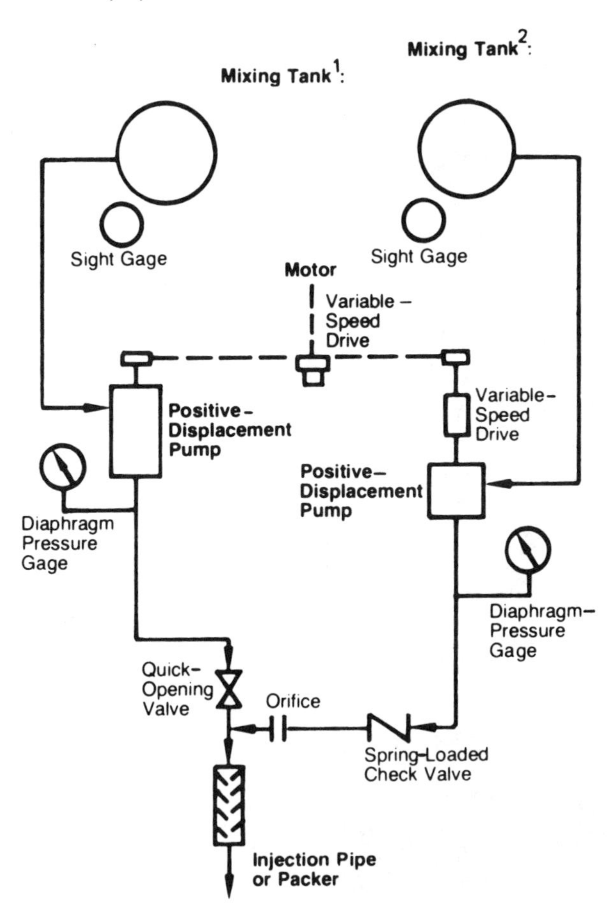

FIGURE 6.3 Components of a metering system.

from the discharge to the suction end. By adjusting this valve the output of either pump can be changed independently. Metering is accomplished by adjusting the pump outputs to the desired values or ratios. This specific unit is driven by an air motor. The total output can be varied while maintaining the ratio output of the pumps by regulating the air supply to the motor. A pneumatic controller (the black box with three gages) is used to stay within pressure limitations. A signal from the discharge line operates a valve which shuts off the air supply to the main motor at any preset pressure.

FIGURE 6.4 Chemical grout metering system.

FIGURE 6.5 Chemical grout metering system.

When the line pressure drops to any present lower value, the air supply valve is automatically opened, and pumping resumes. For electrically driven pump units, similar controls are available to shut off and restart the motor.

Figure 6.5 is yet another metering system, rated at 6 gpm and 400 psi. The catalyst pump is a reciprocating single-piston variable-strokelength pump. Pump output can be reliably set and monitored by the strokelength dial, so a separate flow meter is not needed. The flow meter for the larger pump is a dial-type orifice meter. Motive power is air, monitored by a pressure controller.

The pump sets of Figs. 6.4 and 6.5 look different, use different components, and have somewhat different operating characteristics. Many different systems could be assembled from commercial stock components. Although they too would have different operating characteristics, all well-designed metering systems have one feature in common beyond those of other grout plants: the ability to change and control gel times accurately, while pumping grout, by simple mechanical manipulation of the system's controls.

For most grouts, the gel time control does not extend infinitely but instead covers a finite range determined by the concentration of chemicals in the stock solutions and the operating temperature

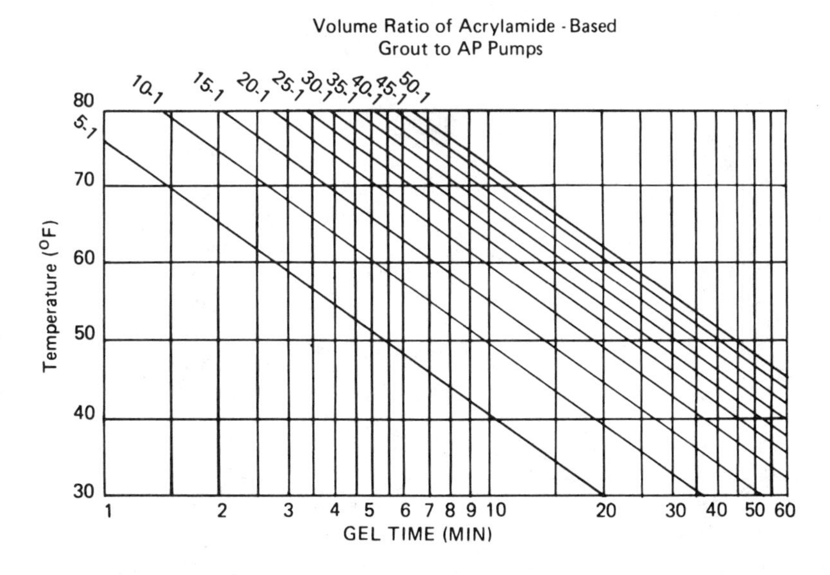

FIGURE 6.6 Metering system—gel time chart. Solution A: 10% acrylamide-based agent, 0.4% dMAPN, and 100 ppm KFE. Solution B: 20% AP.

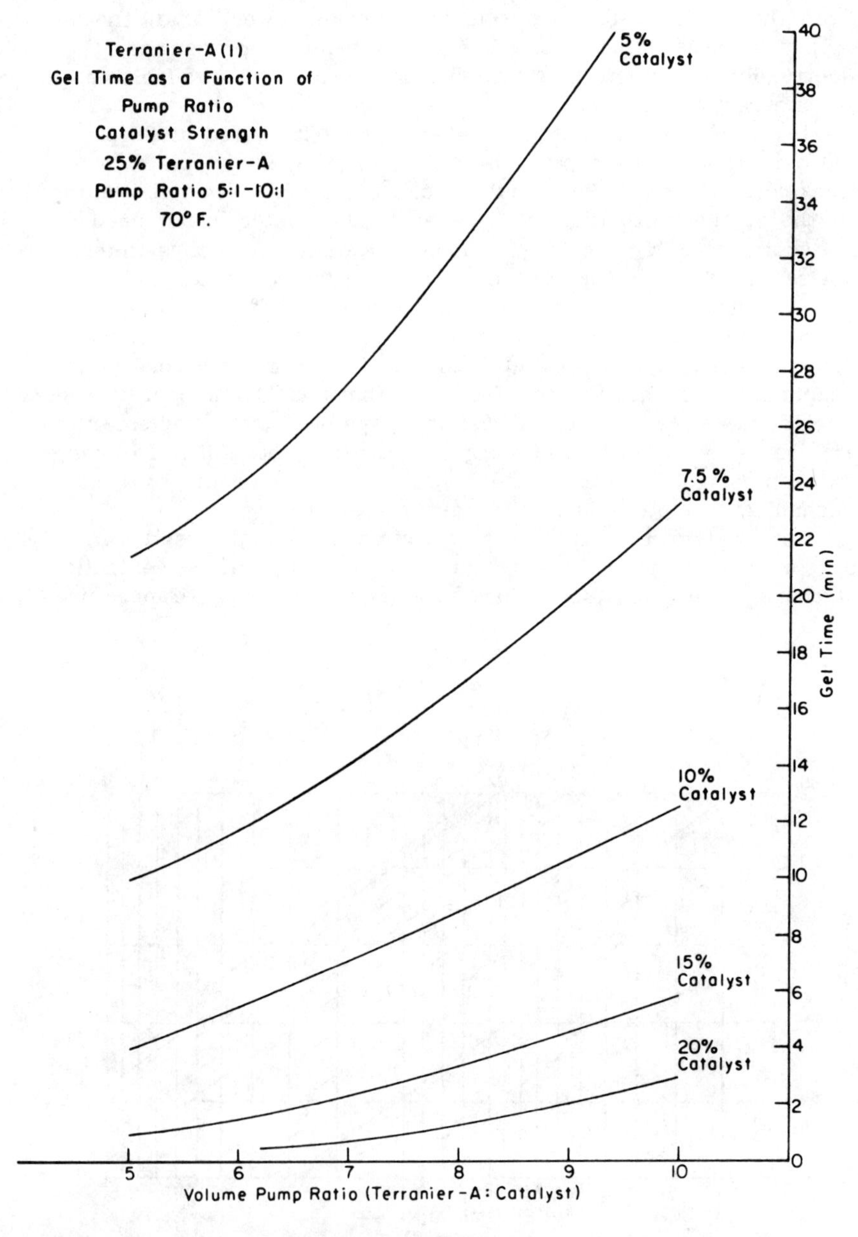

FIGURE 6.7 Metering system—gel time chart. (Courtesy of ITT Rayonier, Inc., Whippany, New Jersey.)

range. The control readings for any pump set may be calibrated directly to specific stock solutions for each different chemical grout. Another more general procedure is to calibrate the pump readings which represent volume output against each other to give a volume ratio. These data can then be used directly with any grout for which volume ratio gel charts are prepared. Figures 6.6 and 6.7 illustrate two such charts, based on different stock solutions, for an acrylamide and a phenoplast grout. For general usage on small jobs, chemical grout plants with 5 to 15 gpm volume capacity and 200 psi to 400 psi pressure capacity are generally adequate. For specific known field conditions, variation may be necessary.

6.6 OTHER CHEMICAL GROUT PUMPS

The criterion for an adequate pumping system for a specific job may be defined as one with adequate volume and pressure capacity and controls which permit full exploitation of those grout properties relevant to the job. For general use on future jobs, it makes sense to design the most flexible system possible, i.e., a metering system. For specific applications it is often feasible to design and use less flexible, and less expensive, systems. Two variations are of interest.

An equal volume system can be converted to a metering system by the addition of a third pump. This should be a low-volume positive displacement type with inherent or added capability for volume control. Either concentrated catalyst or inhibitor can be added to the grout in this fashion, and gel times can thus be changed while pumping.

Another alternative is to use one or more large triplex pumps for the grout and a much smaller triplex for the catalyst. The discharge ends of each pump must be rebuilt to provide a separate discharge for each piston and the piping so arranged that the discharge from each piston may be either sent to the grout pipe or recycled to the solution tanks. The various combinations possible between the grout pistons and the catalyst pistons give a number of different grout setting times for any given tank concentrations.

6.7 COMPONENTS AND MATERIALS OF CONSTRUCTION

By far the most important (and probably the most costly) components of a chemical grouting plant are the pumps themselves. An almost unlimited number of commercial products are available in a wide choice of configurations, operating principles, construction materials, and capacities. Centrifugal pumps generally have too much slippage to work well in a metering system. Rotary, vane, diaphragm, screw, and gear pumps as well as single- and multiple-piston pumps may all be used. Although there is no unassailable proof, many grouters feel that pumps that give a pulsating discharge pressure tend to give better penetration of the formation.

Most grouts are either solids dissolved in water or liquids diluted with water. In terms of their pumping characteristics, they are similar to water; that is, they are nonlubricating liquids of low viscosity, which tend to corrode iron and steel. Many commercial pumps are manufactured specifically to transport heavy petroleum products. These pumps will not function adequately with chemical grouts. There are also pumps designed specifically for high-viscosity liquid or slurries. These, too, should be eliminated from consideration. In terms of materials of construction, stainless steel is the best as well as the most expensive and probably the least available. Construction plastics also generally have excellent corrosion resistance, but pumps made of plastics have comparatively low pressure capacities. Iron and steel will be subject to at least the same degree of corrosion and wear as if water were being pumped. Often, such pumps can be obtained with stainless pistons and valves and ceramic-lined cylinders, which make them totally acceptable for handling chemical grouts.

Pumps for handling catalysts should always be made of stainless steel in the wetted parts. Catalysts are generally very active chemicals and would corrode iron, steel, and bronze very quickly, sometimes in only a few hours. Positive displacement pumps with minimal slippage characteristics should be used, since volume output of the catalyst pump is the only gel time control factor during a pumping operation.

Valves and fittings should be of stainless steel on the discharge side of the catalyst pump. On the suction side, plastic valves and fittings (compatible with the catalyst) may be used. Valves and fittings for the grout pump discharge piping are preferably stainless, but iron and steel fittings may be used if properly maintained and replaced at regular intervals. Brass and bronze fittings are less desirable but may also be used. On the suction side, plastic valves and fittings are adequate.

For use as open–shut switches, quarter turn ball valves are convenient and show by handle position whether they are open or closed. For restricting flow, as in a bypass system, needle valves perform best. Hose of synthetic rubber is available in a wide variety of sizes, pressure capacities, and abrasion-resistant outer casings. Such hose is suitable for both the suction and discharge ends of the pump. On the suction end, noncollapsing hose must of course be used.

Tanks should be of stainless steel, polyethylene, or other suitable plastic. Coated steel tanks are adequate but require constant maintenance. The cost of proper maintenance will usually dictate the use of other tank materials. For some chemicals, aluminum tanks are satisfactory. Under these conditions, aluminum valves and fittings are also adequate; however, they wear more quickly than stainless steel. Except for single-pump batch operation,* large-capacity tanks are not required. The use of four small tanks provides continuous operating capacity. Two of the tanks are connected directly to the pump suctions. The other two are located vertically above the first two. Mixing of solutions is done in the upper two tanks. These solutions are transferred by gravity into the lower tanks as those tanks are emptied by the pumps. The actual size of the tanks need be only large enough to provide mixing time before the lower tanks are emptied. For example, if 5 min are required to mix new stock solutions of grout and catalyst and the grout plant has a maximum pumping capacity of 10 gpm, mixing tanks need not be larger than 50 gal. For most small field jobs, tankage of 30 to 60 gal capacity is adequate. For large jobs, it is often financially expedient to design equipment and tankage for each specific job. Under these conditions, large tanks may offer advantages.

Measuring devices required on grout pumps are for monitoring volume and pressure. Commercial flow meters are available in adequate size and accuracy to cover grouting needs. Accuracy of measurement should be 2% or better. Vertical glass tube meters are the most accurate, can be visually checked for performance, and are also most readily subject to breakage. This type of meter is shown in Fig. 6.4. Indirect measurement of flow through the use of calibrated orifices, or imparted energy, is also accurate provided the function is not disturbed by solid particles lodged in critical places. These types of meters, such as shown in Fig. 6.5, should be checked regularly in order to ensure that they remain in calibration. Since the grout and the catalyst physically contact the operating parts of most flow meters, the materials of construction must be compatible with the chemicals. One source of trouble often overlooked is the use of glass flow meters with silicates. Etching of the glass, of course, occurs.

Pressure gages are also readily available commercially in adequate capacities and accuracies for grouting purposes. Easy-to-read gages are usually 3 to 4 in. in diameter. Accuracy should be 2% or better. Gages should be oil filled and fitted with a protector which keeps

---

*As previously discussed, not recommended for use with chemical grouts.

the liquid from entering the gage internals. The grout plant gages, like the flow meters, must be located in the piping system before the grout is catalyzed. Flow meters (or other flow measuring devices) are needed in both the grout and the catalyst piping. A pressure gage in the catalyst system is desirable but not necessary.

Metering system controls have three separate functions: (1) volume control, (2) pressure control, and (3) gel time control. The volume control is most readily accomplished by a variable-speed operating directly on the main drive. On a gas engine, the built-in throttle serves this purpose. Variable-pitch pulleys may be added to extend the range of operating speeds. (These pulleys may also be added to air and electric motors for the same purpose.) The speed of air motors may be regulated by adjusting the pressure and volume of air supply. A valve and an air regulator are needed in the air supply for this purpose. Electric variable-speed drives (SCR systems) are readily available commercially. All three types of drive systems, properly sized, perform the volume control function adequately. Gas engine drives give complete independence from outside power supply. Air drives are the least hazardous. Electric units are the quietest and most accurate.

The pressure control function is primarily used to prevent inadvertent overpressuring of a formation. During grouting, pressure is most expediently kept at or near the desired operating level by control of the volume pumped (these two parameters are inversely related). When pumping by design or necessity at or near the maximum allowable pressure, protection against damage that might occur from unexpected pressure surges is needed. Such protection can be provided by adjustable blowoff relief valves placed in each discharge line. These valves are generally piped to the pump suction or the solution tanks to prevent wasting chemical solutions. The main problem associated with the use of blowoff relief valves is due to the difficulty of setting two valves to open at exactly the same pressure. Therefore, either the grout valve recycles before the catalyst valve, resulting in some catalyst being pumped into the grout pump, or the reverse happens. In either case, chances are that gel will form in some of the valves, fittings, hoses, and pump components. This will mean downtime before grouting can be resumed. However, the formation is protected from overpressure, and there is no other instrumentation that will function better on a gasoline engine. For air and electric motors commercial instrumentation is readily available which accepts a pressure signal from the discharge lines and shuts off and restarts the main drive as preprogrammed. Shutting the main drive stops both pumps simultaneously, so there is no chance for catalyst to flow into the grout line or vice versa.

6.8 PACKERS

Where the grout hoses meet the hole, pipe, drill rod, or casing through which grout will be placed, a positive leakproof joint is required. Direct connection through pipe fittings can be made to threaded pipe, drill rod, and casing. No such connection is possible to a hole drilled in rock, soil, or concrete, and other devices must be used. These devices are variously called packers, balloon packers, downhole packers, stuffing boxes, isolators, and pressure testers.

A stuffing box is shown in Fig. 6.8. This device threads onto the exposed end of a (large) pipe or casing, and it has a hole through which a smaller pipe can be inserted. A seal is made around the smaller pipe by packing, which is tightened by turning a threaded gland support. Obviously, stuffing boxes are designed for specific sizes of pipe and casing. Their major utility is in making a quick seal for a small pipe entering a much larger pipe.

A typical mechanical packer is shown in Fig. 6.9. By rotating the handle, the flexible rubber sleeve is compressed, and it expands to touch and make a seal against the walls of the pipe or hole or at any desired distance along the hole. Sections of standard pipe, drill rod, or casing may be connected to locate the packer away from the hole collar. Such packers are often referred to as downhole packers.

FIGURE 6.8 Stuffing box. (Courtesy of Sprague and Henwood, Inc., Scranton, Pennsylvania.)

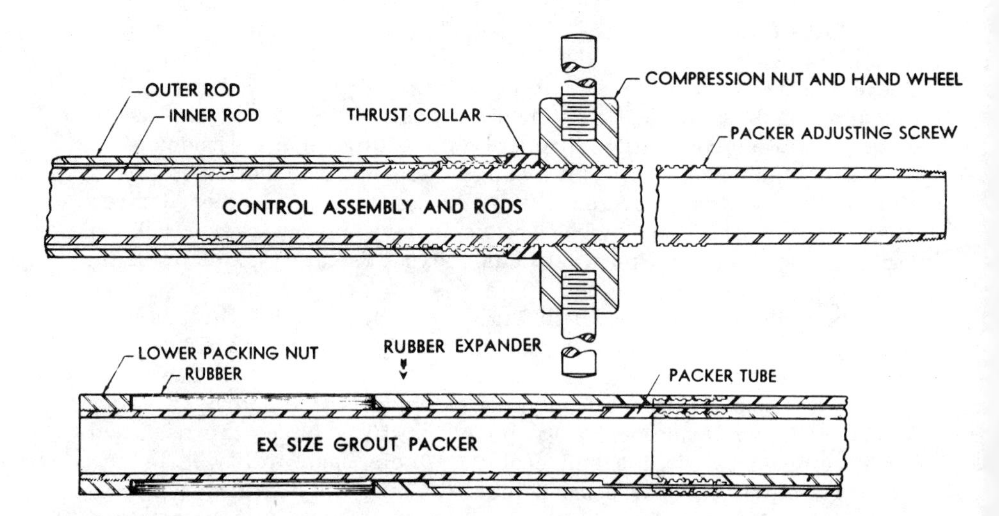

FIGURE 6.9 Mechanical packer. (Courtesy of Sprague and Henwood, Inc., Scranton, Pennsylvania.)

A typical pneumatic, or balloon, packer is shown in Fig. 6.10. A pneumatic packer may be placed at the collar of a hole or moved downhole by the addition of sections of drill rod or pipe. Only one pipe is needed, in contrast to mechanical packers which require a pair of concentric pipes. Pneumatic packers are therefore placed more quickly. In vertical holes, or holes close to vertical, pneumatic packers may be lowered on a wire line which makes placement very quick and simple compared to mechanical packers.

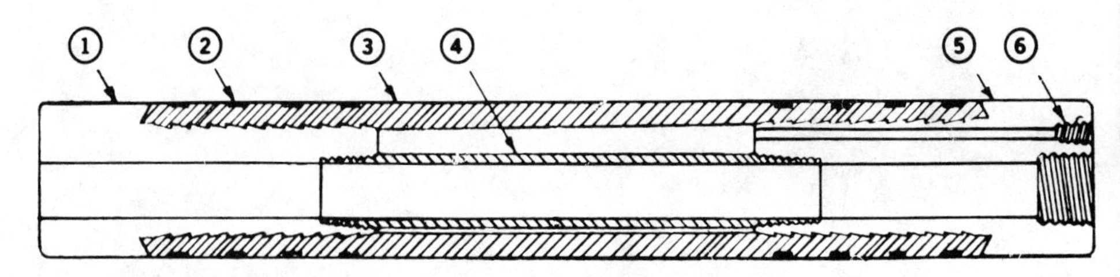

FIGURE 6.10 Pneumatic packer. 1, lower head; 2, punchlock clamp; 3, rubber sleeve; 4, grout tube or packer stem; 5, upper head; 6, air supply.

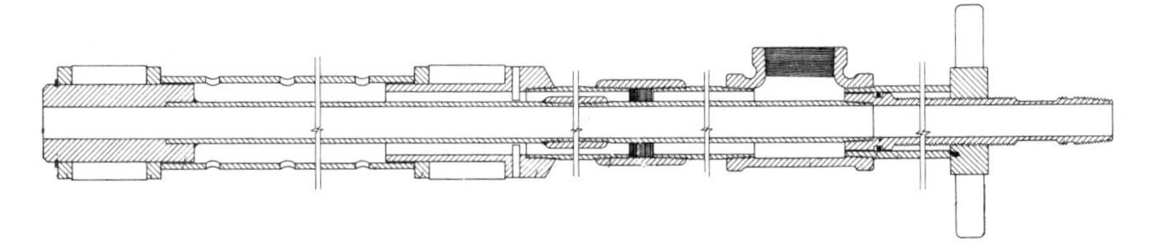

FIGURE 6.11 Mechanical double packer. (Courtesy of Sprague and Henwood, Inc., Scranton, Pennsylvania.)

When a packer is placed and sealed, the entire length of hole between the packer and the bottom or end is open to the flow of grout. There are many times when it is more desirable from an engineering point of view to isolate the grout flow to a specific short length of hole or pipe. Two packers placed close to each other in the same hole may be used for this purpose. Figure 6.11 shows a mechanical arrangement to isolate any desired length of grout hole. This is variously known as a double packer, an isolator, or a pressure tester.

6.9 GROUT PIPES

Grout holes, the final means by which grout is fed from the mixing tanks to the desired location within a soil or rock mass, can be placed by drilling, driving, or jetting. Except for elatively deep placement in loose granular deposits, jetting generally disturbs the formation surrounding the pipe to an extent that makes grouting ineffective. In soft and hard rock, as well as cemented sand and siltstone and concrete, grout holes must be placed by drilling. Tools for drilling holes can range from simple pneumatic jackhammers through sophisticated truck- and trailer-mounted rigs. Beyond the jackhammer stage, drilling is best left to specialists, and process and procedures are not discussed here.

In most soils, except for those which contain more than isolated cobbles and boulders, grout pipes can be placed by driving with simple equipment such as jackhammers and cathead-operated falling weights. For very shallow depths, even sledge hammers may suffice. Standard water pipe in 0.5 and 0.75 in. sizes have been used successfully on many small jobs. These pipes, however, do not stand up long under driving stresses. In the long run, drill rod and casing will prove easier to handle and more economical as well. Drill rods and casing are commercially available in standard

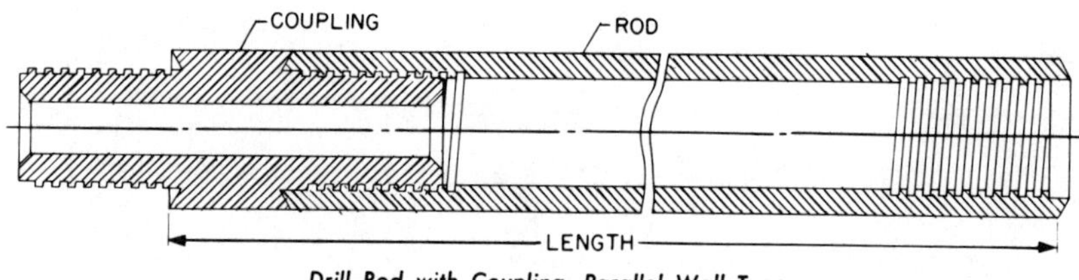

Drill Rod with Coupling, Parallel Wall Type

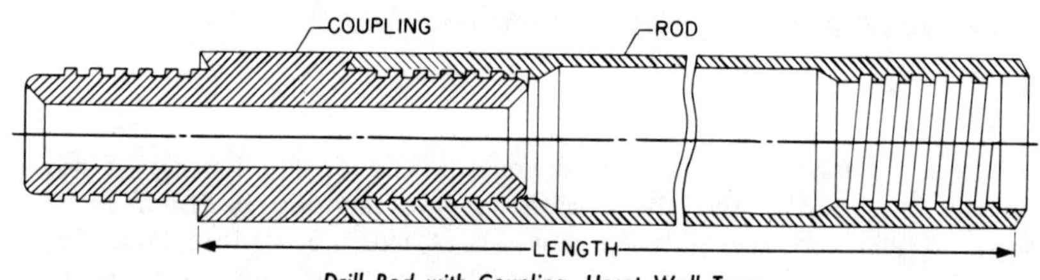

Drill Rod with Coupling, Upset Wall Type

DRILL RODS

Size of Drill Rod		E	A	B	N
Specifications	Outside Diameter	1-5/16"	1-5/8"	1-29/32"	2-3/8"
	I.D. of Parallel Wall Rod	13/16"	1-1/8"	——	——
	I.D. of Upset Wall Rod	——	——	1-13/32"	2"
	Bore of Coupling	7/16"	9/16"	5/8"	1"
	No. of Threads Per Inch	3	3	5	4

FIGURE 6.12 Standard drill rod. (Courtesy of Sprague and Henwood, Inc., Scranton, Pennsylvania.)

sizes from many sources. Dimensions are shown in Fig. 6.12 for drill rods and Fig. 6.13 for casing.

Grouting through pipe, drill rod, and casing is often done through the open end, without the need for a packer. Pipe is generally driven to the maximum depth or distance to be grouted and kept empty during driving by a bolt or rivet taped to the end to keep soil from entering. This end plug must be pushed out by pump pressure when grouting begins. Drill rods and casing may also be drilled into place, with the drill cuttings removed by continual flushing with water. This process keeps the rod or casing from being plugged by the formation.

In the procedures just described, all the grout flows out of the open end of the grout pipe. However, it may enter the formation

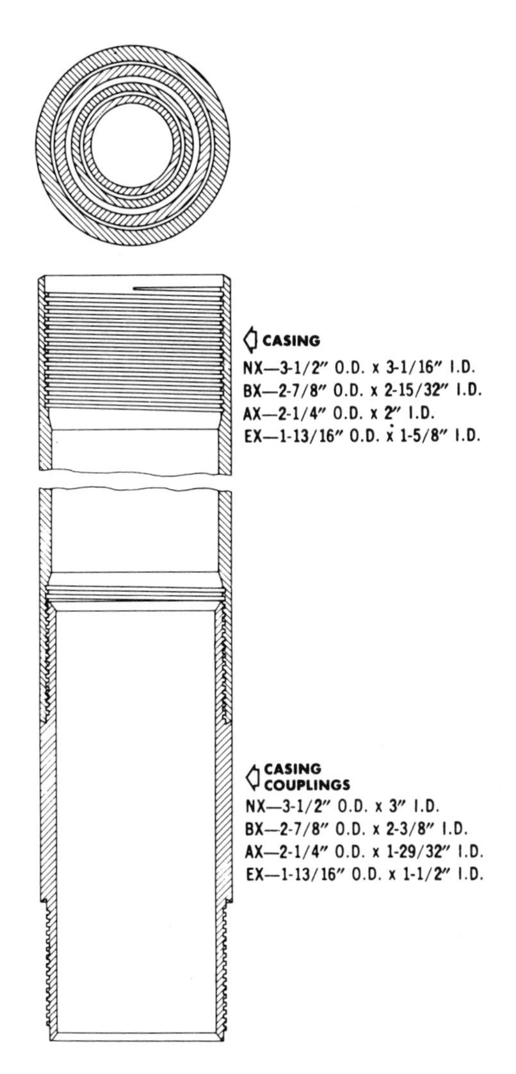

FIGURE 6.13 Standard casing sizes. (Courtesy of Sprague and Henwood, Inc., Scranton, Pennsylvania.)

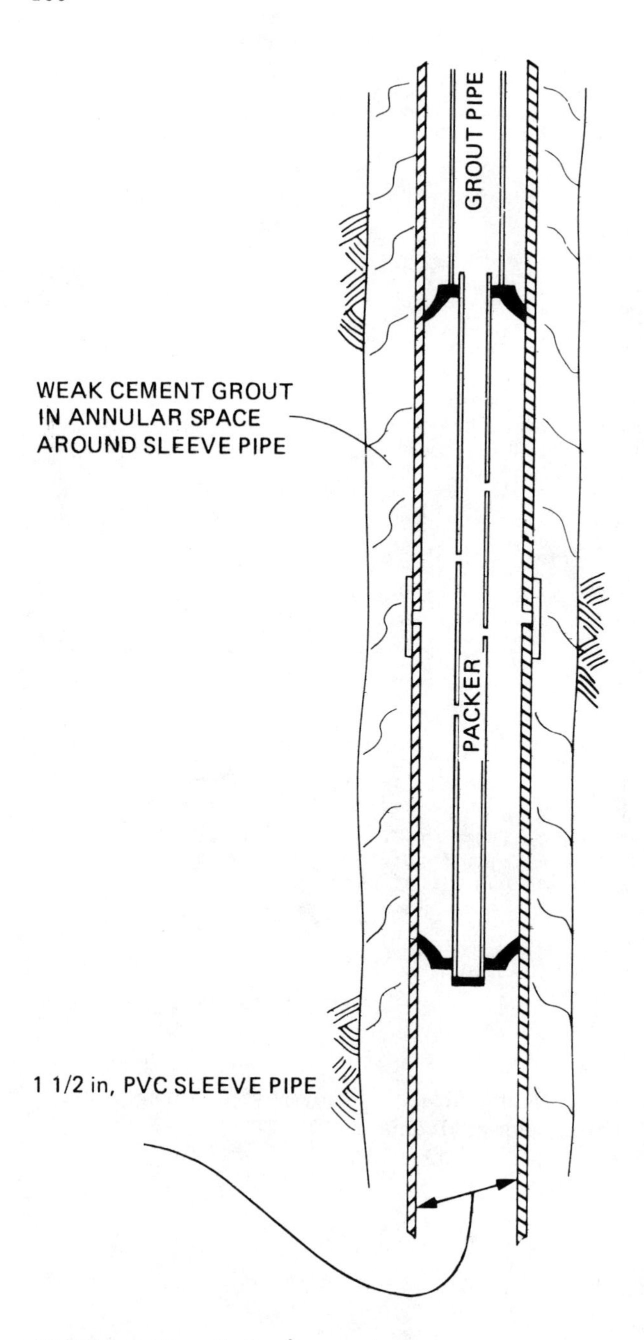

FIGURE 6.14 Tube à manchette.

at any point between the bottom of the grout pipe and the bottom of the hole (after withdrawal of the pipe begins). If the bottom of the hole is in a very porous zone, all the grout may go into this zone regardless of how much the pipe is pulled. A more sophisticated system which gives the grouter complete control over where the grout enters the formation was developed in Europe and is now used on larger jobs in the United States. Called a *tube á manchette*, it is actually a combination of equipment and procedure. The grout pipe itself is basically one in which small holes are placed at regular intervals and covered on the outside by a short piece of rubber tubing. This assembly is placed in a predrilled hole, and the annulus between the formation and the assembly is filled with a slurry of neat cement and bentonite, which is permitted to set (and forms a weak, brittle grout). In use, the holes at any desired depth are isolated by a double packer, and grout is pumped first at high pressure to expand the rubber sleeve and fracture the brittle grout annulus and then at normal pressure as the grout begins to flow into the formation. Figure 6.14 illustrates a section of a tube à manchette.

6.10 INSTRUMENTATION

Every chemical grout plant must contain accurate devices for measuring the grout flow rate and pressure—while grouting. In addition, there must be a fail–safe system for stopping or reducing the grout flow if the pressure exceeds predetermined allowable values. Pressure and flow instruments of many different types are commercially available. Any direct-reading instrument of adequate accuracy and sturdiness is suitable. Control of pumping rate and pressure may be completely in the hands of the operator, or may be handled automatically by sensors that send appropriate signals to the power plant. Suitable electronic and pneumatic sensors are readily available commercially.

In addition to the "must contain" items, other controls are desirable. Most important of these is a systems for controlling and changing the gel times, preferably while the plant is operating. Universally, gel time control is accomplished by varying the quantity of catalyst or inhibitor blended into the grout (see discussion of metering systems, Sec. 6.5). The volume of catalyst in the grout may be controlled by using valves to divert a portion of the total flow to the holding tank, or the pumping rate itself may be varied by either changing the pump speed or the piston stroke length (for positive displacement piston pumps). Again, many commercial products are available for both methods. It is also highly desirable to have thermometers in the system, and a location near the grout pipe where catalyzed grout can be sampled to check gel times.

If manifolding is to be used, more extensive instrumentation is required. Each grout pipe must have its own flow meter, and should have its own pressure gage. In addition, a shut-off valve is needed at each pipe. To avoid the necessity of having a man stationed at each pipe, sensors may be used to display signals on a control panel at the pump, so that one operator can read and control the ongoing flow of grout to many pipes.

It can be seen from the variety of suitable commercial equipment, it isn't practical to specify a "standard" instrumentation set. Figures 6.4 and 6.5 show one possible configuration for a grout plant control panel to handle all expected field conditions.

6.11 SUMMARY

As in other areas of technology, chemical grout pumping systems (in the early days of chemical grouting) suffered from the legacy of cement grouting practice.

It is now recognized that batch systems are ineffective and inefficient for pumping chemicals into a formation. Modern chemical grout pumping plants are sophisticated systems which control the gel time precisely through mechanical functions and which record accurately the ongoing parameters, so that the grouting engineer can modify those parameters to best suit local conditions.

A wide variety of commercial products is available from which suitable pumping systems can be assembled. Equal volume and proportioning systems can be built according to the component diagrams in Sec. 6.5 to cover a wide range of volume and pressure requirements, generally with off-the-shelf industrial products. There is no "standard" chemical grout plant.

Grout pipes often are standard components of the drilling industry, either drill rods or casing. Even the more sophisticated rods such as tube à manchette are available from several commercial sources.

While it is no longer necessary for a grouter to build his own special equipment for placing chemical grouts, more grouters assemble their own grout plants using commercial components and instrumentation.

7

Field Procedures and Tests

7.1 INTRODUCTION

From the time that a decision is made to consider chemical grouting as a possible solution to a specific field problem, many other decisions must be made and acted on. While the decision to "consider" may be made without conscious deliberation, the decisions which follow should be made with deliberation and in a logical sequence. Foremost among all the questions that arise is whether the formations are groutable. The answer to this question becomes the first step in the process of verifying the utility of grouting. It may also be the last step if the formation turns out to be ungroutable.

7.2 DETERMINATION OF GROUTABILITY

A broad determination of groutability can be made on the basis of grain size: Medium to coarse sands can always be readily grouted. Dense fine sands and loose silts can usually be grouted but may cause difficulty. Dense silts should be expected to cause difficulty. Silty clays and clays cannot be grouted. Since natural soils often contain a wide range of grain sizes, another broad criterion is that difficulty in grouting all well-graded materials should be expected when the silt content (particles smaller than the No. 200 sieve) exceeds 20%.

To know the grain size (grading) of soils, it is necessary to take soil borings. This is a specialized field and best left to those who practice it regularly. The grouter, however, must know how to read and interpret a boring log and should have a general knowledge of whether a specific soils investigation is adequate for determining the feasibility of grouting. Often an adequate soils investigation coupled with the grouter's own experience is sufficient for a reasonable determination of groutability. On small jobs, however, there may be no available soils data, and other methods must be used.

7.3 FIELD PUMPING TESTS

Grain size data can, of course, be used as an index of groutability because it is related to the more direct factor of permeability. A rather broad group of criteria based on grain size has in fact been presented in Fig. 4.2 and may be interpreted in terms of permeability.

k* = 10^{-6} or less: ungroutable
k = 10^{-5} to 10^{-6}: groutable with difficulty by grouts with under 5 cP viscosity and ungroutable at higher viscosities
k = 10^{-3} to 10^{-5}: groutable by low-viscosity grouts but with difficulty when μ is more than 10 cP
k = 10^{-1} to 10^{-3}: groutable with all commonly used chemical grouts
k = 10^{-1} or more: use suspended solids grout or chemical grout with a filler

When reliable data on grain size or permeability are not available, it is more direct and often less expensive to run field pumping tests rather than take borings.

Permeability is also an index, actually a closer one than grain size, to the actual desired datum: groutability. It is therefore feasible to bypass the computation of a permeability number and determine directly if a formation will accept grout. This is done very simply by placing a grout hole into the formation and trying to pump liquid into it. For such a field test to produce meaningful data, several criteria must be met. Most obvious, and most important, the point at which grout enters the formation must actually be in that formation. Thus, the pipe must be placed with the same care and skill as if it were an actual grout pipe. The liquid pumped into the formation should have the same viscosity as the grout planned for eventual use. For acrylate-based grouts, water can be used. For other grouts, it may be necessary to artificially increase the viscosity by adding soluble products such as starch.

The volume and pressure at which liquid is pumped into the formation must not exceed either the capabilities of the grout plant

*All values of k are in cm/sec.

or the limitations imposed by safety considerations, specifications, or other reasons. Further, pumping should be continued until the volume placed exceeds that proposed for the actual grout or until equilibrium conditions are established. As can be deduced, pumping tests are best conducted with the grout plant itself.

7.4 FIELD PERMEABILITY TESTS

Permeability tests are often run in the laboratory on soil samples taken in the field. Since so-called "undisturbed" samples cannot be taken of granular soils, the reliability of laboratory test data depends on the accuracy with which the natural soil density and stratification can be reconstituted. Laboratory permeability numbers can readily be in error by an order of magnitude, an amount sufficient to differentiate between successful and unsuccessful grouting jobs. When properly carried out, field permeability tests will give much more reliable data for evaluation of formation groutability. Such tests are performed by either putting water into a formation or withdrawing water from a formation. The actuating forces may be limited to gravimetric, or pumps may be used. In the latter case, the tests are often called *pumping tests*.

Figure 7.1 shows the parameters pertinent to a field pumping test for permeability determination. One test well is required, and two observation wells are needed, both within the drawdown curve and at different radial distances from the test well. The test well is pumped at some constant rate until equilibrium elevations are

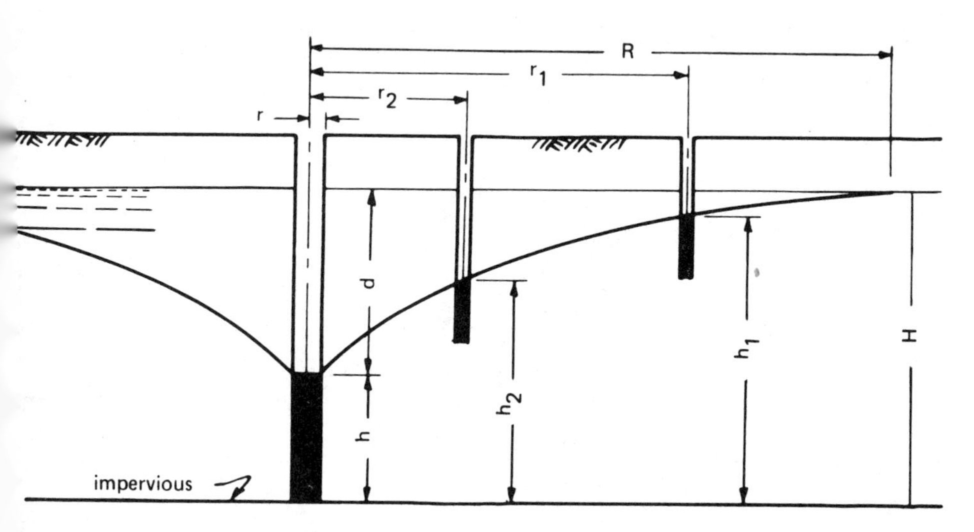

FIGURE 7.1 Field pumping test. (From Ref. 1.)

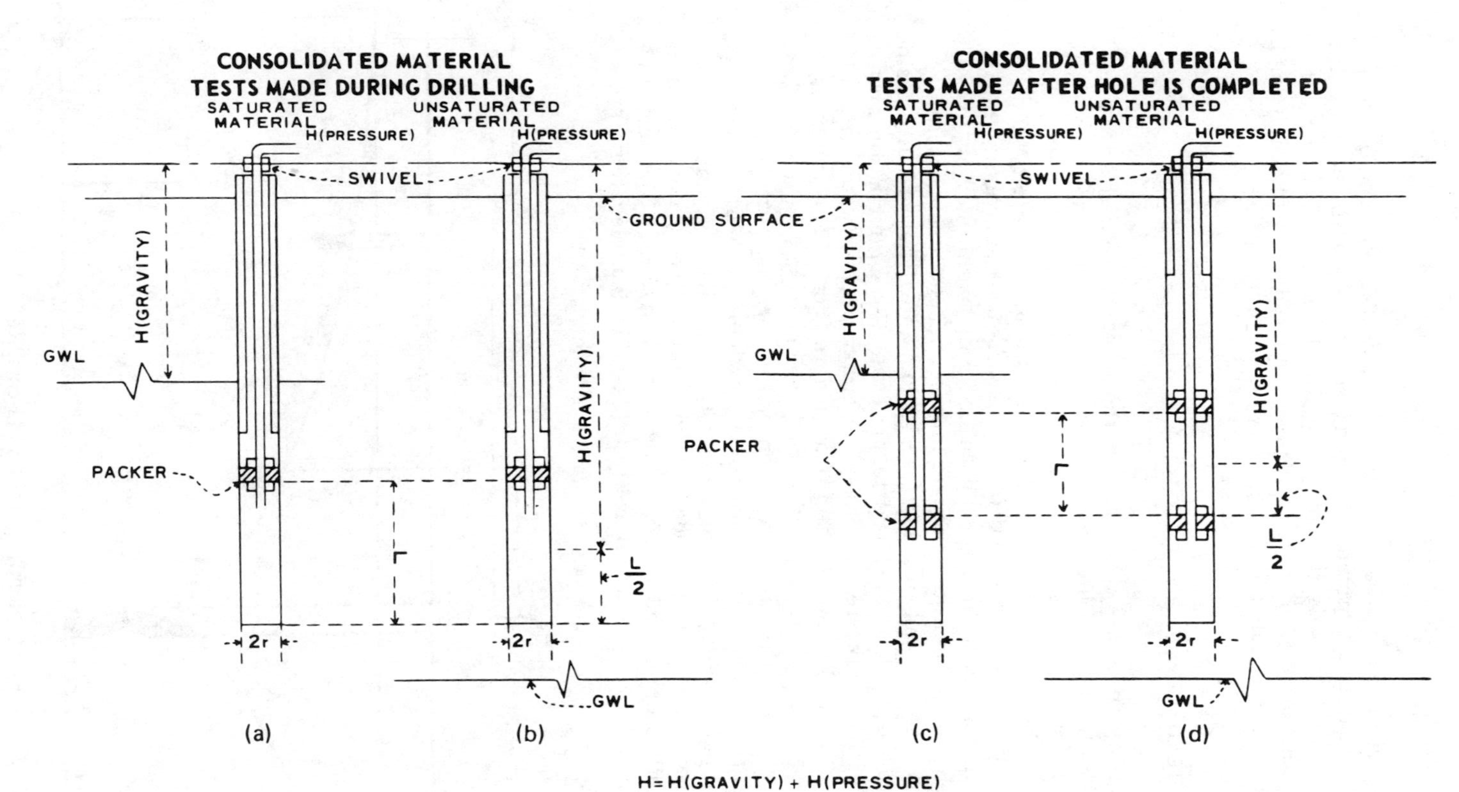

FIGURE 7.2 Packer tests for permeability. (From Ref. 1.)

attained in the observation wells. Field measurements of Q, r_1, r_2, h_1, and h_2 are taken. Permeability is computed from

$$k = \frac{Q \log_e(r_1/r_2)}{\pi(h_1^2 - h_2^2)}$$

The test well should be carried down to an impervious stratum. If this is impractical, then it should be sunk to considerable depth below the water table.

For most accurate results, the observation wells should be located well within the drawdown curve, at significant distances from both the test well and the radius of zero drawdown. An approximate value of k can be found by letting the locations of the observation wells approach these limits. If one location is chosen on the periphery of the test well, r_2 becomes r, the radius of the test well. If the other location is chosen at the point of zero drawdown, r_1 becomes R, the radius of the drawdown curve. R will always be several hundred times the value of r, and therefore the quantity $\log_e (R/r)$ will not vary greatly with small changes in R.

Thus, a value of R can be assumed, which eliminates the necessity for observation wells and yields an approximate value for k:

$$k = \frac{Q \log_e(R/r)}{\pi(H^2 - h^2)}$$

Tests may also be made in portions of a drill hole isolated by one or more packers, as shown in Fig. 7.2. This figure and the procedure described are taken from the Bureau of Reclamation's *Earth Manual*, 1st edition [1].

Figure 7.2 shows a permeability test made in a portion of a drill hole below the casing. This test can be made both above and below the water table provided the hole will remain open. It is commonly used for the pressure testing of bedrock using packers, but it can be used in unconsolidated materials where a top packer is placed just inside the casing.

The formulas for this test are the following:

$$k = \frac{Q}{2\pi LH} \log_e \frac{L}{r}, \quad L \geqq 10r$$

$$= \frac{Q}{2\pi LH} \sinh^{-1} \frac{L}{2r}, \quad 10r > L \geqq r$$

where

k = permeability

Q = constant rate of flow into the hole

L = length of the portion of the hole tested

H = differential head of water

r = radius of hole tested

$\log_e$ = natural logarithm

$\sinh^{-1}$ = arc hyperbolic sine

These formulas have best validity when the thickness of the stratum tested is at least 5L, and they are considered to be more accurate for tests below the groundwater table than above it.

For convenience, the formulas can be written as follows:

$$k = C_p \frac{Q}{H}$$

where k is in feet per year, Q is in gallons per minute, and H is the head of water in feet acting on the test length. Where the test length is below the water table, H is the distance in feet from the water table to the swivel plus applied pressure in units of feet of water. Where the test length is above the water table, H is the distance in units of feet of water. For gravity tests (no applied pressure) measurements for H are made to the water level inside the casing (usually the level of the ground).

Values of C_p are given in Table 7.1 for various lengths of test section and hole diameters.

The usual procedure is to drill the hole, remove the core barrel or other tool, seat the packer, make the test, remove the packer, drill the hole deeper, set the packer again to test the newly drilled section, and repeat the test (see Fig. 7.2a). If the hole stands without casing, a common procedure is to drill it to final depth, fill with water, surge it, and bail out. Then set two packers on pipe or drill stem as shown in Figs. 7.2c and d. The length of packer when expanded should be 5 times the diameter of the hole. The bottom of the pipe holding the packer must be plugged, and its perforated portion must be between the packers. In testing between two packers, it is desirable to start from the bottom of the hole and work upward.

When gravity alone is used to gather data on water inflow or outflow from a drill hole, the procedure is generally called a percolation test. Such tests may be made by either measuring percolation through the bottom of the hole or through the sidewalls. If a

TABLE 7.1 Values of C_p

Length of test section, L	Diameter of test hole			
	EX	AX	BX	NX
1	31,000	28,500	25,800	23,300
2	19,400	18,100	16,800	15,500
3	14,400	13,600	12,700	11,800
4	11,600	11,000	10,300	9,700
5	9,800	9,300	8,800	8,200
6	8,500	8,100	7,600	7,200
7	7,500	7,200	6,800	6,400
8	6,800	6,500	6,100	5,800
9	6,200	5,900	5,600	5,300
10	5,700	5,400	5,200	4,900
15	4,100	3,900	3,700	3,600
20	3,200	3,100	3,000	2,800

Source: Ref. 1.

tube can be sealed tightly into a formation so that no flow occurs along the tube sidewalls, measurement of the rate of rise will yield a value of k. The constant C and the definition of the parameters are shown in Fig. 7.3.

All measured values should be in inches for

$$k = \frac{H \log_e(h_1/h_2)}{C_t}$$

This equation tends to accentuate the k value in the vertical direction.

If it is possible to drill an open hole into a formation, a simple method of approximating the horizontal permeability may be used. The depth of the hole must be large in relation to its diameter for this method to apply. In this method, the rate of rise in the hole is measured at several elevations and the permeability computed by

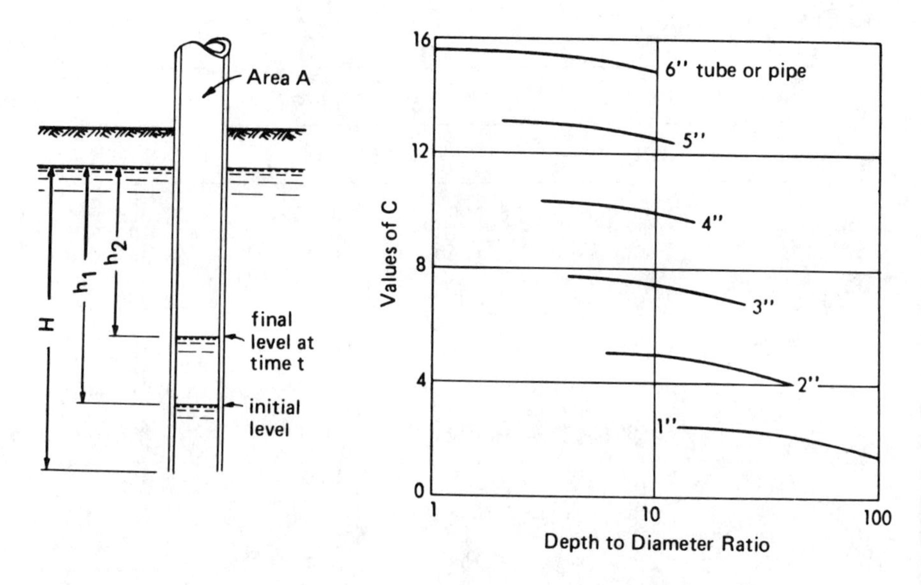

FIGURE 7.3 Parameters for percolation test. (From Ref. 1.)

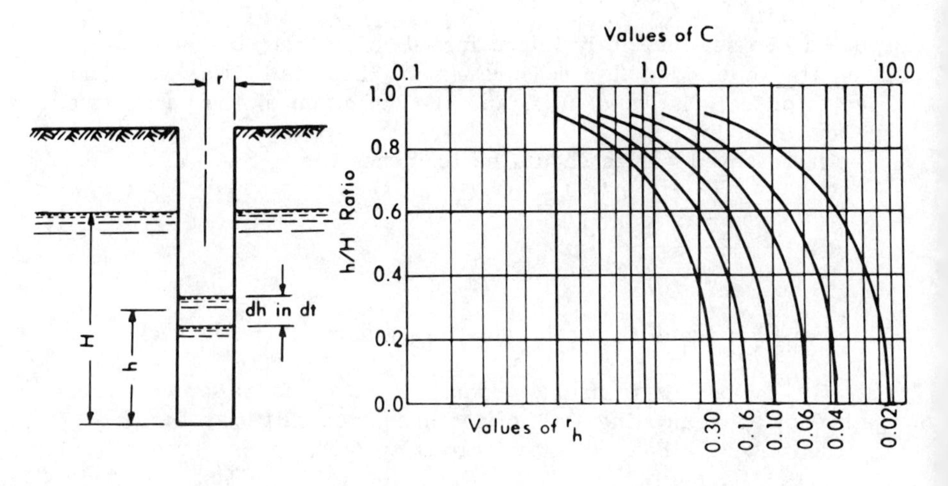

FIGURE 7.4 Parameters for percolation test. (From Ref. 1.)

$$k = 0.2Cr^2H \frac{dh}{dt}$$

The definition of the parameters and the values of C are shown in Fig. 7.4. In all the methods discussed, the wells or holes should be pumped and filled several times prior to taking test measurements in order to minimize the disturbance to natural conditions caused by placing the holes. It should be remembered that uncontrollable or unknown conditions in the field can cause the results of these methods to be quite approximate; even so, they may still be used as a measure of the effectiveness of a grouting operation.

Additional information on pumping test methods can be found in Ref. 2.

7.5 USE OF TRACERS

Seepage channels through soil and rock formations can be delineated through the use of materials called tracers, which are introduced into channels and made to move through them. Radioactive tracers have limited use, primarily due to economics, although in theory radioactives can trace the entire seepage channel, while most other tracers identify only two points on the channel. Organic dyes are the most commonly used tracers for grouting tests. Fluoroscein, marketed under several different trade names, is one commonly used product.

On many jobs it may be necessary to know if the grout to be pumped through a specific hole will reach ground surface or some structural discontinuity such as a tunnel face or wall. It may also be desirable to measure the flow rate for establishing gel times. Colored water may be pumped into the grout hole for this purpose.

Almost any dye which will color water so as to make the water identifiable at low dye concentrations may be used for field pumping tests. Many commercial dyes meet these requirements and are inexpensive and innocuous.

Dye tests, in order to yield useful data, must be closely controlled. Most commercially available dyes are greatly diluted with water for field use. Such solutions have properties (viscosity and density) essentially the same as those of water. Hydraulic data obtained with these dye solutions are directly applicable only to grouts with similarly low viscosity and density. To use dye test data directly with silicate or resin grouts, dye solutions must have their viscosity artificially increased (with a nonlubricating material) to match the higher viscosities of these grouts.

The dilution factor is of major importance in determining whether specific grout holes may be used effectively. For example, suppose a solution using a fluoroscein dye at 100 ppm is prepared. This solution is pumped into a grout hole and subsequently identified as issuing from a surface leak. Since such dyes are visually identifiable even under poor lighting conditions at concentrations of 10 ppm or less, the dye solution may have been diluted with 10 or more times its volume of water before issuing from the leak. No chemical grout could withstand such dilution. If chemicals were pumped, it is most probable that they would not set up due to excessive dilution. This dye test is, therefore, not satisfactory for designing a grout procedure. The example discussed establishes the criterion for dye concentration: When using dyes for grouting design purposes, the maximum concentration of dye should be such that it becomes unidentifiable at dilutions in excess of those which would prohibit grout gelation.

Most chemical grouts are used at field concentrations that preclude gelation at all or at effective gel times when groundwater dilution exceeds 100%. Thus, dye concentration should be such that one-to-one dilution renders the dye unidentifiable. This kind of close control cannot be exercised except by careful measurement of dye and water weights or volumes. The practice of dumping a little dye from its container into a drum of water until the color "looks right" is totally inadequate for good field practice.

Dilution is, of course, also a function of the pumping rate and possibly of the pump characteristics. For this reason dye tests should be conducted with equipment of similar characteristics to that which will be used for grouting, preferably with the chemical grout plant itself.

It may often be desirable to use dye in the grout solution. When this is to be done, a field gel check should be made to determine the effects on induction period. Organic dyes may act as inhibitors or accelerators, particularly at higher concentrations. If dye tests had been run prior to the actual grouting, it is necessary to use a different color dye in the grout itself.

Radioactive tracers are often proposed for locating flow channels within a soil or rock mass. There are several practical limitations. To begin with, radioactive tracers present a health hazard. Also, unless the signal is very strong, it will be lost when the channel moves away from the open soil or rock face. Since the tracer may be optically indistinguishable from water, it may be necessary to use it in conjunction with a dye tracer in order to find outlets quickly. Thus, there appear to be no advantages over the use of dyes alone.

7.6 ADDITIVES

The penetrability of a grout is a function of its viscosity. Any additive to a grout which either increases the viscosity or makes the grout a suspension will decrease the penetrability. In general, if a chemical grout with an additive such as portland cement can be pumped, the use of cement grout by itself may be a more economical solution and should certainly be considered. The exception to this statement is when the chemical grout is needed for gel time control, and the cement is needed for strength. Solid additives used to *extend* the grout (generally defined as giving more grout volume at a lesser cost) will generally prove to be non-cost-effective. The same is true for additives used to increase the viscosity of a grout. The initial use of less expensive, higher-viscosity grouts is a better procedure. At low working temperatures, additives may be needed to prevent freezing of the grout solutions.

Additives may affect grout gel times by acting as accelerators or inhibitors, and their effects must be checked prior to field use. One possible source of (inadvertent) additives often overlooked is the formation to be grouted and its groundwater. Salts from these two sources may have an effect on the grout. Such effects can generally be canceled by using the site groundwater as the source for mixing the grout.

Materials commonly used as grout additives, and their effects and purposes, can generally be found in the technical literature available from the grout manufacturer.

7.7 PUMPING RATE

The rate at which grout may be placed within a formation is one of the more significant factors in the job costs. Therefore, it is generally best to place grout as rapidly as is consistent with safety. The major factor which limits pumping rate is the related pumping pressure (see Sec. 7.8).

In general, when allowable pumping rates decrease toward 1 gpm and less, grouting tends to become an uneconomical method for solving a field problem. On the other end of the scale, pumping at 10 gpm and more generally is not feasible in sands and silts.

The use of volume control to keep grouting pressure under a selected maximum value is common practice in Europe and has been used recently on a major field grouting experiment in the United

States [3]. In theory, the system should work well. It doesn't, because field conditions are never uniform and change continuously as grout sets up in the ground. The method is used despite its drawbacks because good volume control equipment is easier and less expensive to build than good pressure control equipment. However, the result is a fluctuating pressure which often exceeds the desired maximum. In the case of the experiment cited, where the expressed design criterion was to keep grout pressures from exceeding 85% of the fracturing pressure, the actual pressure did in fact often exceed the fracturing pressure by significant amounts. This made the data useless for comparison with results where fracturing pressures were deliberately exceeded.

7.8 PUMPING PRESSURE

The rate at which grout enters a formation increases as the pumping pressure increases. From an economics point of view, it is therefore desirable to pump at the highest possible pressure. However, the possibility of damage to the formation and adjacent structures limits the pumping pressure to specific values for each individual project.

Many grout jobs involve placing grout in zones where there are no adjacent structures and where overpressuring can only cause damage to the formation itself. Under such conditions it has become common practice to grout at a pressure less than 1 psi per foot of depth below ground surface. Only in the recent past has this rule of thumb been questioned and in some cases openly ignored. There is little doubt that this rule of thumb is quietly ignored by field crews, when grout cannot be placed rapidly enough by following it. The basis for the 1 psi per foot limitation stems from rock grouting practice and the general desire to prevent uplift of a formation. To "pick up" a formation by hydraulic pressure, the pressure has to exceed the formation weight. Since rock will weigh about 165 lb/ft^3 its weight per square inch is a little more than 1 psi. Saturated soils will approach 144 lb/ft^3 and thus the rule of thumb is transferred. Actually, the calculation ignores the fact that an isolated soil or rock mass 1 ft square and many feet high cannot be lifted by itself because of the shear resistance at the vertical faces. Only when a very large horizontal zone in a formation is subjected to hydraulic pressure does the possibility of uplift actually exist. This requires a large volume of liquid grout in the formation. Such conditions can exist when grouting with cement. When grouting with chemicals, particularly at short gel times, there is relatively little liquid grout in the formation at any given instant, and the possibility of uplift is generally very remote. This applies to grouting in soils more than to grouting in rock.

When grouting in stratified soils, where vertical permeability is far less than horizontal permeability and the shear strength is higher, if a failure occurs due to overpressure, it will probably result in uplift. In soils whose permeability and strength differences between vertical and horizontal planes are not significant, failures due to overpressure will commonly result in vertical fissures extending in a horizontal direction. Such failure is more properly called fracturing and can be induced by pressures of one-third to one-half those which would be required for uplift. Field research to define magnitudes of fracturing pressures in granular material is very limited [4]. Current indications are values between 2 and 3 psi per foot of depth. Of equal interest is the fact that fracturing pressures can be determined for a specific site by implanting geophones in the formation and listening for the readily identifiable sounds which accompany fracturing.

Several unanswered questions remain in regard to fracturing and allowable pumping pressures. Measurements of pressure are generally taken at the pump and sometimes where the grout pipe enters the formation. The losses can be guessed at, and most practitioners consider them to be small. Nonetheless, until we have research data which actually measure pressures in the formation at the points where fracturing is about to occur, we cannot be sure of the true value of fracturing pressure. In the field, grouting pressures often increase gradually and then suddenly show a drop accompanied by an increase in pumping rate. This is often taken to mean that fracturing had occurred. This may be true, but the pressures assumed to represent fracture were those measured above ground, not in the formation. For additional data on fracturing, see Ref. 5.

When a formation is deliberately fractured by grouting pressure, horizontal and/or vertical cracks open, which fill with liquid grout. Thus, a much larger area exists from which grout can penetrate the voids of a formation. Grouting can obviously proceed at a much more rapid pace than before fracturing. This makes the process of fracturing very attractive to the grouting contractor. Until the very recent past, fracturing had always been considered poor practice, to be avoided if at all possible. In the past several years contractors and engineers alike have begun to question whether fracturing actually is detrimental to a grouting operation. It seems obvious that grout-filled cracks within a stabilized soil mass, particularly with the lower-strength chemical grouts, represent planes of weakness that decrease the overall shear strength. It seems equally obvious that a grouted mass containing grout-filled cracks is stronger than the same zone was before grouting. Thus, it would appear that in soil masses where fracturing would have no detrimental side effects, such as leading to an open surface or beneath a building floor or foundation, the decision as to whether or not to permit fracturing should be based on the actual strength increase needed and whether a fractured formation would still yield such strength. If the grouting is for settlement control, the same considerations apply. If the grouting is for water shutoff only, fracturing should have no detrimental effects.

There is at present very little field data for designing grouting projects at pressures above those which would cause fracture. Until much more is known about this phenomenon, it is suggested that the following guidelines be considered to determine field grouting pressure limits:

1. Keep grouting pressures below 1 psi per foot of depth, or half the fracturing pressure if it is known, when (a) high pumping rates and long gel times are used together, (b) working close to building foundations and underground walls and floors, and (c) the grout take is adequate for the job economics.
2. Keep grouting pressures below 2 psi or the fracturing pressure if it is known when working at short gel times.
3. Permit the grouting pressure to exceed the fracturing pressure by a small amount when grout takes are otherwise less than 2 gpm and the danger of structural damage due to fracturing is negligible.

7.9 GROUTING IN PIPES AND HOLES—MANIFOLDING

When grouting through an open-ended pipe, the point at which all the grout enters the formation is known. In contrast, when grouting at the collar of an open hole drilled in rock or soil which will stay open, grout can enter the formation along the entire hole length, and the actual distribution of grout is not known. To have better control of the grouting operation, packers can be used to isolate short portions of the hole. This reduces the grout take and therefore may not be a feasible procedure in areas of low permeability. To be able to work in small stages, even though the takes are low, it is possible to grout through several holes simultaneously. This process is generally called manifolding.

There is no limit to the number of grout holes that may be manifolded. The number most appropriate is the smallest that permits the grout plant to operate at maximum volume and within job limitations. This will seldom exceed 10 and may be as few as 2.

Manifolding permits the isolation of short portions of each hole, so that the grouting operation can be closely controlled. This requires flow meters, pressure gages, and valves at each grout pipe so that the distribution of grout cannot only be monitored but controlled. Such equipment is an essential element of the process and must be of good quality and kept in continuous operation. The records for each hole of grout take and pressure provide the data from which later judgments are made regarding the possible necessity for regrouting.

On large grouting jobs, and many small ones as well, economy distates continuous rather than intermittent placing of grout holes or pipes. A possible problem which must be avoided, whether grouting one pipe or many, is the sealing of adjacent grout holes and rendering them useless. For this reason, grout pipes must be kept sealed until ready for use. This is automatically accomplished by the tube â manchette, or sleeve pipe, which also permits regrouting at every depth through the original grout pipe. Pipes which are used open ended are left with the bottom seal in place until grouting through the pipe begins. Such pipes are generally pulled at regular intervals and must be redriven to do additional grouting in zones previously treated. There is no way to keep uncased (open) holes from contamination by adjacent grouting. If open holes are to be used, each should be drilled and grouted prior to drilling the next hole.

7.10 USE OF SHORT GEL TIMES

The positive effects of short gel times have been discussed in previous chapters. However, in the field, mental inhibitions to the use of gel times shorter than pumping times still persist—a remaining legacy from cement grouting experiences and from batch systems. This is partly due to the fact that many field problems best solved by the use of short gel times can also be solved by using gel times longer than the pumping times (generally at greater expense).

In fact, the use of batching persists because there are times when even batch systems will solve a problem (almost surely at greater cost). Further, when high pressures and low pumping rates are required (such as when working in deep shafts), small batches of grout pumped successively have been effective. This process (in which a finite number of small grout batches is placed in time succession, so that each batch reaches its gel time also in time succession) approaches two-pump system in which infinitely small volumes of grout reach the gel stage in time succession.

Laboratory work with short gel times started with acrylamide about three decades ago, and was applied in field work almost immediately (see Refs. 2 and 3 of Chap. 5). At that time, it was thought that the use of short gel times would be limited by grout strength, and would not be applicable to the silicates. Later, it was determined that the critical factor was the use of proportioning pumps, and was relatively independent of gel strength and grout viscosity.

In the early attempts to measure viscosity changes during the induction period, use was made of measuring instruments that rotated continuously inside a container of grout. This process lengthened the gel time, and with acrylamide (and now with acrylate), was ascribed to the inhibiting effect of air entrainment. More recently, the same phenomenon has been noted with the silicates. It is now apparent that for some silicate formulations, the setting time of the solution can be extended by vigorous agitation for as long as the agitation continues. Depending upon the degree of agitation, there are some grout formulations that will not set up while in motion. This obviously has application to field work at short gel times.

Experiments can be performed to determine whether or not a specific grout formation will set up while the grout is in motion through a soil or rock formation. Reference 3 details one such experiment. There are three possible modes of action:

1. Motion through a formation destroys the catalyst systems and delays the gel time or prohibits gelation totally.
2. Motion delays the mechanics of bond formation, and the reaction is delayed until the degree of motion falls below some critical level, and
3. Motion has little or no effect on the gelation process.

It is most probable that only numbers 2 and 3 are pertinent to the field applications of commercial chemical grouts. For the grouter, the implication is clear: If a grout does not set up while in motion, the beneficial effects of using gel times shorter than the pumping time are lost. On the other hand, the entire volume of grout placed will probably set up simultaneously, shortly after pumping (and travel within the formation) stops. It is important to know the most probable action of the grout in the formation, in order to design the operation most effectively. New grout formulations should be tested prior to field use to determine if they will set up while in motion.

When working at gel times of 10 min and longer, it may not make much difference if the grout sets in 9 min or 11 min. If one is working at gel times of 2 min, a deviation in setting time of plus or minus 1 min can be crucial. The major factors that can change the formation gel time from the tank gel time are temperature and groundwater chemistry. Temperature differences can be monitored, and controlled through simple measures such as external refrigeration or heating. Groundwater chemistry, on the other hand, is time consuming to analyze and would be difficult to compensate. A simple solution to the groundwater chemistry problem is to use the groundwater itself to mix the grout. Although this may cause the grout setting time to be very different from that shown in supplier's data, the desired setting time can still be preset, and the effects of groundwater chemistry are eliminated.

7.11 SUMMARY

Field tests often give more direct data to the grouter than soil exploration and laboratory tests. This is because the essential bit of data needed by the grouter is different from the data needed in other aspects of geotechnical design. The grouter needs to know the details of how a formation will accept grout, i.e., the rate at which grout can be placed with adequate margin against formation or structural failure. Often, the formation groutability is inferred from other data, such as grain-size analysis, relative density, permeability, and porosity. However, the pertinence of such data is often lost in the translation from sampling through testing through interpretation of data (e.g., is 10% porosity due to three large fissures or coarse strata, or due to an infinite number of fine discrete pores).

On the other hand, simply pumping fluid into a formation bypasses boring, sampling, and permeability testing and establishes directly whether or not a formation will accept grout. (Field tests can also establish permeability values, if such data are needed, with greater accuracy than laboratory tests.)

Field pumping tests must be carried out within the limitations that will be imposed by the job itself, the grout to be used, and the pumping equipment. Thus, they should be done with the grout pumps that will be used on the job, and with a fluid whose viscosity matches that of the proposed grout.

Field testing is often done in order to locate existing or suspected flow channels, and to establish the locations of their exposed terminals. Such tests are an important phase of seepage control work, and are almost always performed with organic dyes (see Chap. 8). Dye testing, as well as field pumping tests, should be done with the grout plant to be used on the project.

Field tests should always include verification that the catalysed grout is gelling as anticipated. For this purpose, sampling valves, located just before the grout enters the pipe, should be installed. Samples should be taken at regular short intervals (5 to 15 min).

REFERENCES

1. *Earth Manual*, 1st Ed. U.S. Bureau of Reclamation, Department of the Interior, Denver, 1960.

2. S. M. Lang, Pumping Test Methods for Determining Aquifer Characteristics, Paper presented at the Sixty-Ninth Annual Meeting of the ASTM, Atlantic City.

3. P. Daniele, J. Hutchinson, R. H. Karol, L. Ospitia, and B. Reim, Gelation of Chemical Grouts While in Motion, *Geotechnical Testing Journal*, Philadelphia, June 1984.

8

Grouting to Shut Off Seepage

8.1 INTRODUCTION

Grouting with cement for control of groundwater became an accepted construction procedure in the late nineteenth century in Europe and, at the turn of the twentieth century, in the United States.

Chemical grouting became an accepted construction procedure between 1920 and 1930, with the successful completion of field jobs using sodium silicate. The modern era of chemical grouting, which saw the introduction of many new and exotic products for field use, began only 30 or 40 years ago, making chemical grouting a relatively new technology.

Procedures and techniques used with cement grouts in the United States were developed primarily by the large federal agencies concerned with dam construction: The Corps of Engineers, The Bureau of Reclamation, and the Soil Conservation Service. Predictably, each of these organizations developed its methods unilaterally, resulting in major areas of difference in philosophy and execution.

It remains difficult, if not impossible, to assess the effects of these differences on the success of field work. This is partly because each field project is unique, and records for two similar jobs, done by different approaches, do not exist. Primarily, however, almost all cement grouting is done to increase the safety factor against some kind of failure. There are generally no precise methods of measuring the safety factors before and after grouting. By way of contrast, remedial grouting (often done for seepage control) is aimed at a specific problem, where failure or incipient failure on a limited scale is recognizable. Grouting either corrects or fails to correct the problem, and the benefits of grouting, as well as the specific procedures used, are directly measurable.

By any reasonable yardstick (volume, cost, man-hours, etc.) the grouting experience of the federal agencies is overwhelmingly in the use of cement. The standards and practices used in cement grouting quite naturally were carried over into chemical grouting usage. Some of these procedures were totally inappropriate and severely limited the success of some of the early chemical grouting experiences. However, these philosophical difficulties have by now been largely overcome. Chemical grouting is accepted as a valid and valuable construction procedure, and the concepts of short gel times, accurate control of gel times, and sophisticated multipump systems and grout pipes have been integrated into practice.

There are two major purposes for grouting, and any field job can be classified in terms of its purposes: (1) to shut off seepage or to create a barrier against ground water flow and (2) to add shear strength to a formation in order to increase bearing capacity, increase stability, reduce settlements and ground movement, and immobilize the particles of a granular mass.

The term *seepage* is difficult to define quantitatively. The ASCE Glossary of Terms [1] states "the flow of small quantities of water through soil, rock or concrete." This definition, of course, depends on the interpretation of the word *small*. Five gallons per minute of water entering the bottom of a deep shaft is a small amount. The same quantity entering a domestic basement is a large amount. Seepage, then, is better defined in terms of the procedures used to eliminate it, rather than by job or quantity of water involved. In contrast to grouting for other purposes, seepage control generally does not require complete grouting of a formation.

8.2 TYPES OF SEEPAGE PROBLEMS

During the construction phase of a project, water inflow is considered a problem (and dealt with) only when the inflow halts or retards construction. However, the same amount of inflow which is tolerable during construction may not be tolerable during the operational phase of the structure. Seepage may also begin after construction is completed, because the elements of the structure modify the normal groundwater flow and/or because of faulty constructions and/or because of foundation movements due to consolidation, earthquakes, and general slope instability. The need for seepage control may be apparent in the design state. If so, remedial measures can be integrated into the overall construction process. However, seepage control procedures are generally carried out after the structure is completed.

Creating a barrier against water flow may be done by grouting specific individual flow channels or by constructing a grouted cut-off, or grout curtain through some of all of a pervious formation. If only a limited number of channels is available for water flow, it is feasible and usually preferable to use procedures aimed at individual channels. These are discussed in this chapter.

If many channels are available for water flow, but flow is occurring only through a few of them, seepage control procedures often result in shifting the water flow from grouted channels to previously dry channels. If, in the end, a large number of channels must be grouted, other procedures will probably be more cost effective than treating flow channels one at a time.

If many flow channels are available and flowing, such as in the case of granular soils, or severely fissured rock, procedures are generally used which attempt to impermeabilize a predetermined volume of the formulation. These are discussed in Chap. 9.

Typical seepage problems include infiltration through fissures in rock, such as the tunnels in Figs. 8.1 and 8.2, and through joints and porous zones in concrete, as shown in Fig. 8.3. Figure 8.1 shows a drift (tunnel) in a copper mine in Canada, 250 ft below ground surface. Many thousands of feet of such drifts are required even in a small mine to provide access from the main shafts to the ore bodies. Typically, the drifts will intersect numerous minor fault zones as well as some major ones. Some of the cracks and fissures will conduct surface water from rivers, lakes, and precipitation into the drifts. In Fig. 8.1, individual leaks cannot be seen, but their aggregate readily shows as several inches of water on the floor. Water from this and every other wet drift must be collected and pumped out of the mine. As the ore zones are developed, total seepage increases. Eventually, it becomes more economical to shut off the seepage than pay continuously increasing pumping costs.

Figure 8.4 is a geologic map of the 250 level containing Fig. 8.1, which was taken looking toward point b. Figure 8.5 is an enlarged view of the zone, showing the cracks and faults as plotted by the mine geologists. Also shown are a number of grout holes, which were drilled into fault zones identified on the geologic map. Grouting through these holes proved totally ineffective in reducing the seepage.

Figure 8.2 is a slope entering a coal seam near Pittsburgh. The photo shows a location several hundred feet downslope. Groundwater entered at many points, generally as drips and occasionally as a very small steady flow. Most of the year, the water was of no consequence. In winter, however, icicles formed as shown, interfering with coal car movement along the tracks and also making the area hazardous for personnel. It was possible to seal the leaks completely by chemical grouting.

Figure 8.3 is a vehicular tunnel in Baltimore. Water is shown entering in small amounts through a construction joint in the concrete.

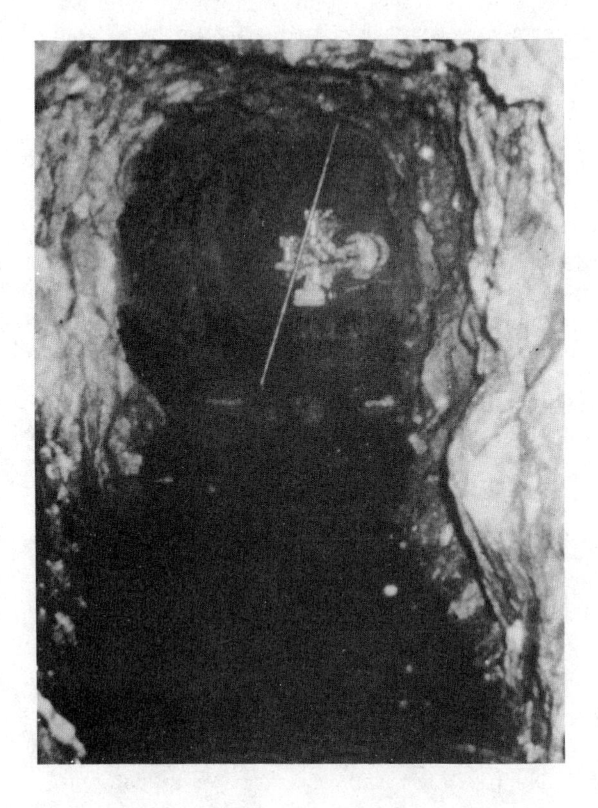

FIGURE 8.1 Seepage into mine drift.

FIGURE 8.2 Freezing of coal slope seepage.

FIGURE 8.3 Concrete construction joint seepage.

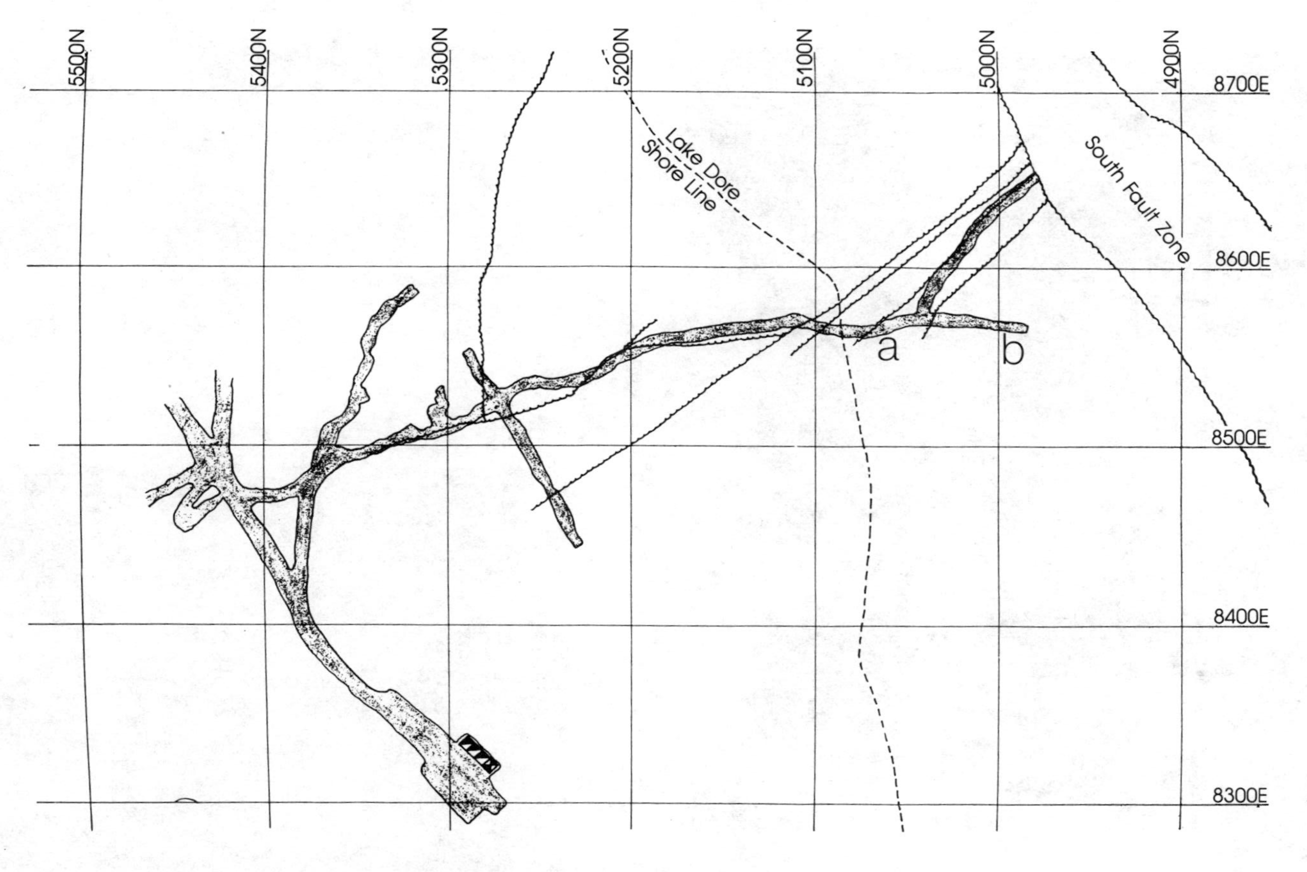

FIGURE 8.4 Geologic map of the drift shown in Fig. 8.1.

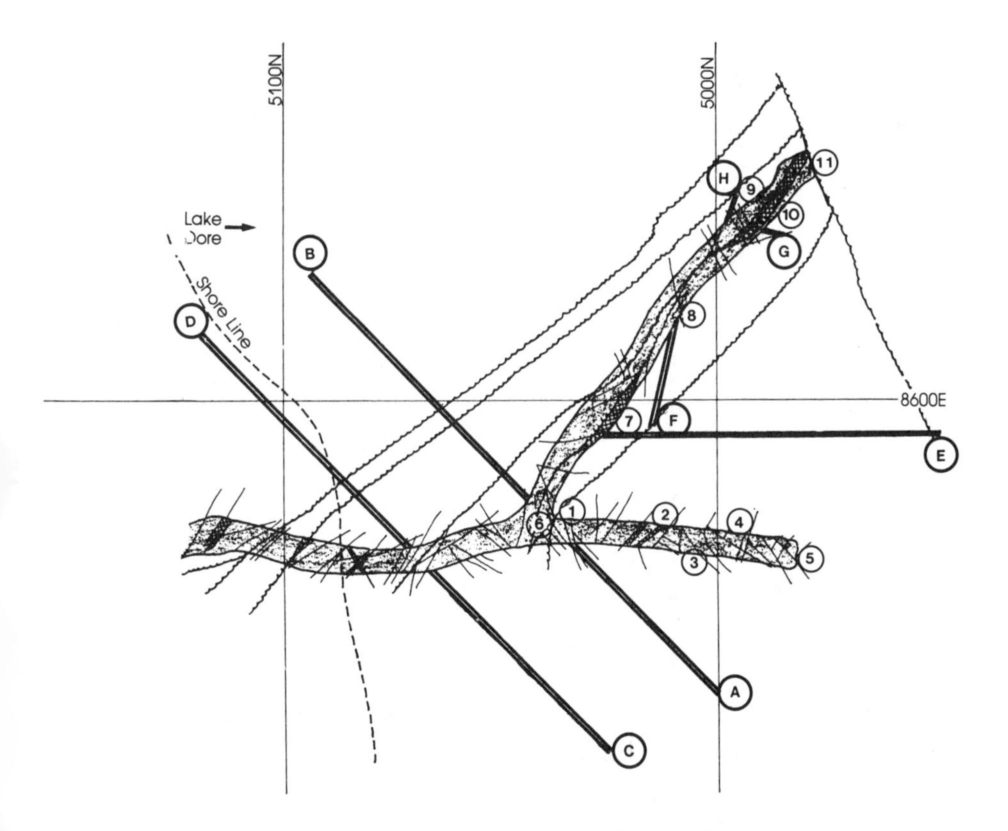

FIGURE 8.5 Enlarged view of a portion of Fig. 8.4.

The volume of water was insignificant, but tiles could not be placed over the wet zone. Grouting with chemicals successfully stopped the seepage.

8.3 LABORATORY STUDIES

The reason the coal slope and tunnel jobs were successful, as opposed to the mine work shown in Fig. 8.1, is that different techniques were used. These were developed by field and laboratory research to determine why the grouting in the copper mine was ineffective. The original field work dates back to 1957 and is one

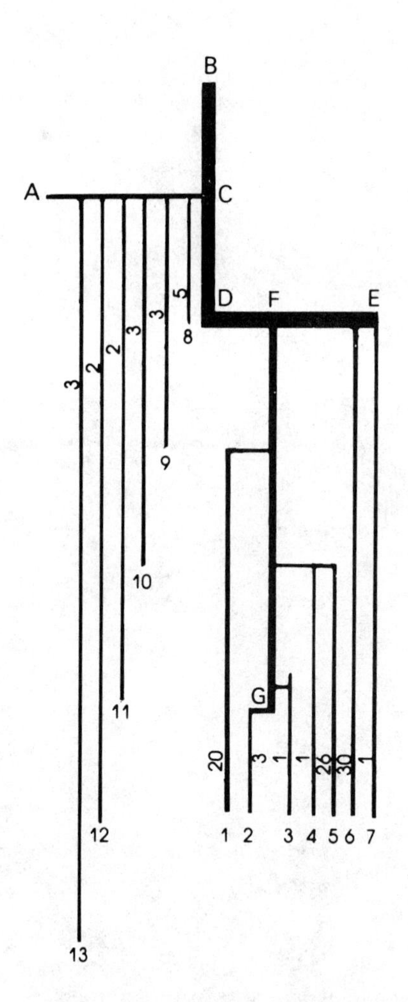

FIGURE 8.6 Model for seepage study.

of the earliest attempts to apply chemicals (other than silicates) to mine grouting. This work indicated clearly that the route grout would follow (from its injection point to its exit point or final location) could not be determined with any accuracy by interpretation of the geologic map coupled with visual site examination. In fact, the complexity of seepage through a fissured rock mass virtually precludes the effective preplanning of a grouting operation.

Following the initial mine grouting, a series of laboratory experiments were set up to simulate field seepage conditions [2]. Lucite tubing was used to represent seepage channels and drill holes. In the model shown in Fig. 8.6, BCDE is 3/8 in. ID; AC and FG are 3/16 in. ID; lines 1, 5, and 6 are 1/8 in. ID; and the rest are 1/16 in. ID. ACDE represents a fault zone, and BD represents a drill hole. Lines 1 through 13 represent seepage channels leading from the fault zone to the mine drift.

In one experiment point B was shut and water was pumped through point A at a constant rate until equilibrium was reached. The numbers on the lines in Fig. 8.6 show the percent of total flow going through each line. When the volume pumped through point A was changed and equilibrium again established, the percent of total flow going through each line varied as shown in Table 8.1.

Other experiments verified the conclusion that can be drawn from this one, that percent flow is not a direct function of total flow and that the variation was much higher in pipes with low percentage flows than in those with high percentages.

In another series of experiments, dyed water was pumped in at both points A and B to study the paths and mixing of grout and groundwater. After equilibrium had been established with a static head of red-colored water at A causing seepage in each of the 13 pipes, blue-colored water was pumped in at B. At a pumping pressure less than the static head at A, no flow from B occurred. As the pressure at B was increased to a value above that at A, flow in the BC direction started. Mixing occurred at point C, with all the fluid from B (blue) moving in the CD direction. As the flow from B increased, the color in line CD became predominantly blue, and equilibrium could be established with lines 1 through 7 flowing blue and lines 8 through 13 flowing red. This condition persists over a considerable pressure increase at B. Only when the volume entering at B became very large compared to that at A did any of the blue fluid begin to enter the pipes 8 through 13. It was found by experimentation that if the larger openings such as 1, 5, and 6 were plugged, much less flow was required at B to cause blue fluid to move into pipes 8 through 13.

A number of different models were used in the experiments partially described above. It was found after working with any given model with a number of different conditions that sufficient data

TABLE 8.1 Two-Minute Flow-Through Model

Pipe No.	cc	%	cc	%
1	305	20-	169	20+
2	43	3-	22	3-
3	21	1+	8	1-
4	15	1	9	1+
5	395	26-	232	28+
6	455	30+	210	26-
7	12	1-	5	1-
8	78	5	44	5+
9	48	3	32	4-
10	46	3	37	4+
11	35	2+	20	2+
12	34	2+	7	1-
13	39	3-	33	4
Total flow	1526		828	

were available so that the seepage system could be sealed with a preplanned procedure. The required data for any system could be summarized by the following descriptions:

1. The required pumping pressure to fill all the seepage channels with groundwater dilution held to negligible proportions
2. The time lapse from the start of pumping until the pumped fluid reaches the end of each seepage channel
3. The volume of pumped fluid required to fill all the channels at the required pumping pressure

Such data can be obtained only by a pumping test. It cannot be obtained by visual examination even for very simple seepage systems, because the addition of external pumping pressure, which will be required for grouting, changes the characteristics of the seepage system.

Lab work and field work both indicate conclusively that when pressure conditions within a seepage system are changed, the very small leaks are much more affected than the large ones. Thus, it should be expected that a field pumping test will reveal leaks which were not flowing under normal static head conditions.

The laboratory studies showed clearly that at some point within a seepage system the external fluid pumped mixes with the internal fluid (in field work, this means grout would mix with groundwater). At small pumping volumes it should be expected that somewhere within the seepage system grout will be diluted with groundwater to the extent that it will not gel. As the pumping volume is increased, the detrimental dilution will decrease. Under both conditions, the grout tends to flow toward and into that portion of the seepage system farthest from the source of groundwater supply.

The laboratory studies pointed the way to techniques, refined by field experimentation, which work well in stopping seepage in fractured rock and which are also useful in treating fractured and porous concrete. The technique consists of first drilling a short hole that will intersect water-bearing fissures and cracks. Holes are drilled in the simplest fashion, often by jackhammer. Dry holes are generally worthless and should be abandoned. Wet holes (holes that strike water) are generally useful and are dye-tested as soon as completed.

The dye test is done with the grout plant by pumping dyed water through a packer placed at the collar of the wet hole just drilled. The dye concentration must be carefully controlled, as discussed in Sec. 7.5. Pumping pressures should be kept well below the pump capacity. If these criteria limit the pumping rate to less than a gallon per minute, it may be best to abandon the hole and drill a new one. (This criteria applies to work with low-viscosity grouts. When working with polyurethane, cost-effective pumping at much lower rates is possible.) The rate noted is approximate and may well vary from job to job. At some low rate for each specific job it will become economically more feasible to drill new holes to find higher takes rather than to treat tight ones. Job experience will soon dictate which wet holes need not even be tested.

When the dye test begins, the adjacent wall area is carefully watched for evidence of dye. When dye is first seen at any point, the time since the start of pumping is noted as well as the pumping pressure and rate. Dye tests may be stopped when dye appears at one point or may be continued until a number of different locations show dye. (Generally, the points where dye appears are in areas which are already wet or flowing. This is the assumed condition in the discussion which follows. If dye appears only in areas which were dry prior to the dye test, the hole should be abandoned.) For each point, time, pressure, and rate are recorded.

Every time dye appears, this indicates that an open seepage channel exists between the packer and the point where dye appears. The exact location of the channel within the rock mass is not known, but if the drill hole being tested made water, then the established channel must reach into the water source somewhere. Therefore, the hole is worth grouting.

The time recorded for the appearance of dye is the maximum gel time that can be used to seal that particular channel. If longer gel times are used, the grout will run out of the leak before it sets. If the time is short, of the order of 5 or 10 sec, this may be too short a gel time to handle readily. The effective time can be lengthened by lowering the pumping rate. This is a necessary step when attempting to seal a number of zones from one hole and one of the zones has a very short return time.

If the return time is long, say, 5 min or more, it can be shortened by increasing the pumping rate. This will generally raise the pumping pressure and may therefore not be feasible if allowable working pressures would be exceeded. It is usually not productive to treat holes that show return times of 10 min and more when holes with shorter return times are available.

Polyurethane use is increasing for seepage control in fractured rock and concrete, as well as for sealing concrete construction joints. These grouts will probably not be effective in openings less than 0.01 in. unless very high pumping pressures are used. Field procedures are the same as for the less viscous chemical grouts, althrough shorter grout holes and higher pressures are generally used.

8.4 FIELD WORK

Except for very limited special cases, all of the grouts used must be permanent. They must have adequate strength and imperveousness, and these properties must not deteriorate with age or by contact with ground or groundwater. The grouts must have controllable setting times over a wide range, be acceptably nonhazardous to humans and ecology, and inexpensive enough to be competitive with other construction alternatives. Finally, they must have a low enough viscosity or particle size to permit placement at acceptable economic rates and safe pressures. Domestically, most seepage-control work currently uses the acrylics, with the use of polyurethanes on the increase, and the use of silicates waning (see also data on microfine cements).

Metering pumps are highly desirable for dealing with seepage problems for both the dye tests and the actual grouting. (For water-reactive polyurethanes, single pumps can often be used effectively.) Once a hole has been drilled and tested and it has been determined that grouting is in order, the pump suction lines may be switched directly from the dyed water tanks to the grout tanks. The use of valved quick-couplings at these points saves a good deal of time. Dye may also be used in the grout, a color different from that used for dye testing. Proportions of this dye must also be carefully controlled, so that it is not identifiable at 100% dilution.

When pumping begins, the pumping volume should be brought as quickly as possible to that used during the tests, and the grout itself should have a gel time of about three-quarters the previously recorded return time. The pumping pressure is monitored to make sure it does not exceed the allowable, but otherwise no attempt need be made at this stage to keep the pressure at dye test values. The leak is watched closely. It should begin to seal at about the recorded return time. If this does not happen and dye (of the color used in the grout) does not appear at the leak, then dilution beyond the ability to gel has occurred. (This would normally mean too high a dye concentration was used in the dye tests.) If dye does appear but the leak does not seal, then dilution of the grout has extended the gel time beyond the return time. This may be counteracted by decreasing the gel time (easy to do with metering equipment but very difficult with equal volume equipment) or by decreasing the pumping rate (easy to do with either kind of equipment).

As the leak begins to seal, the pumping pressure will rise, particularly if a single channel is being treated. If the rise is rapid or reaches high enough values, it will blow out the seal just made and reopen the leak. Therefore, it is important, as the leak begins to seal, to keep the pressure from rising by reducing the pumping rate. It is desirable at this time to continue pumping and if possible to place additional grout, since the grout now being pumped is most probably going directly into the source of the seepage. If field conditions and experience offer no clue to the additional volume to be placed, pump an amount that does not exceed that pumped up to the time the leak sealed.

Once gel begins forming in the seepage channel, the entire seepage net is altered, and all previously gathered seepage data may become totally unreliable. This is the primary reason why holes should be drilled, tested, and grouted one at a time. For holes which feed more than one leak, considerable change in return times will occur once sealing of one leak begins. For such holes, sealing of additional leaks becomes a trial-and-error proposition, with a good chance of blowing out earlier seals with the higher pressures which may be needed for other leaks.

The process described can be readily used to seal one leak or zone at a time, and if the total number of leaking zones is small, the technique is economical for complete seepage control. (Actually, 100% shutoff can be obtained for individual leaks but is often not economically feasible for a system with many leaks. For example, it may cost as much to shut off the last 10% as it does the first 90% of the total seepage.) However, if the number of leaks are many, the method becomes very time consuming, and it becomes necessary to shut off leaks by shutting off the source of seepage. Grout pumped through the hole after all the external leaks have closed is very effective for this purpose.

The application of the method discussed is illustrated by a seepage problem in Pennsylvania. Note in this and in other case histories which will appear later that field personnel often deviate from practices described in the text as most desirable. Near Pittsburgh, a slope to a coal vein was excavated about 300 ft from a river. Water problems were anticipated, and pregrouting with cement to a depth of 20 to 30 ft was completed prior to the start of excavation. This was effective in the treated zone, and excavation showed rock fissures up to 4 in. wide that were full or partially filled with cement grout. When excavation reached river level and continued lower, small seepages were encountered. At first, these were not severe enough to stop the concrete casing operation. (Weep holes were placed at the bottom of the casing to control pressure buildup.) At a depth of about 65 ft below riverbed and 200 to 250 ft downslope, a water channel flowing at 25 to 30 gpm was intercepted. This flow was sufficient to halt the excavation and concreting, and the contractor turned to chemical grouting to control water inflow.

The following quotations from the grouting engineer's job report indicate the field procedures used:

> GROUTING PROCEDURE
>
> (A) Drilling
>
> Holes were drilled into the rock around the main flow according to Figure 1 [see Fig. 8.7]. Holes were drilled from 10 feet to 14 feet deep, about 2 feet to 4 feet apart. The purpose of drilling this way was to attempt to instersect the main channel of flow. This was accomplished with Holes No. 2, No. 4 and No. 5. Hole No. 6 also showed a little moisture but the remaining two holes, No. 3, No. 7 were dry. The main flow was through Hole No. 5 and Hole No. 4. The total flow increased after drilling from 25 to 30 gallons per minute to about 35 gallons per minute.

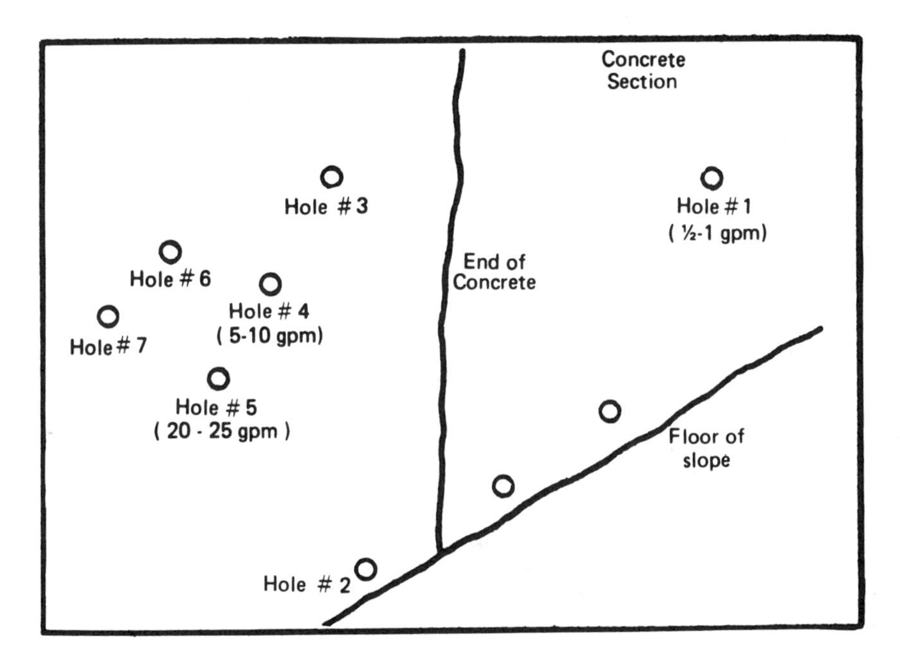

FIGURE 8.7 Grout hole location [noted as Figure 1 in the quotation in text].

(B) Pumping of Dye

Fluoroscein dye was used. The dyed water solution was pumped into Hole No. 1 on the north wall of the slope. This was one of the weep holes which had a 2-inch pipe in it, so no packer was needed. Water was coming from this hole at 1/2 to 1 gallon per minute.

It was decided to pump into this hole because the dyed solution would flow down slope which would make it possible to learn if this flow connected into the main flow at the bottom. Pumping here would also indicate how fast the water in the channel was flowing and therefore determined how fast a gel time was needed. The dye solution was placed through the grout pump. Pumping rate of dye was 5.5 gallons per minute. The dye first showed up in Hole No. 5 in two minutes, ten seconds. It then appeared in Hole No. 4 in two minutes, 30 seconds; in Hole No. 6 in 3 minutes, 30 seconds; in Hole No. 2 and the pool of water at the bottom of the slope in 5 minutes, 15 seconds. Holes No. 3 and No. 7 were dry. Since all the holes except the dry ones showed traces of dye, it indicated that they were all connected by a common channel. Pressures encountered during pumping of the dye were negligible.

The information obtained here indicated that possibly one injection could control the whole problem at the base of the slope. As it turned out, however, three were necessary.

PUMPING OF GROUT

It was decided to pump the grout in the vicinity of the main flow, since the closer the application is to the main flow, the better chances are of putting the grout where it would be effective.

Hole No. 4 with a flow of 5–10 gpm was chosen for the first injection because the flow rate and pressure were not sufficient to hinder placement of a packer. It also was close to the main flow which appears in Hole No. 5 with a strong rate of 20–25 gpm.

Shot No. 1 in Hole No. 4

Pumping Rate—5-1/2 gallons per minute
Gel Time—3-1/2 minutes at 40°F (temp. of solutions)
Pumping Time—7 minutes
Grout Placed—35 gallons
Max. Pressure—125 psi (hole gelled up, pumping stopped)

Shot No. 2 in Hole No. 5

Average Pumping Rate—6.3 gallons per minute
Gel Time at Start of Pumping—3 min., 30 sec at 40°F
Gel Time at End of Pumping Cycle—about 1 min., 30 sec
Catalyst pumps at full capacity
Pumping Time—15 minutes
Grout Placed—95 gallons
Maximum Pressure—125 psi.

Grout was first pumped into the hole at 5-1/2 gallons per minute. The pressure was not high (30 psi.) so the rate was increased to about 7-1/2 gallons per minute. The dyed grout started to push out through the rockface and out of Holes No. 6 and No. 3 (dry hole); it also started to show along the floor of the slope. At this point the catalyst pumps were opened to full speed. The pressure went up to 70 psi., then to 125 psi. indicating the hole was gelled. Pumping was stopped.

Shots No. 1 and No. 2 stopped 90% or more of the water coming in at the bottom of the slope through the rockface on the north wall. There was still a small flow of water coming in at Hole No. 2 and behind the concrete wall. To stop this, Shot No. 3 was placed in Hole No. 1.

Shot No. 3 in Hole No. 1

Gel Time—3-1/2 minutes at 40°F.
Pumping Time—8 minutes
Grout Placed—15 gallons

Maximum operating pressure allowed behind concrete casing was reached, 35 psi. Pumps started to recycle, hole gelled, pumping stopped.

Shots No. 1, No. 2, No. 3 controlled the water problem 100% at the base of the slope.

When seepage occurs through widely spaced cracks or construction joints on a smooth surface, it is feasible to treat each crack or joint separately, often directly at its exposed location. Such was the case in the following job description, which was done by grouting within a concrete dam, part of an electric generating facility in New Jersey.

A passageway, approximately 45 ft below the level of the lower reservoir, provides access to draft tube and waterwheels. The passageway, 5 ft × 7 ft × 100 ft, is fully and smoothly concreted. Three openings spaced equidistant along the 100 ft length lead directly to these draft tubes and waterwheels.

Leaks occurring at several points in the three access ways were a continual annoyance to maintenance and inspection personnel. They were found mainly along construction joints in the concrete structure and to a lesser extent in honeycombed areas in the concrete. Water at 35°F flowed from these points in well-defined streams, in some cases at 2 to 3 gpm.

Since the walls were smooth and the construction joints straight, it was feasible to use a pressure plate, or surface packer, fitted with an 0.5-in.-thick sponge rubber gasket around its perimeter. Details of the packer are shown in Fig. 8.8. Such packers may be made to any length. However, the longer they are, the more force is required to hold them in place. Normally they are braced against an opposite wall. Surface packers can also be used to stop seepage from porous concrete, as described from a portion of the engineer's report on this field work.

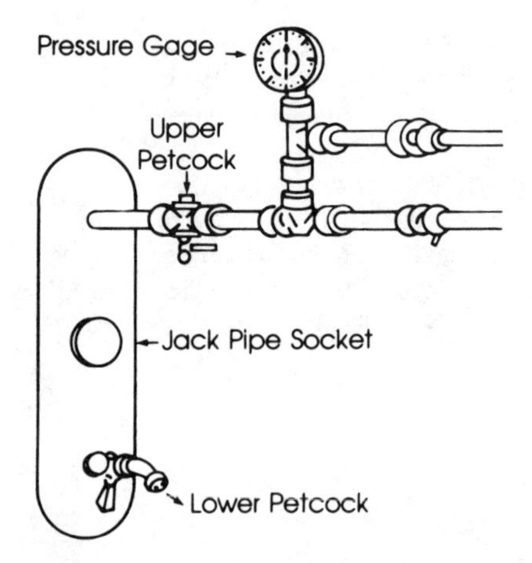

FIGURE 8.8 Surface packer details.

> The pump hoses were connected and the pump actuated. When the tracer dye in the grout was observed coming out of the lower petcock, this was shut off. Pumping was continued and when tracer dye flowed from the upper petcock to indicate that air and ground water had been eliminated from behind the plate, that petcock was closed. With these two exits closed, the grout could then be forced into the honeycombed concrete. Dye eventually was seen seeping from fissures well beyond the area covered by the plate. By pumping slowly, gel was made to form in these fissures, thus extending the effectiveness of the pressure plate.

Grouting with the plate, as just described, differs from the packer method in that the grout must enter the voids against the flow of water rather than with it; thus, additional pressure is required.

Good penetration was achieved, and the leaks covered by the plate were stopped, as were some extending a considerable distance beyond the plate. The tracer dye, mixed in with the grout, offered a ready means of detecting the travel of the grout through the fissures.

Sealing isolated cracks and construction joints is more generally done by pumping grout through packers placed in drilled holes. The following excerpt from the engineer's report describes typical field procedures:

> *Drilling Holes /Placement of Packers*—One-inch holes were drilled in the vicinity of the leaks to the point of intersection of the

construction joints, 8 to 10 inches from the surface. An electric handheld diamond coring drill was used. Next, packers were inserted in the holes and tightened to provide a pressure-tight seal.

After placing a packer, it was necessary to effect a seal around the construction joint over an area sufficient to enable the grout to travel and subsequently gel without becoming dissipated. The seal was achieved by forcing lead wool into the joint. In addition, a 2 × 4 × 48-inch plank, covered on one surface with a half-inch layer of sponge rubbrer, was placed against the joint and firmly held by 2 × 4's braced against the opposite wall.

Pumping Grout/Dye Tracer—A hand-operated, equal volume, dual piston pump was connected to the previously secured packer. Grout was formulated to give a gel of about 12 seconds.

Red dye mixed with the chemical grout served to trace the flow and indicated the time required for the grout to travel through the confined area and out to the surface. This determined the duration of effective pumping time, which proved to be only a few seconds.

Slow, steady pumping at less than 1.2 gallons per minute proved most effective. Some chemical was lost due to imperfections in the seal along the joint but this decreased as the solution began to gel.

Pumping was continued until it took more than a reasonable effort to operate the pump handle—about 20 pounds. At this point, after a 30-second waiting time, the pump was disconnected and the sealing board and packer removed. The leakage had been arrested.

Sealing seepage through a finite number of channels is possible and generally practical when the channels exist as cracks and fissures in an impermeable medium. When the channels exist as preferred flow zones in a generally permeable medium, it must be recognized that sealing the flow zone will only divert the flow to a new channel. Thus, when dealing with porous media such as sandstone and cinder block (as well as the more obvious granular deposits), the job economics should be based on total stabilization, not just stabilization of those zones which are wet prior to grouting.

The necessity for postconstruction water proofing often occurs due to changes in the water table. These changes may be due to natural causes, but more often they are induced by other construction projects. Such was the case at a missile complex in the western part of the United States.

The underground installations of the complex are connected by 2200 lineal feet of corrugated metal tunnels. Nominal IDs of the tunnels vary from 7.5 to 9.5 ft. The larger tunnels consist of 5 gage corrugated plate, and the smaller are of 7 gage. All tunnels consist of 8-ft-long sections made by joining five plates longitudinally by a double row of 0.75 in. bolts, staggered 6 in. on centers. Sections are joined by a circumferential row of 0.75 in. bolts on about 9.5 in. centers.

Preconstruction exploration found no natural groundwater within 200 ft of the surface, in deposits consisting of 3 to 5 ft of silty sand, 15 ft of silty sand with white caliche beds, and 20 ft of shattered basalt underlain by competent, dense columnar basalt. During this study it became known that the site was within a planned federal irrigation project. Rather than move the site to nonirrigable land, it was decided to provide waterproofing to the underground structures.

All the tunnels were founded on or in the shattered basalt. Most of the tunnel footage was placed in trench-type (open) excavations. Backfill consisted of a mixture of the two upper strata, with embedded rock fragments in some areas, compacted to 90% to 95% modified AASHO maximum density. Waterproofing provisions included external membranes at the junctions of tunnels with other facilities, lead washers for all tunnel bolts, welding of all laps, and, on the exterior tunnel surface, a 1/8-in.-thick asbestos-fibered asphalt mastic placed on an asphalt prime coat.

Irrigation began in the area after all tunnels were completed and backfilled. Some 2.5 months later, seepage into the tunnels was first detected. Within another month and a half, leakage was occurring throughout the entire complex, not only in the tunnels but in other facilities as well. The major source of seepage was the bolt holes in the tunnels. At the peak of water inflow, the total seepage exceeded 150,000 gpd.

The seepage fell off as irrigation ceased in the fall, and an intensive repair effort was mounted, primarily by welding bolt heads to the tunnel plate. Prior to the start of the next irrigation season, seepage had been reduced to between 2000 and 3000 gpd, an amount that could be handled by the regular sump pump system. It was known, however, that seepage would increase greatly as the new season's irrigation began.

While the repair effort was going on, a 50 ft section of tunnel was set aside as a test section to evaluate chemical grouting as a solution to the seepage problem. Five days of work in the tunnel demonstrated that grouting with acrylamide-based materials was an economically feasible method for sealing existing leaks and for handling anticipated seepage as the water table continued to rise. The test program demonstrated the presence of continuous channels at the plate laps along the bolt lines through which grout would travel, effectively sealing the bolt holes. Tests also showed that grout hole spacings as much as 4 ft apart along the laps were adequate and that grout volumes of 10 to 15 gal per hole were needed. Both lateral and circumferential joints could be treated in this fashion. Based on the test grouting, specifications were prepared, and a grouting contract was let at about the time the second irrigation season would start.

Grout holes were placed by drilling a 3/8 in. hole through the inner plate at the lap and either threading the hole or cementing a square threaded tab over the hole. Spacing of circumferential holes was 5 to 6 ft, longitudinal spacing was 4 ft. Grout plant hoses would be attached to the threaded holes through a nipple and valve placed as soon as each hole was finished. Holes were placed and water-tested prior to grouting, and the results of the water test (take, pressure, and time of movement to observation holes) were used to set the pumping parameters and the gel times. Grouting pressures were not permitted to exceed 45 psi. In general, the volume per hole was 5 gal or less at gel times shorter than a minute, and pumping rates were around 0.5 to 0.75 gpm.*

Grouting was carried on over a 4 month period in rolled steel structures, neoprene joints, and concrete to steel joints as well as in the tunnels. Grouting work extended through and was completed about a month after the completion of the second irrigation season. In the tunnels themselves, 4400 holes were grouted with 16,200 gal of grout, averaging 3.7 gal per hole and 7.6 gal per linear foot of tunnel (many holes did not accept grout). At the completion of grouting, practically all seepage inflow had been sealed.

Problems requiring seepage control may occur very gradually over a period of many years, or even because our perception of personal safety (or legal liability) change with time. Such was the case in Milwaukee, where a 130-year-old cave system and brewery museum, owned by Miller Brewing Company, had to be closed. Leaking groundwater had caused a slippery, unsafe condition on the slate floors [3].

The arched caves are 150 ft long and as deep as 60 ft below ground surface, with walls up to 44 in. thick. The walls were constructed of large cut-limestone blocks faced with a brick lining. The soil outside the tunnel caves is clay. A layer of deteriorated block and granular soil surrounds the tunnels. The caves were used to store beer until 1906, when they were abandoned in favor

---

*The grout plant used was a specially designed equal volume system with an output of 0.2 to 1.2 gpm. Gel times were controlled by varying the catalyst percentage in the solution tank.

of the (then) new refrigeration systems. The caves were made into a brewery museum, and opened to the public in 1950. Annually, some 150,000 visitors are attracted. Seepage through the walls had made the tunnel floors increasingly slippery and unsafe, and finally necessitated the closing of the museum.

Numerous attempts were made to seal the tunnels by working from the surface. None of these attempts was successful. Work was then started from the inside in 1983. Grout holes were placed by the grouter, McCoy Contractors (Milwaukee, Wisconsin) on 18 in. centers, through the walls into the surrounding backfill. Using AC-400 at gel times varying between 30 sec and 4 min, 3000 gal of grout were injected into about 900 holes (representing about 45% of the total job). Grout pressures were kept below the overburden pressure. Figure 8.9 shows details of the grouting operation. A year of monitoring indicated a 95% seepage cut-off. The rest of the work was then completed with 5000 gal through 1200 holes. The museum has since re-opened.

When grouting cracks whose exposed openings exceed 1/16 in. (1.5 mm), it is often worthwhile caulking those openings prior to grouting. This reduces loss of grout, permits working at longer gel times, and provides a pressure backing to permit grout to flow into the crack against the seepage flow. A useful discussion of case histories where such procedures were used is in Ref. 4.

Cracks of this and larger dimensions are readily sealed with polyurethane grouts, which provide greater resistance to extrusion than the acrylic based materials.

It is sometimes possible to use judiciously placed drain holes to create a preferred flow path for subsequent grouting. Such procedures were used to control seepage into the underground portion of a grain silo, in a job done by Denver Grouting Services, Inc.

In Wray, Colorado, a grain elevator pit had been built 12 ft deep in free-flowing sand, in a zone where the water table often rose to 4 ft below ground surface. Some time after construction, every construction joint and crack began leaking, necessitating the installation of a sump pump. Shortly, this unit was in continuous operation at 25 to 36 gal per min. In order to eliminate this continual expense, as well as to avoid the contamination of several thousand bushels of grain if pump failure occurred, remedial measures were required. Three options were available:

1. Replace the elevator pit with a new one, properly sealed during construction.
2. Install an expensive fail-safe pump system with battery backup.
3. Seal the existing structure.

FIGURE 8.9 Grouting from the inside of a tunnel.

The owner elected the third option, and had tried unsuccessfully to pump sealers around the outside of the pit, prior to contacting a grouting company.

Holes were drilled on a regular pattern through the walls and floor. Each hole was closed with a valve as soon as sand was seen running in. Then vertical grout holes were placed on about 4 ft centers about 12 ft from the structure perimeter. Sodium silicate grout, catalysed with ethyl acetate and formamide, with gel times ranging from 18 to 25 min were pumped through the grout pipes, and the valves inside the pit were opened to draw the grout toward the pit. By controlling the gel time and the pumping rate, grout was made to set up at the pit walls, effectively sealing the pit one section at a time. After 10 years of exposure, the pit is still completely dry.

Previous discussion of grouting materials (see Chap. 4) had indicated a growing core of data mitigating against the use of sodium silicates for seepage control. Much of this data stems from recent laboratory studies, most of which used a catalyst system composed of Formamide as reactant and ethyl acetate as accelerator. The successful 10-year field performance of the particular grout used in Colorado might be related to its specific composition. Additional laboratory studies are needed to investigate this possibility.

8.5 SUMMARY

Seepage occurs because nature or construction results in an open channel between a water source and a point at lower potential. The location of the channel in rock and soil is unknown and cannot be deduced with accuracy from geologic data.

The most effective procedure to control seepage would be to plug the seepage channel at the water source. This is almost always impractical, and field procedures are designed to plug the seepage channel at the point of seepage. In practice, short holes are drilled to intercept (by trial and error) the seepage channel 5 to 10 ft from the point of seepage. This establishes a circuit containing part of the original seepage channel. By proper gel time and volume control, the circuit can be gelled, stopping the seepage.

Seepage through cracks and joints in concrete can be treated similarly and may also be grouted directly through a surface packer. A comprehensive study of tunnel sealing procedures appears in Ref. 5.

REFERENCES

1. Committee on Grouting, Preliminary glossary of terms related to grouting, *J. Geotech. Eng. Div. ASCE, 106*(GT7):803–815 (July 1980).

2. R. H. Karol, Study of Mine Seepage, Technical Report, American Cyanamid Company, 1958.

3. *Civil Engineering*, ASCE, New York, Sept. 1984, page 14.

4. S. T. Waring, Chemical Grouting of Water Bearing Cracks (Paper at ACI 1985 Convention, Chicago, Illinois).

5. Bruce La Penta, Tunnel Seepage Control by the Interior Grouting Method, Special Project Report to the Department of Civil Engineering, Rutgers University, New Brunswick, NJ, July, 1989.

9

Grout Curtains

9.1 INTRODUCTION

Grout curtains, also called grouted cutoffs, are barriers of groundwater flow created by grouting a volume of soil or rock of large extent normal to the flow direction and generally of limited thickness in the flow direction. Typically, a grout curtain could be used alongside or underneath a dam to prevent drainage of the impounded water. Curtains may also be placed around construction sites or shafts to prevent water inflow. Where the required service life of a cutoff is of limited duration, well points or other construction methods often prove more practical. For long-term shutoffs, where the zone to be impermeabilized is close to ground surface, slurry walls have in the past decade become very competitive with grout curtains. Where the treatment is deep or below a structure which cannot be breached, grout curtains remain the most practical and sometimes the only solution.

9.2 SELECTION OF GROUT

All the common, commercially available grouts, if applied properly, will reduce the permeability of a granular formation to values similar to those of clays. (See Chap. 4 for a discussion of the permanence of silicates under high hydraulic heads.) This is more than adequate for water cutoff. Strength is not a critical factor, since the formations to be grouted are generally stable in their natural conditions. Resistance of the grout to extrusion from the soil voids is of course of importance. However, even the so-called "weak" grouts such as polyphenols can resist hydraulic heads of several hundred pounds per square inch for every foot of curtain thickness (in sands and silts; in coarse sand and gravels extrusion resistance is lower). Since the minimum practical thickness of a grout curtain is 5 ft or even more, adequate resistance to extrusion will always exist. Consolidation of the grout is sometimes thought to be a possible problem. However, grout gels, like clays, will consolidate only under mechanical pressure. The erection of a grout curtain causes no significant change in mechanical pressure on the gel. The curtain, if it functions properly, does create a hydraulic gradient from one side to the other. However, hydraulic pressures will not cause consolidation, and as previously noted, adequate resistance to extrusion is always present. In view of the foregoing discussion, the only factor of importance in selecting the grout is its ability to penetrate the formations to be grouted (Secs. 7.2, 7.3, and 7.4). If more than one grout meet this criterion, then economic factors enter into the selection.

Grout curtains are generally large jobs in terms of time and material involved. For large jobs of any kind, there is often economic merit in the use of more than one grout, using a less expensive material for the first treatment (cement, clay, and bentonite should be considered if they will penetrate coarser zones) and following with a (generally) more expensive and less viscous material to handle residual permeability.

An interesting discussion of the grout selection process appears in the paragraphs quoted below in reference to grouting for a dam in Cypress [1].

> Before it was possible to select a proposed hole layout it was necessary to choose a chemical grout with properties acceptable to the Engineer. In relation to gel permeability, permanence and grout viscosity, several grout systems could be considered, namely chrome Lignin/dichromate (e.g., TDM and Sumisoil), resorcinol formaldehydes (e.g., MQ14), acrylamides (e.g., AM-9), phenol formaldehydes (e.g., MQ4) and silicate based grouts. When considering potential hazards from the materials, however, the choice narrows further, since the acrylamide AM-9 is neurotoxic and the chrome lignin/dichromate systems, whilst less toxic, are dermatitic. The phenol fomaldehydes by comparison are non-toxic, and tests to date on resorcinol formaldehydes show that there is little danger if the materials are ingested. Silicate based systems are also non-toxic and have been used successfully on other dams and are less expensive than any of the other chemical grouts mentioned above. It was on this basis that sodium aluminate/sodium silicate grout was proposed to the Engineer.

Extensive tests were carried out on a sodium aluminate/ sodium silicate grout which had been successfully used at Mattmark Dam in Switzerland [1a] to establish the suitability of this type of grout for Asprokremmos Dam. The mix tested comprised of:

Solution (1)	Sodium aluminate	16 parts/wt
	Water	320 parts/wt
	(i.e., 4.8% Sodium aluminate)	
Solution (2)	Sodium silicate (M75)	522 parts/wt
	Water	284 parts/wt

and from the results obtained, the following conclusions were drawn:

The sodium aluminate/sodium silicate mix tested had an initial viscosity of 3.6cP and gel time of approximately 33 minutes at 20°C when tested under laboratory conditions. The gel time of this mix could be extended to approximately 80 minutes by reducing the sodium aluminate content of solution (1). A reduction in sodium aluminate content, however, resulted in a weaker gel and at concentrations less than 3.6% in solution (1), the gel strength was judged to be too weak to be of any value where "permanence" is required.

The gel time could also be extended by increasing the total water content of the mix. However, for gel times greater than 60 minutes, the degree of dilution required is excessive (see Graph 3).

The viscosity of a sodium aluminate/sodium silicate grout solution increased gradually with time and a mix with an initial viscosity of 3.6cP and gel times 73 minutes at 20°C remains below 5.0cP for 50% of its life and below 10cP for 75% of its liquid life.

The sodium aluminate/sodium silicate reaction is temperature sensitive and a rise of approximately 8°C will halve the gelation time.

Further tests carried out by an independent body, as directed by the Engineer, also showed that the chemical grout proposed was satisfactory with regard to permanence and that there was very little likelihood of any leached material acting as a dispersive agent on the clay core.

9.3 GROUT CURTAIN PATTERNS

The pattern for a grout curtain is a planview of the location of each grout line or row, and every hole in each row. The sequence of grouting each hole should be noted on the pattern. It has been shown previously (Sec. 5.8) that in order to approach total cutoff a grout curtain must contain at least three rows of grout holes and that the inner row should be grouted last. The distance between rows, and the distance between holes in each row, is selected by balancing the cost of placing grout holes against the cost of the volume of grout required. For any selected distance, the spread of grout horizontally must be at least half the spacing. The volume of grout required for a 6 ft spacing of holes is 4 times the volume required for a 3 ft spacing. Thus, while the cost of drilling decreases as a linear function, the cost of grout increases as a quadratic function. For any specific job the actual costs of drilling and grout can be readily computed for several different spacings to determine the specific spacing for minimum cost. This generally turns out to be in the 2 to 5 ft range. The process is graphically illustrated in Fig. 9.1.

Determination of optimum spacing is a mathematical process suitable for computer solution. A program (written in BASIC) to achieve this end appears in Appendix C. This program was written in 1985 by Keith Foglia, then a student at Rutgers University. Slight revisions were made in 1987. The use of the program is illustrated in the example which follows.

Assume a grout curtain is to be placed between 20 and 70 ft below grade to protect a dam, and will extend from the side of the dam for 100 ft to reach a rock face. Tube à manchettes will be used for grouting, and it is estimated to cost $5.00 per ft to place the grout pipe through overburden, and $20.00 per ft through the zone to be grouted. The grout is estimated to cost $1.50 per gal in place, and the grouted zone has a groutable porosity of 35%. Feeding this data into the program yields the printout shown in Figure 9.2.

In addition to optimum spacing, this program also gives the total job cost. Thus, the cost effectiveness of other types of grout and grout pipes may be quickly evaluated. Also, since some of the cost factors are estimates, it is possible to evaluate the effects on job cost of possible variations in unit cost of pipe placement and grout-in-place. Note that costs entered are in-place (i.e., labor is included).

The length of a grout curtain is often determined by the physical parameters of the job. A cutoff between two foundations obviously

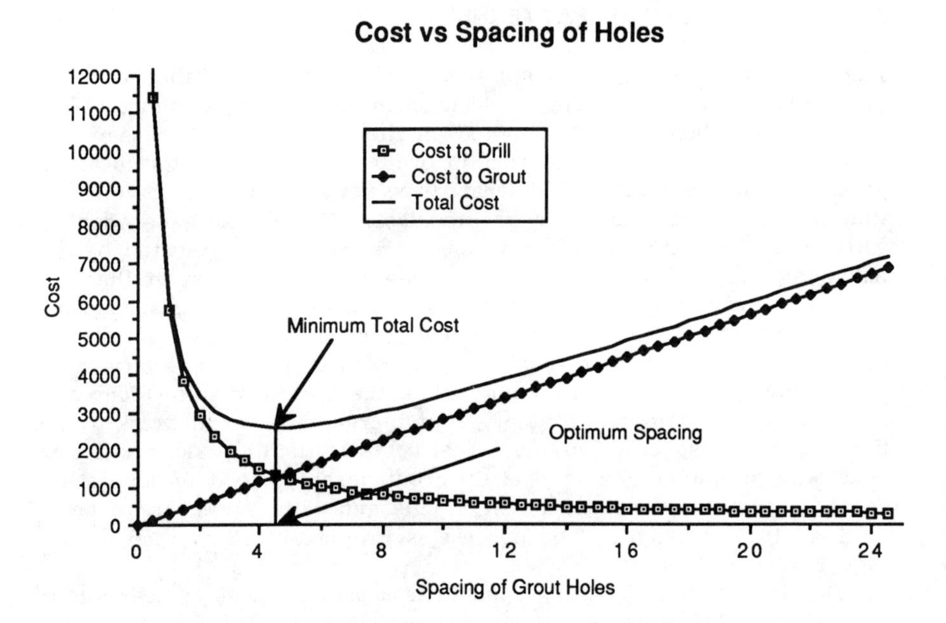

FIGURE 9.1 Optimum spacing of grout holes. (Courtesy of Keith Foglia, Rutgers—The State University of New Jersey, New Brunswick, New Jersey.)

has a length equal to the distance between them. A grout curtain on one side of a dam, however, need not extend indefinitely or to the closest impervious formation. Such curtains function by extending the otherwise short flow paths far enough so that flow is reduced to tolerable amounts. The length may be extended to where more permeable zones terminate or may be designed on hydraulic principles alone.

The depth of a grout curtain is determined by the soil profile. Unless the bottom of the curtain reaches relatively impervious material, the curtain will be ineffective if shallow and very expensive if deep.

FIGURE 9.2 Computer printout for problem in grout hole spacing.

GROUT CURTAIN DESIGN

```
This is a Basic program to compute the most economical spacing of
chemical grouting holes for a Grout Curtain, which will cut off the
flow of ground water.

It also computes:    Number of Holes to be Drilled
                     Volume of Grout to be Placed in each hole
                     Cost of Drilling and Placing Pipes
                     Cost of Grouting
                     Total Cost


(Hold down the shift and press the PrtSc key to print any screen.)

                      Enter M for Methodology

                      Press enter to continue?
                        METHODOLOGY


This Program correlates drilling costs and cost of grout in order
to determine the optimum spacing of holes for minimum cost.

Cost of Grouting = Volume of Grout x Cost per Unit Volume

Cost of Drilling = Number of Holes x Cost to drill each Hole

Number of Holes = (3 x Length of Curtain divided by Spacing) + 1

Total Cost = Cost of Grouting + Cost of Drilling

The minimum Total Cost can therefore be determined by taking the
derivative of the Total Cost equation and setting it equal to zero.

Solving the result for Spacing gives the Theoretical Optimum Spacing.


                      Press any key to continue

Enter Porosity of soil to be treated. If known? n=? .35
NOTE:   If porosity is unknown but void ratio is known,
        just press return to go to the next step.
Enter Void Ratio of soil to be treated. If known? e=?

Porosity will be taken to be n=0.350

Will job require drilling through Over Burden Soil
to get to the area to be grouted?  (Y/N)? y

Estimate Cost (in $ Per Foot) of drilling through Over Burden soil.
Be sure to include cost of labor and materials.

Enter Over Burden Drilling Cost (in $ Per Foot)? 5
Enter depth of Over Burden soil to be drilled through in ft. ? 20

Estimate Cost (in $ Per Foot) of drilling through soil to be treated.
Be sure to include cost of labor and materials used and placed.

Enter Drilling Cost (in $ Per Foot)? 20
Enter depth of soil to be treated in ft. ? 50

Enter Cost of Grout in $ per gallon? 1.5
```

(a)

```
Based on three rows of holes in an off-set pattern.  Like...
```

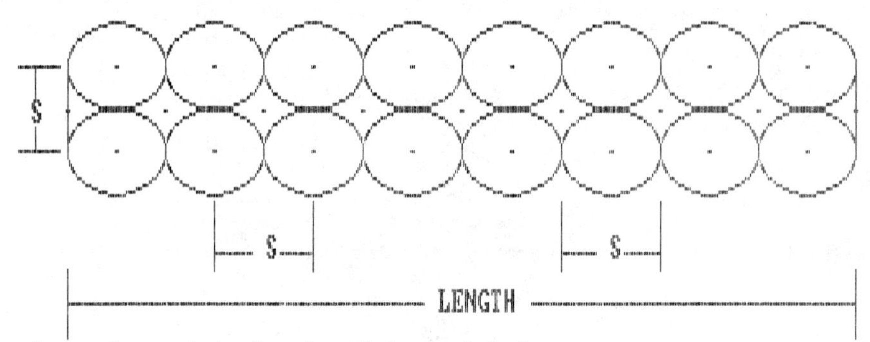


```
The volume of grout to be placed in each hole:
                                   In each outer hole=944.064 gal.
                                   In each inner hole=257.956 gal.

Total volume of grout to be used=   70820.760 gal.
Total cost of grouting=    $106231.10

Total cost of the job=     $216231.20
                         (Press any key for diagram of your situation)
```

(b)

```
Your 100 hole pattern should look like...
```

```
                     Spacing of Grout Holes= 3.030 ft.

Enter  Q  if you want to Quit.  Just press ENTER to Run again?
```

(c)

FIGURE 9.2 (Continued)

9.4 DESIGN OF A GROUT CURTAIN

The first step in the design of a grout curtain is the spatial definition of the soil or rock volume to be grouted. The design then defines the location of grout holes and the sequence of grouting. For each hole the volume of grout per lineal foot of hole is determined, based on the void volume and the pipe spacing, to allow sufficient overlap between grouted zones. The intent is to form a stabilized cylinder of a desired specific diameter along the length of pipe. The diameter is selected so that stabilized masses from adjacent grout holes will be in contact with each other, and overlap slightly.

In practice, it is difficult to synchronize the pumping rate and pipe pulling (or driving) rate to obtain a uniform grout placement rate along the pipe length. It is common practice to pull (or drive) the pipe in increments and hold it in position for whatever length of time is required to place the desired volume of grout. If small volumes of grout are placed at considerable distances apart, the obvious result is isolated stabilized spheres (or flattened spheres). As the distance between placement points decreases, the stabilized masses approach each other. The stabilized masses will also approach each other, if the distance between placement points remains constant but grout volume increases. Experiment and experience have shown that the chances of achieving a relatively uniform cylindrical shape are best when the distance the pipe is pulled (or driven) between grout injections does not exceed the grout flow distance normal to the pipe. For example, if a stabilized cylinder 5 ft in diameter is wanted, in a soil with 30% voids, 45 gal of grout is needed per ft of grout hole. The pipe should not be pulled more than 30 in. At 30 in. pulling distance, 112 gal should be paced. (The grouting could also be done by injecting 77 gal at 18 in. intervals, etc.)

Even when the proper relationship between volumes and pulling distance is observed, nonuniform penetration can still occur in natural deposits when these are stratified. Degrees of penetration can vary as much as natural permeability differences. Such nonuniformity has adverse effects on the ability to carry out a field grouting operation in accordance with the engineering design.

It would be of major value to be able to obtain uniform penetration regardless of permeability differences in the soil profile. In assessing the cause for penetration differences, it becomes apparent that the grout that is injected first will seek the easiest flow paths (through the most pervious materials) and will flow preferentially through those paths. To modify this condition, other factors must be introduced. If the grout were made to set prior to the completion of the grouting operation, it would set in the more open channels where it had gone first, and force the remaining grout to flow into the finer ones. Accurate control of the gel time thus becomes an important factor in optaining more uniform penetration in stratified deposits. Just as in controlling the detrimental effects of groundwater flow, more uniform penetration in stratified deposits also requires keeping gel times to a minimum.

The operating principles can be summarized as follows:

1. The pipe pulling distance must be related to volume placed at one point.
2. The dispersion effects of gravity and groundwater should be kept to a minimum.
3. Excess penetration in coarse strata must be controlled to permit grouting of adjacent finer strata.

The first criterion requires only arithmetic and a knowledge of soil voids. Isolated stabilized spheres will result if the distance the pipe is pulled between injection is greater than the diameter of the spheres formed by the volumes pumped. Experiments at decreasing pipe-pulling distances readily show that the stabilized shape begins to approach a cylinder as the distance the pipe is pulled approaches the radial spread of the grout, as discussed previously.

The second criterion requires that grout be placed at a substantially greater rate than the flow of groundwater past the placement point, and that the gel time does not exceed the pumping time. In the formations where chemical grouts would be considered—those too fine to be treated by cement—pumping rates more than 1 gpm are adequate to prevent dispersion under laminar flow conditions. (Turbulent flow does not occur in such soils other than at surfaces exposed by excavation.) The control of gel times not to exceed the pumping time is readily done with dual pumping systems but is difficult and frustrating with batch systems.

The third criterion requires that the gel time be shorter than the pumping time. (The alternative is to make additional injections in the same zone after the first injection has set. This would require additional drilling and would certainly be more costly.) This process is feasible with chemical grouts but obviously cannot work with a batch system. Dual pumps and continuous catalysys are required.

9.5 CONSTRUCTION OF A GROUT CURTAIN

When a grout curtain is built for a dam prior to the full impounding of water behind the dam, there may be no way to evaluate performance of the curtain for a long time after its completion. Even when performance can be evaluated quickly, there is often no way to relate poor performance to faulty construction openings or windows in the curtains that were not grouted. Thus, complete and detailed records are vital for each grout hole. Adequate records will indicate the location of probable windows and permit retreatment of such zones while grouting is still going on.

In contrast to the seepage problems discussed in Chap. 8, grout curtains cannot economically be constructed by trial and error in the field. The entire program of grouting must be predesigned, based on data from a soil study and an adequate concept of anticipated performance. The procedure is illustrated by a case study of a small grouted cutoff required to facilitate construction of a dam. In this job, there were other methods of solving the problem which may appear more practical. Nonetheless, the field engineer selected chemical grouting. The job is of interest because it was fully documented, to the point where graphic reproduction of the grouting operation was possible.

In the province of Chiapas, Mexico, an earthfill dam was under construction. The river to be dammed had been diverted upstream of the dam site. At the core wall location, the channel which the river had previously cut through bed rock was 25 to 30 ft wide and full of riverbed sand and silt deposits to a depth of about 5 ft. When these deposits were excavated in the core wall area, it was found that significant flow was still occurring through the sands, even though the river had been diverted. This flow formed a standing pool in the excavation, which could not be emptied by pumping.

The first field expedient tried was excavating upstream from the core wall site. Here, too, a standing pool was formed. It was hoped that pumping from the upstream pool might lower the water level in the core wall zone. This did not occur.

A wall of riverbed deposits, about 10 ft wide at its top, separated the two pools. It was decided to place a grout curtain in this wall. To facilitate grouting, a concrete cap about 6 ft thick was cast on the wall, using sandbag forms. This cap also covered two rather large boulders, which had been left in place. Several random holes were drilled and cement grout was pumped. However, the riverbed deposits would not accept cement, so chemicals had to be used.

The total amount of seepage was not known, but as noted previously, it was great enough to prevent adequate drying of the core wall area. The visible seepage through the dam was estimated at 50 to 60 gpm. Most of this seepage was concentrated in one major channel. It seemed most likely that this flow channel had formed between the concrete cap and the sand.

Since the grout curtain was to be placed in sand, it was obvious that closure of the flow channels alone would only result in the formation of new channels. Therefore, it was determined that a complete curtain must be placed. Since the geometry of the curtain could not provide confining effects (see the last paragraph of Sec. 5.8), a three-row pattern was dictated.

Data from which to compute an in-place void ratio were not available. Using average values for void ratios of sands and data from Fig. 5.32 on grout spread, it was estimated that 25 gal per vertical foot of hole would give radial flows of 20 to 25 in. Accordingly, a spacing of 1 m was selected for drill holes in each row and for the spacing between outer rows. The third row of holes would be located in the center of the squares previously formed. Grout volumes were to be held at 25 gal per vertical foot, where sufficient penetration of grout pipe was obtained to warrant several lifts.

One row of holes had already been placed at about a meter apart prior to the planning of a grout program. These were the holes through which cement grout had been tried. This row became the upstream row of the pattern. The spacing of holes in the latter two rows was more closely controlled than in the upstream row. A plan view of the location of holes in the completed pattern is shown in Fig. 9.3.

The numbers near the drill holes represent the sequence in which the holes were grouted. It was the original intention to stagger holes in the outer rows, then fill in the outer rows, and then do the center row on a stagger system (Sec. 5.8). Deviations from the ideal sequence were made in the interest of expedience to avoid long delays in completing the pattern. In the center row, holes were grouted several times; the numbers indicate the sequence of the first treatment.

The pumping unit used for all the work was a medium-pressure proportioning system composed of two equal volume piston pumps

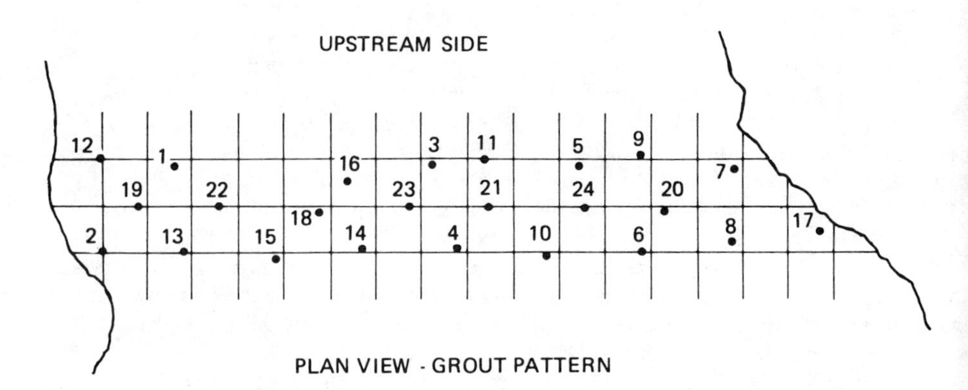

FIGURE 9.3 Plan view of grout hole locations.

and a small variable-volume piston pump. The small pump was used intermittently to add concentrated accelerator, as required. Four 35 gal plastic tanks were used for mixing. Chemicals were mixed in 25 gal batches in the upper tanks and hand-dumped into the lower tanks for pumping. An acrylamide-based grout was used throughout the work.

Gel times were controlled by the amount of catalysts used during mixing. Further control was mechanically available by feeding accelerator with the variable-volume pump. Dye was used in all grout mixes for quick identification of leakage.

Grout pipes were placed by jetting through holes drilled in the concrete cap.

In accordance with good engineering practice, detailed records were kept for each grout hole. These records are summarized in Table 9.1.

It is possible from these records (and from the detailed log of each hole, which is not reproduced here) to plot the most probable location of the grouted mass resulting from each injection. By doing this, the complete grout curtain may be reconstructed. In plotting the shape of the grouted mass for each injection, use is made of the following data:

1. Depth of the injection pipe
2. Number of lifts
3. Volume pumped at each lift
4. Average radial spread of grout in sand
5. Proximity and shape of previously grouted masses
6. Physical effects such as seepage started or stopped, extrusion of material from other grout holes, deviations from average pressure and volume, etc.

The use of very short gel times helps to ensure that the grout remains near the spot where it entered the formation (Sec. 5.5). This is assumed to be so in plotted grout location. The shape of stabilized masses resulting from a sequence of lifts during grouting has been established by laboratory tests under many different conditions (Chap. 5).

Figures 9.4, 9.5, and 9.6 represent sections taken vertically through the grout holes in the three rows of the pattern. These sections show the concrete cap, the two large boulders which were left in place when the cap was poured, the sand through which seepage is occurring, and rock walls and the bottom of the riverbed. Dimensions for plotting were recorded during the drilling and grouting operation. Numbers above the cap represent the sequence in which holes were grouted. The vertical lines show the deepest penetration of the grout pipe at each hole location. The shaded

TABLE 9.1 Field Data

Hole no.	Depth treated m	Depth treated ft	Total volume (gal)	No. of lifts	Max. pressure (kg/cm^2)	Max. pressure (psi)	Average gel time (min)	Average pump rate liters/min	Average pump rate gpm	Min. gel time (min)
1	0.80	2.6	75	3	1.9	25	1.5	—	—	0.7
2	0.45	1.5	100	2	1.9	25	1.5	—	—	0.7
3	1.25	4.1	100	4	7.7	100	1.0	25	6	—
4	1.08	3.5	100	4	3.9	50	1.0	25	6	—
5	0.50	1.6	100	2	3.9	50	1.0	25	6	—
6	0.50	1.6	100	2	3.9	50	1.25	23	5.5	—
7	0.00	0.0	100	0	7.7	100	1.25	42	10	—
8	0.00	0.0	100	0	—	—	1.5	44	10.5	—
9	0.00	0.0	25	0	—	—	1.5	—	—	—
10	0.40	1.3	100	2	2.7	35	1.6	32	7.7	0.7
11	1.00	3.3	150	4	—	—	1.5	25	6	—
12	1.15	3.8	150	4	—	—	0.7	29	7	0.1
13	0.25	0.8	100	1	5.8	75	0.7	37	9	—
14	0.53	1.7	75	1	5.8	75	0.7	39	9.3	—
15	0.90	2.9	100	4	3.9	60	—	27	6.6	0.1
16	1.50	4.9	125	4	7.7	100	1.5	23	5.5	—
17	0.95	3.1	75	3	—	—	1.6	35	8.3	0.1
18	0.37	1.2	75	0	—	—	2.0	21	5.0	—
19	0.95	3.1	110	3	2.7	35	1.3	21	5.1	0.5
18	1.17	3.8	40	0	—	—	1.5	42	10	—
18	1.17	3.8	25	0	—	—	—	—	—	—
19	1.40	4.6	75	0	4.6	60	0.7	35	8.3	—
20	0.0	0.0	5	0	4.6	60	—	—	—	—
21	0.0	0.0	100	0	2.3	30	0.8	29	6.9	0.5
20	0.62	2.0	25	0	13.5	175	—	15	3.6	0.2
22	0.0	0.0	55	0	7.7	100	0.8	12	3	0.5
23	0.0	0.0	100	0	3.9	50	0.75	27	6.6	—
24	0.0	0.0	30	0	7.7	100	0.6	15	3.6	—

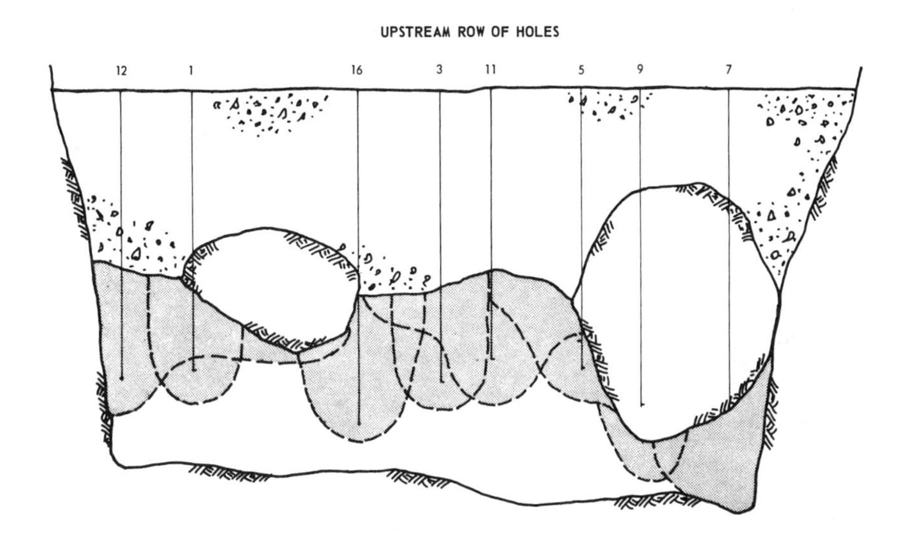

FIGURE 9.4 Vertical section through upstream row of holes.

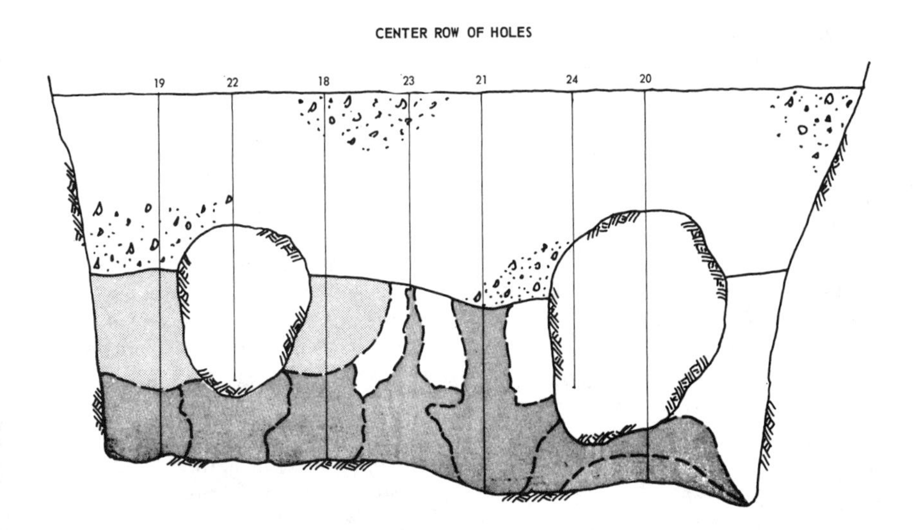

FIGURE 9.5 Vertical section through center row of holes.

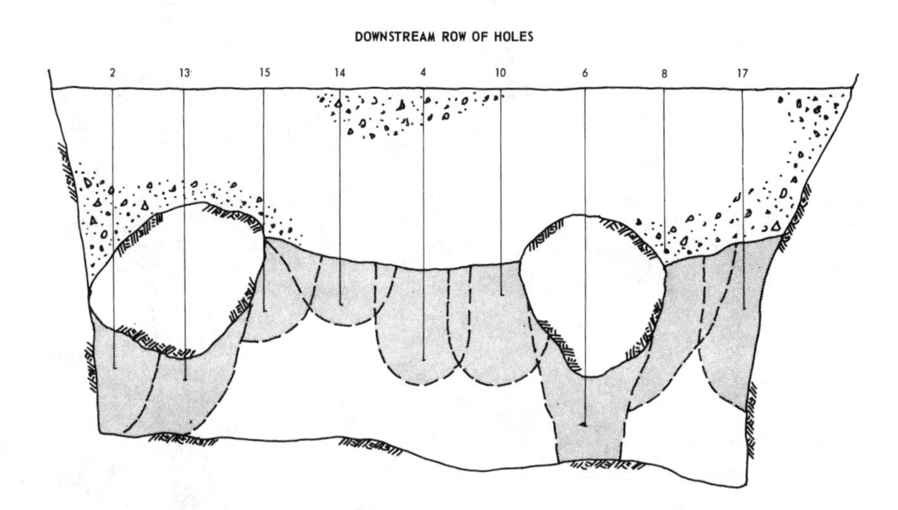

FIGURE 9.6 Vertical section through downstream row of holes.

area represents the grouted area. The dashed outlines around and within the shaded area represent for each hole the most probable grouted shape. These outlines were reconstructed in numerical sequence of grouting using the criteria previously cited.

Sections of the outer rows indicate clearly that a considerable portion of the vertical profile in the sand remained ungrouted. This, of course, was also indicated during the field operation by the records of pipe penetration. For this reason, it was imperative that holes in the center row reach bedrock. Care was taken to ensure that half of them did. In several cases, specifically holes 18 and 19, it was necessary to treat the hole several times, from the top down, in order to be able to place a grout pipe to bedrock. This accounts for the extra stabilized shapes shown for those holes.

The open area shown within the sand mass in the center row represents a soil volume which the grouting records indicate as most probably stabilized by previous injection in the outer rows.

The case history just described represents a successful application in that it made it possible to excavate to beckrock in the core wall area with no water inflow. Nonetheless, there are a number of procedures which might well have done better. To begin with, the job from start to finish took a little over 50 working hours. Of these, about 5.25 hours were actually spent pumping grout. This is a very low ratio and could be easily improved. One-third to two-thirds are reasonable ratios to aim for. On a small job, such as the one detailed, the low ratio is a minor overall economic factor. On large jobs, however, an efficient operation becomes an economic necessity.

Jetting was used to place the grout pipes. However, the jetting procedure resulted in excessive loss of sand, primarily due to poor control of the procedure. In a limited volume such as existed, jetting very often will open new channels. Driving would have been a preferable way of placing the grout pipes.

In several instances leakage past the packer was sealed by the use of fast gel times. In at least one case, this procedure also lost the hole. It is poor practice to depend on the properties of the grout to make up for the deficiencies in the equipment.

On the positive side the grouting operation was planned in full detail prior to the start of field work and carried through essentially in accord with the original planning. Full records were kept, permitting the identification and treatment of deficient zones. The grout curtain was completed to bedrock even though shutoff was attained about halfway through the planned program. Such a procedure is necessary for anything but a temporary shutoff.

9.6 ROCKY REACH DAM

A large, successful, and excellently documented grout curtain was constructed at the Rocky Reach Hydroelectric Project in Wenatchee, Washington, in the late 1950s. In this curtain, chemical grout was used to increase the cutoff effectiveness of a previously constructed clay–cement curtain (approximately 0.25 million gal of chemicals followed 1 million ft^3 of clay–cement grout). The job was reported in ASCE journals [2] and was the subject of a Doctoral thesis [3] that was later printed by the Corps of Engineers [4]. These publications are recommended reading for engineers planning extensive grout curtains.

The dam is on the Columbia River, which at the dam site follows the steep, western wall of a canyon. Between the dam and the eastern canyon wall is an 0.5-mile wide terrace made up of pervious strata (ranging from silts to gravels) deposited by the river in its earlier courses. These ancient riverbeds, some of them directly exposed on the present riverbanks, could readily drain the reservoir. A grout curtain, 1800 ft long, was designed and constructed to prevent such drainage.

A cement–clay grout followed by a cement–bentonite grout was used to fill the coarser voids. Over 1 million ft^3 of cement-based grout was placed in a three-row pattern over the entire curtain length, with two additional rows of holes near the dam. The effectiveness of the cutoff was monitored by piezometers placed

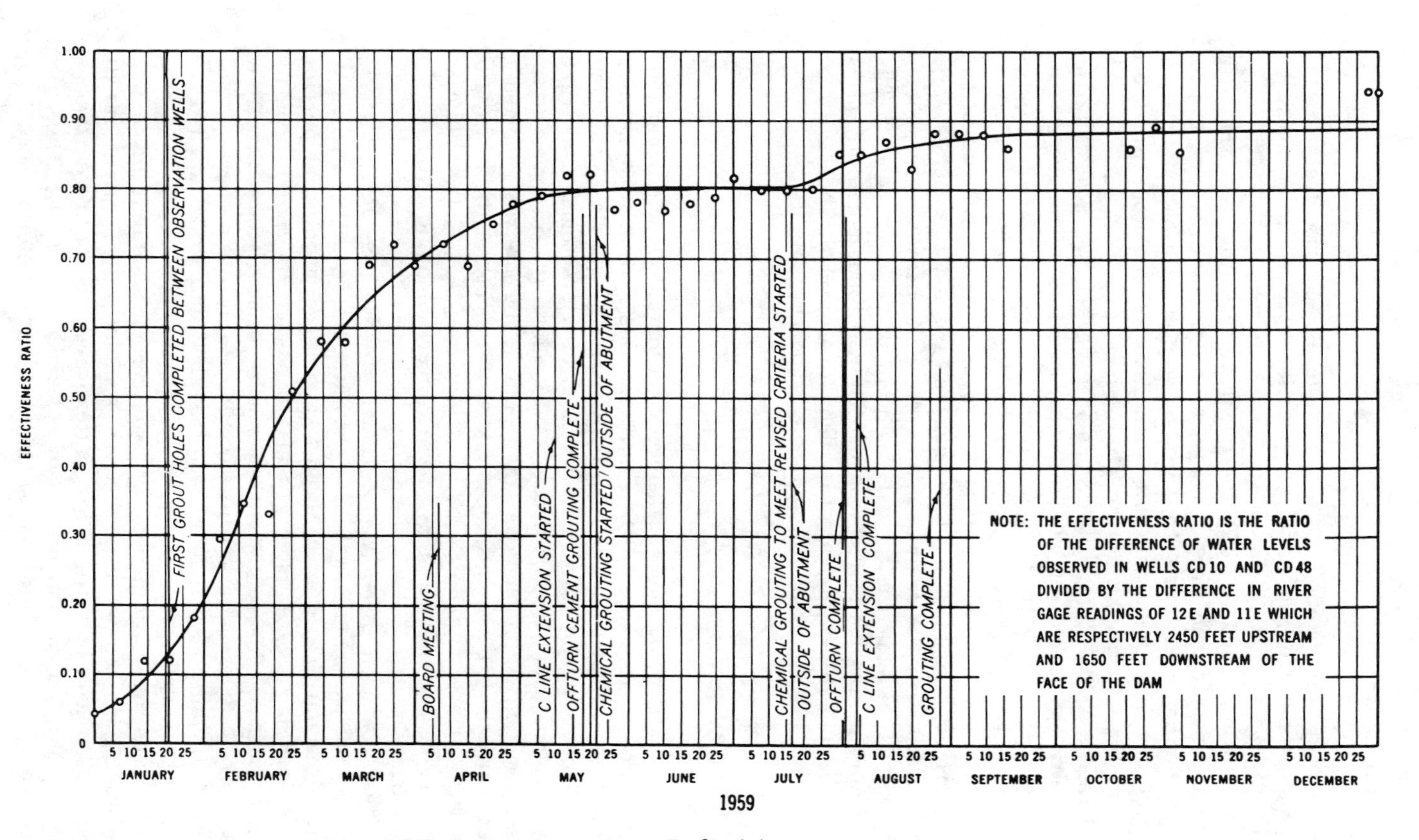

FIGURE 9.7 Grout curtain effectiveness. (From Ref. 4.)

upstream and downstream of the curtain. The difference in elevation between two specific holes, compared with the difference in river levels (actually, headwater and tailwater elevations) opposite those holes, is an index of effectiveness, and this index is an excellent way to assess quickly the effects of grouting. This index for the Rocky Reach Dam, plotted against time, is shown in Fig. 9.7. The chart makes it clear that cutoff effectiveness leveled off at about 80% for cement-based grouts.

Chemical grouting began in an attempt to increase the effectiveness. Original criteria were too restrictive (i.e., allowable grouting pressures were too low) and indicated no benefits. Revised criteria began to give positive results almost as soon as instituted. Chemical grout was placed along the center row of holes for the 1500 ft of curtain nearest to the dam. During a 6 month period, almost 0.5 million gal of acrylamide grout were placed, with pumping rates as high as 12 gpm (pressure refusal was considered 3 gpm at 30 psi). During this time the cutoff effectiveness increased to 89%. The reservoir behind the dam has been full for many years, and all indications are that the cutoff is still functioning at its original effectiveness.

9.7 SMALL GROUT CURTAINS

The design of an electric generating station in Illinois called for the construction of several miles of earth dikes 20 ft high to form cooling water ponds. Borings showed the materials underlying the dikes to consist of 5 ft of gray, silty CLAY to clayey SILT with a trace of sand, 17 ft of gray to brown fine to medium SAND with a trace of gravel, 5 ft of coarse SAND and GRAVEL, and then LIMEROCK.

An impervious barrier was needed between the dike and the limerock to prevent possible draining of the ponds. A slurry trench was used over most of the project to provide the barrier. In one area, along a 100 ft section where the dikes intersected, it was planned to drive sheet piping. After the completion of the slurry walls and the removal of the equipment, when pile driving commenced, it was found that the lower strata of coarse material prevented a satisfactory interface with the limerock. At this point, a grouted cutoff was selected as an alternative.

A grouting pattern was selected with holes in two rows spaced 2 ft apart and a third row of holes in the center of the squares previously formed.

Grout holes were drilled with a hollow stem auger. The grout pipe was placed through the stem, and then the auger was withdrawn. Grout was placed in stages as the grout pipe was withdrawn.

The two outer rows were grouted with SIROC. A concentration of 65% had to be used to counteract deleterious components of the soil formation. One row was grouted completely before any holes in the other row were grouted. (Generally, the grouting sequence is staggered across the rows.) Gel times were 10 to 20 min, the pumping rate was 5 gpm, and pumping pressures ranged from 1 to 1.5 psi per ft of depth. The average volume per hole was 240 gal, or about 9 gal per vertical ft of hole.

On the second of the outer rows, parameters were similar, except for a lower take of about 185 gal per hole, or somewhat under 7 gal per vertical ft of hole.

In the center row, an acrylamide grout was used for penetration into zones that could not be entered by the more viscous silicate grout. Average pumping pressures were higher, since the larger voids were already filled, and volume per hole averaged 160 gal, or about 6 gal/ft of hole.

The postgrouting history indicates that seepage problems did not occur. However, this is one of many grouting jobs which was done as an additional precaution against some type of incipient problem. It is possible that the upper layer of cohesive material would have prevented seepage by itself. There are no data available to justify either the conclusion that the grout curtain functioned properly (because in reality it may not have been needed) or that the grout curtain was not needed (because in reality there may have been pervious windows through the upper cohesive zone).

It is interesting to note that the original engineer's job report omitted data on gel times and pumping pressures. This is typical in many commercial companies to keep from revealing data which may be of help to competitors and is of little interest to top management.

Mud ponds and retention and settling basins are often made by constructing a continuous dike on a relatively impervious formation of a size and shape adequate to retain the desired liquid volume. Such was the case in Delaware, where a dike some 20 ft high was built of fine, granular materials to retain sulfate solutions.

After a period of use, three leaks developed close to each other at the outer base of one of the dike walls. Chemical grouting was selected as the method for sealing the leaks and for preventing the formation of new leaks along the entire length of the dike wall that was leaking.

Several borings were taken from the top of the dike to verify the reported soil formations. pH numbers were determined for each sample recovered. These show very low values below 10 ft, verifying that seepage of the retained acids was occurring. SIROC was selected for this project, and the soil samples were used to check the effects of the low pH on the gel time.

Grout pipes were hand-driven on 8 ft centers along the top of the dike for the entire 280 ft length of the seeping wall. Grout was injected at low pressures as the pipes were slowly withdrawn. Pressures were kept below 1 psi per foot of depth, pumping rates varied from 3 to 6 gpm, and gel times were about 15 min. In areas of high take and in the vicinity of the leaks (grout holes 12 through 15), intermediate pipes were driven and grouted. The work resulted in complete sealing of the leaks.

(Again, the engineer's original job report omitted actual values for pumping pressures and rates.)

Small dams are often constructed by simply building triangular cross-section fill across a valley. Often, the materials used are selected because they are local, and compaction control is lax or nonexistent. As might be expected, such dams often leak. One such case is Ash Basin No. 2, in Snyder County, Pennsylvania [5]. Dimensions are shown in Figure 9.8.

The dam was built in 1955 to store residual fly ash. A clay core was provided. By 1964 the storage basis was full, and the dam was raised 10 ft. In 1971, the enlarged basin was full. The crest would have to be raised another 30 ft to provide additional storage.

In 1982 the owner made the decision to raise the crest. However, numerous water seeps were occurring through the lower portion of the embankment, and several larger leaks also existed. Since these problems would be enlarged by the higher fluid head, it was necessary to eliminate the seepage. The method chosen was a grout curtain.

Pumping tests through drill holes showed permeabilities as high as 10^{-2} cm/sec. It was decided to use either cement grout or chemical grout, based on pumping tests in each grout hole. The contractor (Hayward Baker Co., Odenton, Maryland) used a criterion based on his past experience to grout with cement if the reference number (RN) was over 100, and to use chemical grout for lower numbers. The RN is defined as

$$\frac{V}{Lt} \times f$$

where

V = number of gallons pumped

L = uncased length of hole in feet

t = pumping time in min, and

f = 400 @ 5 psi
200 @ 10 psi
100 @ 20 psi

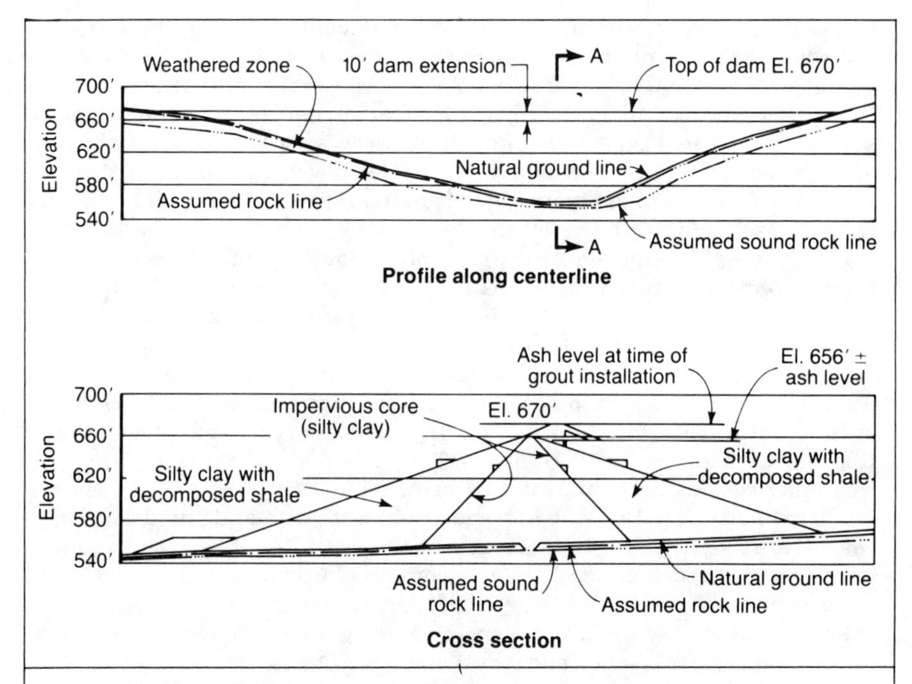

The core of the 870 ft long, 106 ft high dam consists of silty clay materials on native, slightly weathered shale; the remainder of the embankment is made of silty clay with decomposed shale.

FIGURE 9.8 Dimensions of Ash Basin No. 2.

A single row curtain was selected, with primary holes 20 ft apart. Secondary and tertiary holes were used to give a final spacing of 5 ft. The cement grout consisted of a mixture of Type 1 Portland Cement and Type F fly ash, on about a 5 to 2 ratio. The chemical grout selected was AC-400.

A total of 2500 bags of cement, 1200 bags of fly ash and 43,000 gal of chemical grout were used. For most of the job, pressures were kept below 0.5 psi per foot of overburden, and pumping rates were under 4 gpm. Chemical grout gel times were between 10 and 15 min.

A weir constructed prior to grouting had registered a total seepage of 35 gpm. The grout program reduced this quantity to 2 gpm.

Permeability tests, as illustrated in the case history above, are useful in the process of grout selection. Another use of such data is illustrated by the quote below [6].

> Water pressure tests conducted during siting of the landfill indicates that the upper jointed and weathered granodiorite has a permeability coefficient on the order of 5×10^{-4} cm/sec. Laboratory tests indicate a much lower permeability for the intact rock. Using a table relating joint frequency, width and permeability, the observed joint frequency of between 3 and 30 joints per ft gives a range between 0.002 and 0.04 in. for the joint openings. This is consistent with the characteristic jointing pattern of the rock.

Generally, fissures below 0.01 in. are ungroutable with cement. Thus, the tests above indicate that chemical grouts must be used.

The table referred to in the quoted paragraph is from E. Hock and J. W. Bray, *Rock Slope Engineering*, The Institution of Mining & Metallurgy, London, 1974.

Suggested standards for grouting in dams have been proposed by A. C. Houlsby [7]. Figure 9.9 shows data from that paper.

Grouting for dams, whether for foundation support or water control, often involves such large quantities of labor and materials that it may become impossible for manual evaluation of the records in time for adequate field response. Electronic monitoring is an obvious solution. With the recent advent of powerful micro computers, this solution can be realized. Reference 6 contains two papers discussing the grouting of foundations and abutments for large dams. Although most of the data refers to cement grouting, there are some significant references to the use of chemicals, as illustrated by the following:

> A single row grout curtain with a maximum depth of 393 ft (120 m) was constructed under the core of the dam. The grouting was with pressures ranging from 15 psi (0.10 MNm^2) to 300 psi (2.07 MN/m^2).

Closure criterion used was a maximum of 1 Lugeon leakage from water pressure tests in check holes. A Lugeon is defined as a water loss of 1l/min/m of hole per 10 atmospheres of pressure. In order to meet this criterion in the brecciated fault zones, supplementary chemical grouting was performed. The chemical grout used was sodium silicate with additives of sodium aluminate and sodium hydroxide.

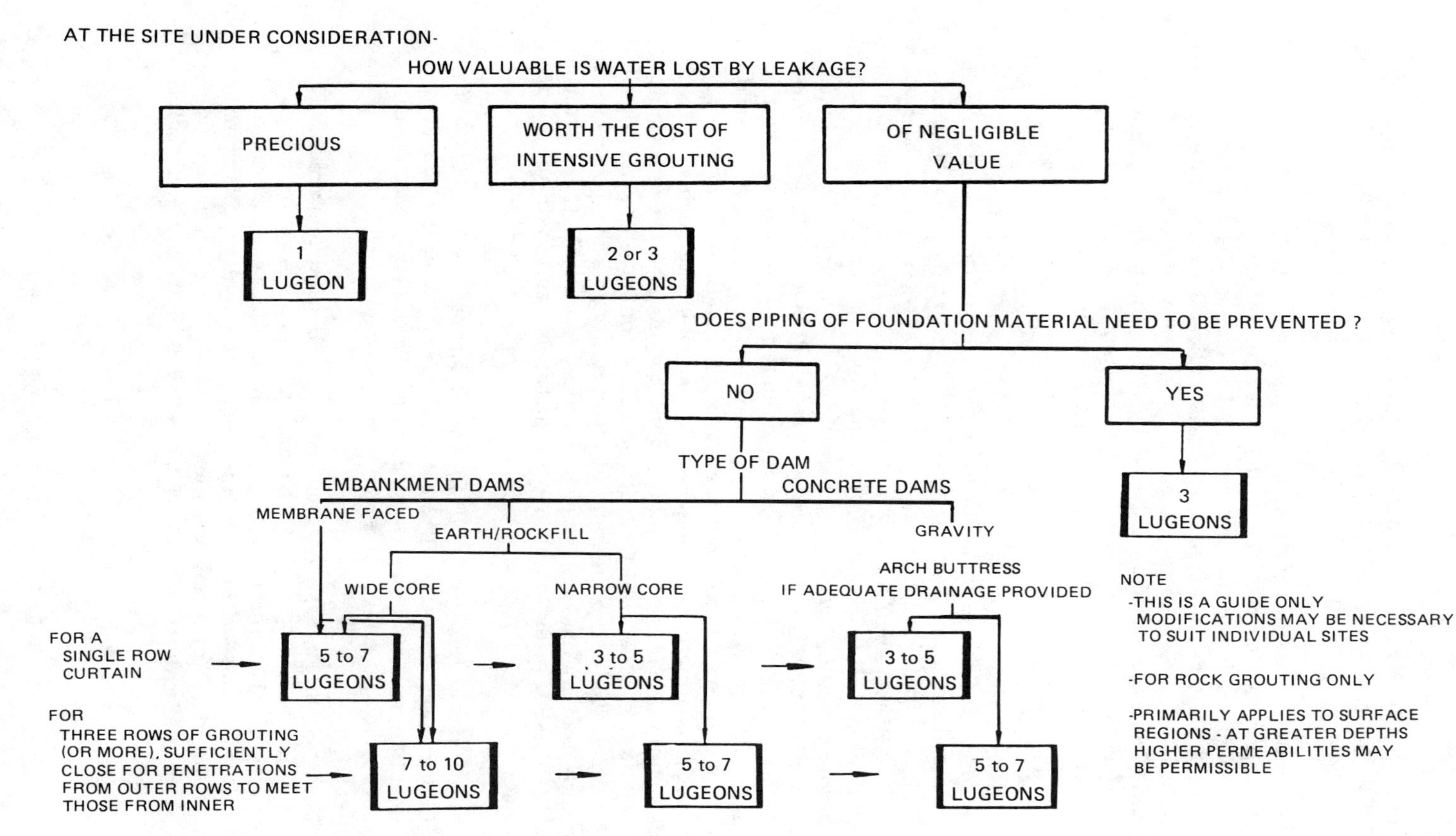

FIGURE 9.9 Suggested standards for grouting. (From Ref. 7.)

Remedial measures for field problems cannot be fitted into a common mold, even for projects such as grout curtains—which normally have many common elements and procedures. An example illustrating this point is the case history below, which basically involves a cement grout curtain functioning close to the border of acceptability.

King Talal Dam is built on the Zarqa River in Jordan to provide irrigation for 59,000 dunams of land along the Jordan River Valley. It is an earth and rockfill dam with a central clay core, about 100 meters high. A cross-section of the dam, shown in Figure 9.10, shows the location of the two major formations, the Kurnub sandstones and the Zarqa limestones, and also the location of two major faults in the left abutment.

Provisions were made in addition to the usual foundation grouting, in order to prevent excessive flow through the left abutment. These included extensive cement grouting at several levels within the abutment and drainage curtains to keep hydraulic pressures low.

Filling of the reservoir started in winter of 1976–1977. Minor leakage of clear water through the left abutment began, and was controlled by additional grouting. At normal operational levels, the left abutment seepage was 35 to 50 L/sec (over 800 g/min). Al-through this quantity is well within acceptable limits, half of it was coming in at one spot—a drainage window (No. 69) in the drainage gallery (Fig. 9.11).

After several years of operation, it was decided to heighten the dam in order to increase its irrigation potential. The crest was to be raised 16 m to elevation 185. This would increase the hydraulic head through the seepage channels, and there was concern that the total seepage volume would increase beyond acceptable limits and/or that the increased flow rates would induce piping. Accordingly, a study was undertaken in the early eighties to assess the problems and propose remedial measures.

The first portion of the study consisted of a detailed review of all the grouting records, as well as instrumentation data and flow records. Based on these data, and numerous on-site visual reviews of existing conditions, a preliminary proposal for testing and remedial measures was written in the field. Excerpts from this proposal appear below (modified for clarity to readers not familiar with all field details).

GROUT CURTAIN CLOSURE STUDY

CONDITION I:

Seepage through window 69 (in Zarqa):

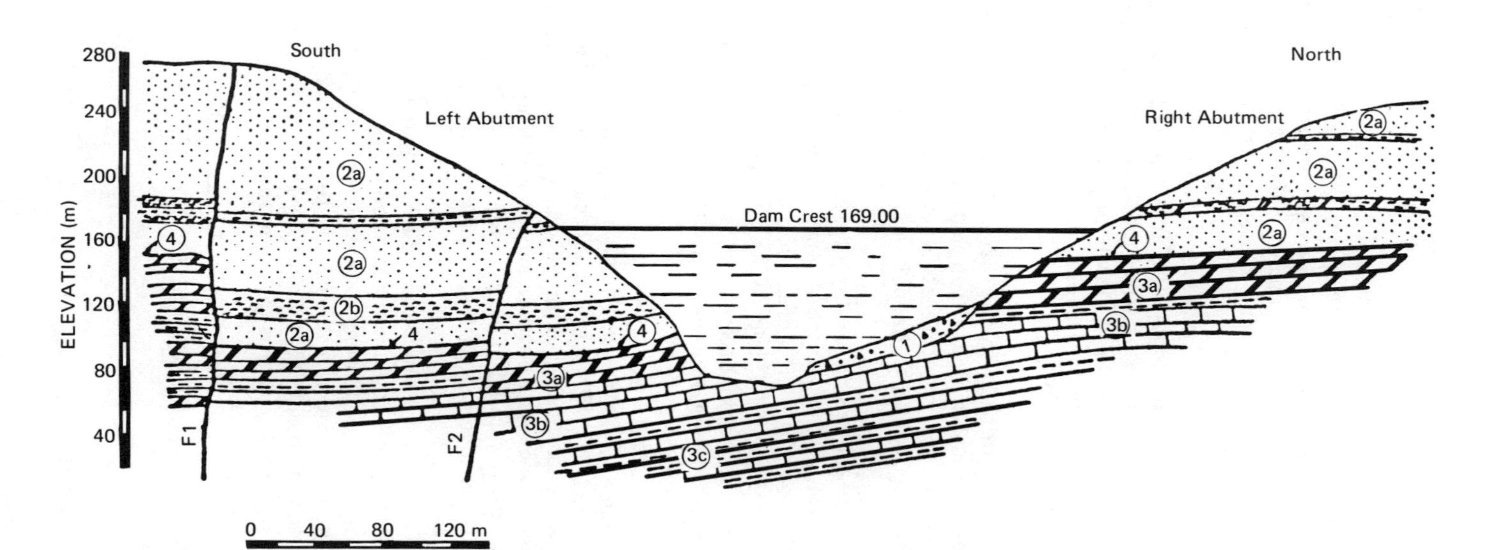

FIGURE 9.10 Geological cross-section along the King Talal Dam's axis. 1 = slope debris; 2a = Kurnub formation sandstone; 2b = Kurnub formation shales with thin layer of dolomite; 3a = Zarqa formation dolomite; 3b = Zarqa formation limestone; 3c = Zarqa formation claystone with marl; 4 = Kurnub-Zarqa contact: F_1 and F_2 = geological faults. (From Y. M. Masannat, *Geotechnical Considerations in the Treatment of Foundations of the Left Abutment of King Talal Dam*, National Research Council of Canada, 1980, pp. 34–43.)

FIGURE 9.11 Drainage window No. 69.

DATA:

1. Water temperature readings suggest source of water is from low reservoir level.
2. Flow first appeared at reservoir level 107.
3. Flow volume reflects reservoir head.
4. Grouting analysis shows probable curtain windows in left abutment.

INFERENCE:

Seepage path goes through or underneath grout curtain, most probably in the Zarqa.

TREATMENT:

Add a second row of grout holes upstream of the curtain in the Zarqa.

DETAILS:

1. Use cement grout with short setting time with provisions for very short setting times, if grout appears at window 69.

2. Cement content, pumping rates and pressures to duplicate previous work, at starting of pumping.
3. Spacing between holes to be two meters. New row of holes offset upstream as far as possible. All holes vertical.

CONDITION II:

Weak, friable sandstone at Kurnub Zarqa contact.

DATA:

1. Poor core recovery in drillholes.
2. Slope exposure show very weak cementation, and evidence of weathering along fissures.
3. Grouted areas in the formation continue to take grout on subsequent treatment.
4. Field permeability ranges around 10^{-3} cm/sec.
5. Lab permeability of intact cores ranges from 10^{-4} to 10^{-6} cm/sec.

INFERENCE:

1. Primary permeability is too low for effective grouting with cement or chemicals.
2. Secondary permeability might be (at least partially) treated effectively with cement.
3. Either: a) seepage pressures are breaking the sandstone particle bonds, with possible movement of grains and formation of pipes, or b) the cement grout was displaced from its desired location prior to initial set.

TREATMENT:

1. Take cores in several zones where regrouting at the contact has shown large takes. Examine drill holes with bore hole camera.
    a. If all fissures are grout filled, grouting was effective, and additional (typical) cement grouting is warranted.
    b. If no fissures containing cement are found, this indicates that all cement was washed away before it could set. Grouting with fast setting cement, or cement-chemical grout mixture is indicated.
    c. If some open fissures, and some filled with cement exist, this suggests the possible formation of piping channels. Such zones should be grouted with low viscosity chemical grouts to seal the pipes. Scheduled checks of new erosion should be established.

2. Grout the lower sandstone at the contact with the Zarqa. Use a three-row curtain with spacing of outer rows as wide as gallery width allows. Angle rows up and down stream to obtain a row spacing at contact (outer to outer) of 3 meters. Type of grout to be determined by test program in the south end of gallery 92, evaluating:
   a. Portland Cement with accelerators.
   b. Portland Cement with Chemical Grout.
   c. Low viscosity chemical grout.

Sleeve pipes or tube à manchette should be used in the entire test section, and at least along the center row of holes in the total curtain.

CONDITION III:

Large grout takes with no evidence of closure.

DATA:

Grouting records do not show a refusal signature.

INFERENCE:

Openings (ungrouted zones) could remain within the grout curtain.

TREATMENT:

Regrout with quick setting cement grout, using new vertical grout holes on upstream side of curtain. Checking continuity of curtain is necessary.

DETAILS:

Use inclined holes in plane of curtain. Measure permeability at regular adjacent intervals, using a double packer.

Based on this proposal, a final report covering all aspects of geotechnical activities was submitted and implemented.

9.8 SUMMARY

Flow of water through large pervious masses can be controlled on a temporary basis by pumping and by well-point systems, and on a temporary or permanent basis by slurry walls and grout curtains.

Where the construction site is available to heavy equipment, and the treated zone extends down from the surface or close to it, slurry walls are often most cost effective. Where the treated zone is far below the surface, or underneath an existing structure, grout curtains are often the only viable alternative for permanent flow control.

Grout curtains (cutoff walls) can be used before, during, and after the completion of a construction project to control the flow of subsurface water. When used after completion of construction, it is usually because visual signs of distress have appeared. Under these conditions grouting can be done virtually by trial and error, since the effects of grouting can be seen immediately. On the other hand, when cutoff walls are grouted prior to or in the early stages of construction, the work must be preplanned in detail and accurate records kept of each grout hole, so that regrouting is done where indicated and the possibility of windows can be kept to a minimum.

REFERENCES

1. L. A. Bell, A cut off in rock and alluvium, *Grouting in Geotechnical Engineering*, ASCE, New York, 1982, pp. 178–179.

1a. H. Einstein and G. Schnitter, Selection of Chemical Grout for Mattmark Dam, *Proceedings of A.S.C.E.*, Vol. 96, SM6, pp. 2007–2023.

2. F. W. Swiger, Design and Construction of Grouted Cutoff, Rocky Research Hydroelectric Power Projcet, Paper presented at the June 1960 National Meeting of the ASCE.

3. J. E. Wagner, Construction and Performance of the Grouted Cutoff, Rocky Reach Hydroelectric Project, Doctoral thesis, University of Illinois, Urbana, Jan. 18, 1961.

4. J. E. Wagner, Miscellaneous Paper No. 2-417, March 1961, Corps of Engineers, Vicsburg.

5. W. H. Baker, H. N. Gazaway, and G. Kautzman, Grouting rehabs earth dam, *Civil Engineering*, New York, Sept. 1984, pp. 49–52.

6. E. D. Graf, D. J. Rhoades, and K. L. Gaught, Chemical grout curtains at Ox Maintain dam, *Issues in Dam Grouting*, ASCE, New York, April 30, 1982, pp. 99–103.

7. A. C. Houlsby, Cement grouting for dams, *Grouting in Geotechnical Engineering*, ASCE, New York, 1982, pp. 1–34.

10

Grouting for Strength

10.1 INTRODUCTION

There are many occasions when it would be desirable to be able to add strength to a soil formation. For clays this can be done by preconsolidating (overloading). For granular materials, other procedures are available, including densification and grouting. If the fluids (air or water) in the soil voids are replaced by a solid, it becomes more difficult for the individual soil grains to undergo relative displacements, thus adding shear resistance (or strength) to the soil mass. This additional strength may be useful in preventing the movement of material from under loaded zones, in increasing the bearing capacity and the slope stability of grouted formations, and in reducing settlements in zones adjacent to or above excavations. Figures 1.7, 1.8, and 1.9 illustrate some of these applications of grouting.

10.2 STRENGTH OF GROUTED SOILS

The shear strength of a granular soil is due primarily to the nesting (or interlocking) of grains and the consequent resistance of the grains to rolling or sliding over each other. Conditions which cause the grains to interlock more strongly increase the soil shear strength. Thus, dense sands are stronger than loose sands, and loaded sand masses are stronger than unloaded sand masses. Since granular materials weigh 100 lb/ft^3 or more (dry), soil deposits are loaded by their own weight, and a sand stratum extending 10 ft below ground surface is considerably stronger at its bottom (where the grains are confined by vertical and lateral pressures) than it is at the top, where the grains are relatively free.

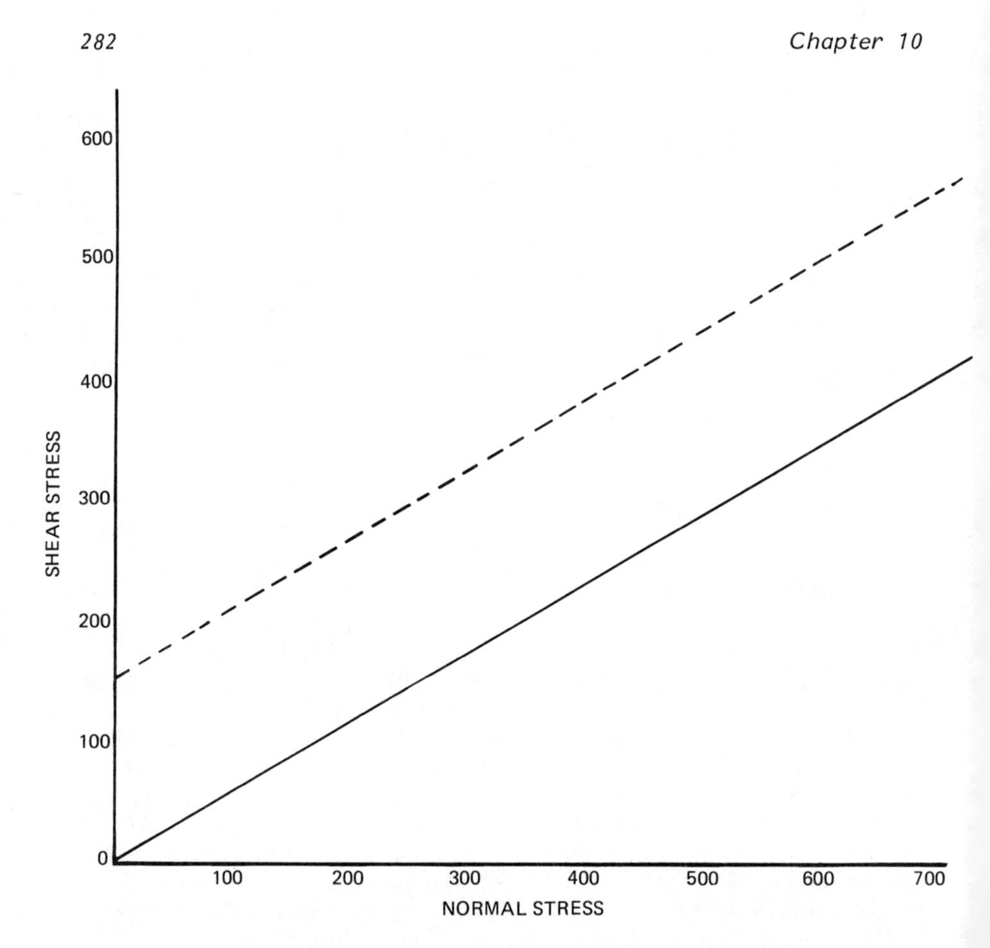

FIGURE 10.1 Representation of shear strength of a granular soil.

The shear strength of a granular material is represented by the solid line in Fig. 10.1 (this is the same as Fig. 4.3 except that the axes are scaled; see also Sec. 4.2 for a discussion of strength). The friction angle shown (30°) is typical. From the figure, assuming the soil weighs 100 lb/ft^3, at 5 ft below grade (equivalent to a normal stress of 500 lb/ft^2) the shear strength is about 290 lb/ft^2 (about 2 psi). Ten feet below ground surface, the shear strength is 4 psi, etc.

Granular soils have no unconfined strength and cannot be subjected to unconfined compression testing. To determine the shear strength of granular materials, artificial confining pressures must be applied, as in triaxial and direct shear testing. See Chap. 9 of

Ref. 1. Grouted soils, however, can be tested in unconfined compression. The unconfined strength added by the grout appears as an intercept of the vertical axis. In clays and clayey soils, this intercept is called cohesion, and grouting may indeed be considered to add cohesion to the soil. Grouting generally has little effect on the friction angle, so the strength of a grouted soil can be represented by the dashed line in Fig. 10.1. In the figure, the grout strength is shown as 150 lb/ft^2 or about 1 psi. As small as this value may be, it still more than doubles the ungrouted soil strength at a 2 ft depth. It is hard to conceive of a grout that does not provide at least 10 psi of cohesive strength. Thus, any grout will make a significant, if not major, increase in the ungrouted strength of granular deposits within a short distance of ground or exposed surface. (This statement does not apply to relatively shallow soils heavily loaded by foundations or otherwise confined.) It is only at great depths (for example, 500 ft, where the vertical pressure due to the weight of the soil generates shear strengths of the order of 200 psi) that the contribution of most chemical grouts to soil shear strength will become negligible. (These comments are based on two assumptions, namely that the grout does not modify the soil friction angle and that the soil is not loosened or fissured by the grouting process.)

Cement grouts are generally thought of as adding significant strength to a grouted formation, while clays (when used as grouts) are generally thought of as adding no strength. All chemical grouts fall between these two extremes. Neat cement grouts can have unconfined compressive strengths of 1000 to 1500 psi. Only some of the resorcinol-formaldehydes approach or exceed these values. (This does not account for epoxies and polyesters, which have been used for rock grouting but due to their high viscosities and costs are not applicable to soils.) Other chemical grouts, including those thought of as "strong," fall far below neat cement in strength. For strength applications, the grouts considered strong, and generally used, include the high-concentration silicates, the aminoplasts, and some of the phenoplasts. The grouts generally not used for strength applications include the acrylamides and acrylates, the lignosulfonates, the low-concentration silicates, and some of the phenoplasts.

It is actually the creep strength of a grouted soil (either in unconfined compression or triaxial compression, depending on the specific application) that should be used for design purposes, not the unconfined compressive strength. In the absence of specific creep data, the value of one-fourth to one-half of the unconfined strength may be used for applications. A suitable safety factor must be applied to these suggested values.

In selecting a grout to be used for any application, it is obviously necessary to select one with sufficiently low viscosity to be able to penetrate the formation. In strength applications, this criterion may eliminate all the grouts with adequate strength. In terms of usable strength (including a safety factor of 2), a range of values for preliminary design use is given as follows (these values must be verified by actual tests prior to final use in design of a grouting operation):

Lignosulfonates	5 to 10 psi
Low-concentration silicates	5 to 15 psi
Acrylamides, acrylates	5 to 20 psi
Phenoplasts	5 to 30 psi
Aminoplasts	10 to 50 psi
High-concentration silicates	20 to 50 psi

10.3 GROUTING FOR STABILITY

New construction in the vicinity of existing structures often causes concern about the possible reduction in bearing capacity under existing footings or foundations. Such was the case illustrated by Fig. 1.7, where a structure was to be placed occupying all the space between two adjacent buildings. The basement excavation would extend 11 ft below the foundation of one of the existing buildings. The supporting soil was a loose, dry sand which, if untreated, would run out from under the existing foundations during excavation.

Sodium silicate grout was used to solidify a soil mass below the existing foundations. Foundation loads could then be transmitted through the treated soil to depths below the effects of excavation. Often excavation can then be carried out adjacent to the treated soil mass without the need for external shoring or bracing.

A problem very similar in nature but much different in scale occurred during construction of a subway system in Pittsburgh [2]. This problem resulted in the largest chemical grouting job in United States history to that date.

The Sixth Avenue portion had to be built under a street right-of-way only 36 ft wide, lined with large buildings whose spread footings were founded well above the subway invert. A 50 ft wide station intruded under the sidewalks and under building vaults beneath the sidewalks. Six of the buildings, of different size and construction, required underpinning to avoid excessive or catastrophic settlements.

The soil profile consisted of 65 ft of granular materials over sedimentary bed rock. Site borings classified the soil as fine to coarse sand and gravel with a trace of silt. Laboratory tests indicated a permeability range from 1 to 10^{-2} cm/sec, and fines

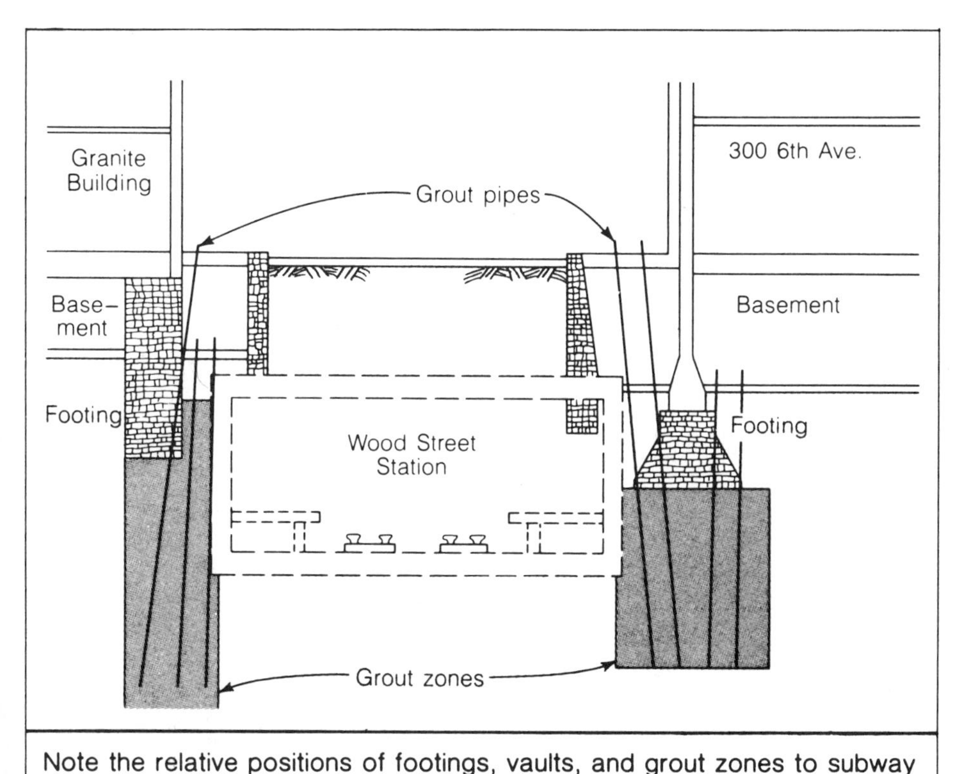

Note the relative positions of footings, vaults, and grout zones to subway station.

FIGURE 10.2 Sixth Avenue subway: cross-section. (From Ref. 2.)

(passing the #200 sieve) of 7% or less. On these bases, chemical grout was approved for underpinning.

Sodium silicate grout was selected and used throughout the project at a 50% concentration, with setting times of 30 to 45 min (grouted soil strengths of 100 psi were required). Grout pressures were generally kept below overburden values, and pumping rates were dictated by the pressure limitation. Figure 10.2 shows a typical operating cross section at the subway station. During a 4 month period, over a million gallons of grout were placed through nearly 4 miles of sleeve-port grout pipes, with injection totals reaching as high as 20,000 gal/day.

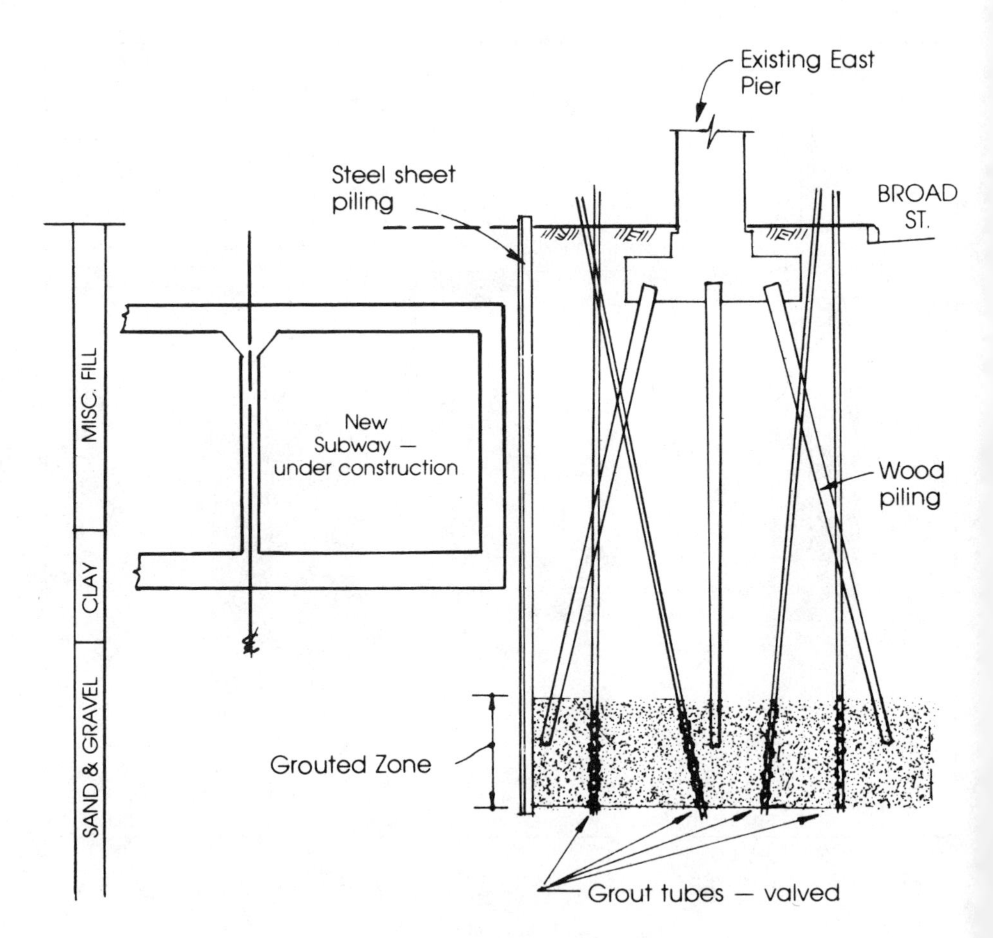

FIGURE 10.3 Proximity of proposed subway and existing bridge pier. (Courtesy of Raymond International, Raymond Concrete Pile Division, New York.)

During the grouting, elevation surveys were done to detect possible surface heave, seismic tests were conducted to verify grouted zones, and block samples from test pits were taken to verify strength values. Subject to positive tests, payment was ultimately made for two items: linear measure of grouted pipes, and volume of grout placed.

Soil movement induced by underground excavation or by vibration due either to construction procedures or facility operations may also be a possible cause of damage to previously placed structures. This potential was recognized in the planning for a subway near a bridge abutment.

In designing the Philadelphia Broad Street subway extension, the reinforced concrete box of the subway passed very close to the pile-supported east pier of the Walt Whitman Bridge approach. Figure 10.3 shows the proximity of the two structures. Although excavation for the subway would stop well above the load-bearing zone of the wood piling, it was feared that vibration due to subway operation might density the granular soils, causing settlement of the pier. To eliminate this possibility, it was decided to grout the granular soils in the load-bearing zone.

The nature of the zone to be grouted required a low-viscosity grout, and a phenoplast was selected for this job. The Stabilator system was chosen for grout platement. Basically, this system uses lightweight casing with cantilever spring valves placed in accordance with job requirements. The casing follows a knockoff drill bit into place. A double packer is used to isolate the desired grouting zone. Grout pressure opens the spring valves. (See Fig. 10.8.)

A grout pipe spacing of 5 ft was selected, giving a pattern five rows wide with 23 grout holes in each row. Grout would be placed in a blanket from 4 ft below the deepest pile to 3 ft above the most shallow pile tip. The two outer rows on each side of the curtain were grouted first using predetermined quantities of grout (quantities selected to give a 5 ft radial flow at the existing void ratio) with a 5 to 15 min gel time. The center rows were then grouted to pressure refusal.

After the completion of the grouting, borings were taken to verify the results. For two specific areas, changes in blow counts are shown in Fig. 10.4. Performance history over the period (several years) since the completion of the grouting indicates no settlement problems. This job is considered a successful grout application by the technical people involved. However, now that no settlements have occurred, it is impossible to prove that they would have occurred without grouting. Unfortunately, the only way to prove that grouting was needed would have been not to grout and risk a failure. In this case, such a risk was not justified.

DEPTH FEET 0	GENERAL DESCRIPTION	STANDARD PENETRATION BLOW COUNT			
		LOCATION #1		LOCATION #2	
		BEFORE GROUTING	AFTER GROUTING	BEFORE GROUTING	AFTER GROUTING
5	MISCELLANEOUS FILL	23/1"	16	17	
		3	4	13	
10		15	16	2	
15		3	9	9	
20	Firm Silty Clay with decayed vegetation	10	7	14	21
					58
					59
25	Dense medium to fine sand with trace fine gravel and occasional silt layers	30	33	27	37
		36			139
					55
			100/2"	30	
30					100/3"
					GROUTED ZONE
		38		40	
		40	67		
35	Dense gravelly sand		97		
			79	28	
		64	69		
			GROUTED ZONE		
40				48	
		46		38	
45		52		20	
		Fig. 2			

FIGURE 10.4 Test boring reports. (Courtesy of Soiltch Division, Raymond International, Cherry Hill, New Jersey.)

Loss of support for a foundation may occur for reasons other than adjacent construction. In the case of a 165-ft-high brick chimney, wood piling supporting the foundation slab was deteriorating and transferring the load to the underlying sand. The sand, of course, was loose due to the pile decay and not capable of carrying the load without excessive settlement. Grouting was selected as the best method of avoiding heavy settlements.

A section through the chimney foundation is shown in Fig. 1.8. A sodium silicate grout of high concentration was selected, since strength was needed. (The specifications for the job required an unconfined compressive strength of 5 tons/ft^2, about 70 psi.) First, grout pipes were placed vertically in a ring around the foundation area and injected with calcium chloride solution. The curtain thus formed would tend to give immediate solidification of any grout solution which tended to flow away from beneath the foundation. Silicate grout with a short gel time was then injected through 0.5 and 0.75 in. pipes driven vertically and horizontally underneath the slab. About 500 ft^3 of soil were grouted.

In sands such as the formation treated (St. Peter sand, in the Minneapolis area), unconfined compressive strengths of stabilized soil samples averaged 10 tons/ft^2. Thus, the project engineers were satisfied that their strength specifications were exceeded. However, at the time this job was done, little attention was paid to the significance of creep of silicate grouted soils. The creep strength of stabilized soil samples was, of course, not determined but would probably have ranged from 3 to 6 tons/ft^2. The chimney did not fall, because of the safety factor built into the requirement of 5 tons/ft^2 in the specifications as well as the probability that the settlements which did occur were uniform rather than differential.

Grouting to increase soil stability and bearing capacity is often done as a remedial measure when a problem has already developed. Sometimes, however, pregrouting is done early enough to be a preventive, rather than remedial, measure. Such was the case in Rumford, Maine, during the planning phase of new storage tanks. The foundation soil to a depth of 10 ft below the proposed foundation mat was a loose, brown silty sand with grading as shown in Fig. 10.5. Below the 10 ft depth, the soil increased in density. The loose material had an allowable bearing capacity of 1.5 tons/ft^2. The mat, 3 ft thick by 36 ft wide by 68 ft long, would transmit 3.5 tons/ft^2 to the soil when the tanks were full. To increase the soil-bearing capacity, a grouting program was designed using cement grout followed by chemicals.

Grout sleeves were placed (vertically) on 2 ft centers throughout the mat-reinforcing steel, and the concrete was poured around them. After the mat had cured, grout pipes were driven through

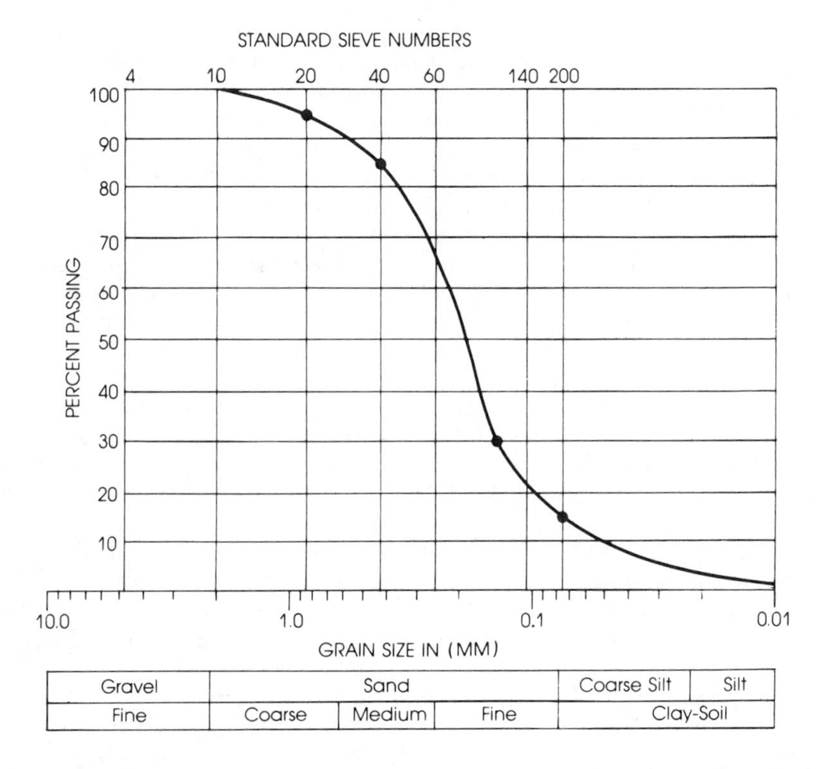

FIGURE 10.5 Graduation curve for soil beneath mat foundation.

the sleeves to a depth of 19 ft below the bottom of the mat. Alternate pipes were first grouted with a thixotropic cement grout (in this case, two bags of cement and 8 lb of a lubricating-type admixture put into 30 gal of water). The design volume to be injected per foot was based on the assumption that half the voids could be filled. Actually, due to excessive pressures, only two-thirds of the design volume was placed. When the cement grouting operation was finished, grout pipes were placed in the alternate untreated locations, and a silicate grout was injected. Again, the design volume was based on filling half of the voids. About 80% of the design volume was actually placed. The entire grouting operation was completed in 33 days. After 3 years of operation, measurements showed that no differential settlements had occurred.

This job was large enough to take advantage of a procedure which always merits consideration: use of an inexpensive, viscous material in the initial part of the grout program to fill the larger voids followed by less viscous materials to seal the finer fissures and voids. In this case, the material costs for the cement-based grout were about one-third the costs of the silicates, and placement costs were similar. (There are many conditions where placement costs for chemicals are far lower than for cements, partially or completely offsetting the difference in material costs.)

The main concern in this job was to increase the bearing capacity and reduce the settlement potential of the foundation soil. This can be done by densifying the material and by adding cohesion. Filling the voids with a grout will add cohesion but will not densify the soil. Fracturing the soil by extruding lenses and fingers of solid grout through the mass will density the soil but may not add cohesion. In this case, where there are no distinct strata of different materials, fracturing will occur primarily along vertical planes. There is little doubt, on this job, that both phenomena occurred. Most probably, the fracturing was primarily due to cement grouting and the void filling to chemical grouting.

There is considerable difference of opinion among practitioners regarding fracturing. Most of the disagreements relate to the use of the weaker grouts. A lens or sheet of a high-strength chemical or cement grout will not weaken the grouted formation. This statement applies to granular soils, not to rock. Lenses or sheets of the weaker grouts, such as the acrylates, chrome-lignins, and some of the phenoplasts and silicates, may quite possibly form zones of weakness through which failure planes can more readily develop. The fracturing process should generally be avoided with these weaker grouts.

Crystal River, Florida, is the site of electric generating facilities of Florida Power Corp. Expansion plans for a nuclear power facility were formulated in the 1960s. The greater part of the facility would be on a mat, imposing loads which range from 2.5 to 7.8 ksf. Primary foundations would be carried 20 to 30 ft below original grade. The site is underlain by solutioned limerock. Extensive cement grouting to fill solution voids was to be carried out to prepare the foundations for the structural loads. Chemical grouting on a lesser scale was done to supplement the cement grouting.

The geologic description of the formation, typical for many large limestone deposits, is excerpted from the engineering report:

> The carbonate rocks of both the Inglis and Avon Park Formations have been subject to past solution activity favoring the primary joint sets of the regional fracture system. The effects of the solutioning have been found to be particularly intense at the intersection of the primary and secondary joint sets within the rocks of the Inglis Formation, forming a network of nearly vertically oriented solution channels

throughout the area of study. Within the immediate plant site, the area of most intensive solutioning appears to correspond to the location of a series of fractures intersecting near the southeast edge of the Reactor Building.

Voids in the carbonate rocks formed by solution processes were found in many instances to be filled with loose to medium dense deposits of sand and silty sand (secondary infilling) transported by groundwater circulation. Within the areas of most intense solutioning, zones of highly decomposed limerock were found to be in close proximity and to be intermixed with the infill sediments. The infill deposits and the associated heavily leached limerock were generally found to extend from elevation +50 to +60 down to elevations from +25 to +35. It is in these areas that the supporting character of the foundation materials was suspect, leading to the use of chemical grouting as a supplement to the scheduled cement grout foundation treatment.

A feasibility study was undertaken to determine if chemical grouting could indeed improve the foundation performance in two questionable areas. A portion of this study was a technical literature research review. Sodium silicate was selected for use as the only grout with performance histories of 30 years or more. (Terranier grout was used on portions of the project as a temporary water barrier to aid the construction process. It was not used for permanent foundation support.) SIROC grout was selected for the advantages of a single-shot system and on the assumption that the end product of SIROC was sufficiently similar to that of the Joosten process so that the Joosten history of permanence was transferable. Available data on SIROC strength and penetrability indicated that the material should meet the job requirements. (These data were acceptable at that time. At present, our current knowledge of creep might cause us to question the ultimate strength value specified as being unnecessarily high, and we might cast a jaundiced eye on the statement that SIROC would readily penetrate materials with up to 20% minus 200 material.)

A field test grouting program was conducted to define optimum production grouting techniques and appropriate acceptance test procedures. The size and scope of the field test, and the applicability of much of it to general field testing procedures, merit detailed description. Much of what follows is condensed from the engineer's report:

Test Areas 1 and 2, as located on [Fig. 10.6], were selected to be representative of the infill and soft limerock deposits delineated by earlier borings, as shown in [Fig. 10.7]. Test Area 1, a 14 by 16 foot block, was used to evaluate SIROC and develop production grouting techniques. Test Area 2, an 8 × 30 foot block, was used to evaluate Terranier "C" grout, to be used for establishing a cutoff wall. Because of the proximity of dewatering wells, Test Area 1 was enclosed by a peripheral grout curtain, and a single row curtain was used at one edge of Test Area 2. These curtains were intended to reduce the hydraulic gradient within the test areas and prevent excessive dilution and loss of grout.

Primary injection holes for consolidation grouting in Test Area 1 were located on an eight feet square spacing with a single injection at the center of the square. Secondary injections were located to complete a four foot grid in the 8 × 8 ft. consolidation test block. In the western half of the test block, a tertiary order of injection was used to split the four feet grid, completing, in this area, a diamond hole pattern with injections spaced 2.8 ft. apart. The grout injection pattern in Test Area 2, excluding the curtain wall, employed primary and secondary holes to effect a three foot closure grid. Grout hole patterns employed for the curtain walls and consolidation areas are shown on [Fig. 10.6], which also shows location of borings for other test purposes.

Grouting of the test areas was accomplished by injecting grout through valved ports isolated in a perforated grout tube by a double packer, similar in concept to the established Tube-à-Manchette system. This system developed by the Swedish Stabilator Company also employed a unique method of grout tube installation, using an air drive "Alvik" drill mounted on an (Atlas Copco) air track. The "Alvik" drill powered an eccentric or expendable concentric drill bit which extended through and slightly in advance of the grout tubing and drilled a slightly larger diameter hole than the 60 mm valve tubing. By this means, the grout tube advanced simultaneously with the drill bit, resulting in a very rapid penetration. During the drilling operation, water was utilized as a circulating fluid to remove the cuttings.

The 60 mm valved tubing used for SIROC and Terranier grout injection contained four rows of grout ports, about 1 cm. in diameter, oriented 90° apart. The spacing of the ports in each row was 28 in. o.c. to achieve a sequence of two injection ports every 14 in. of the tube length. Each port was covered by a spring leaf to form a valve to

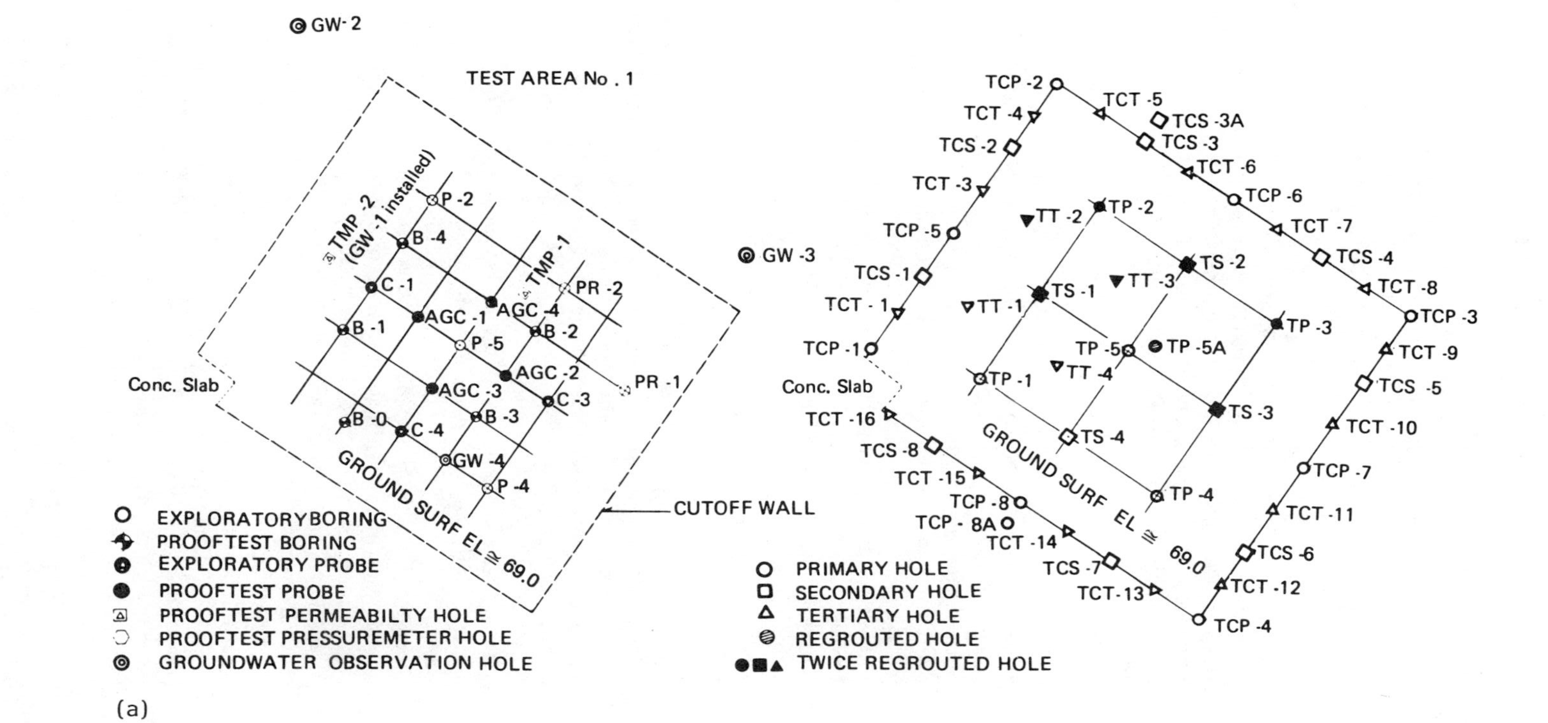
GW-2
TEST AREA No. 1
TMP -2
(GW -1 installed)
TMP -1
P -2
B -4
C -1
AGC -1
AGC -4
PR -2
B -1
P -5
B -2
AGC -2
PR -1
AGC -3
C -3
B -0
C -4
B -3
GW -4
P -4
Conc. Slab
GROUND SURF EL ≅ 69.0
CUTOFF WALL
GW -3
EXPLORATORYBORING
PROOFTEST BORING
EXPLORATORY PROBE
PROOFTEST PROBE
PROOFTEST PERMEABILTY HOLE
PROOFTEST PRESSUREMETER HOLE
GROUNDWATER OBSERVATION HOLE
TCP -2
TCT -5
TCS -3A
TCS -3
TCT -4
TCS -2
TCT -6
TCP -6
TCT -3
TCT -7
TCP -5
TCS -4
TCS -1
TCT -8
TCT - 1
TCP -3
TCP -1
TCT -9
TCS -5
TCT -10
TCP -7
TCT -11
TCS -6
TCT -12
TCP -4
TCT -16
TCS -8
TCT -15
TCP -8
TCP - 8A
TCT -14
TCS -7
TCT-13
TT -2
TP -2
TS -2
TT -3
TS -1
TT -1
TP -3
TP -5
TP -5A
TT -4
TP -1
TS -3
TS -4
TP -4
GROUND SURF EL ≅ 69.0
PRIMARY HOLE
SECONDARY HOLE
TERTIARY HOLE
REGROUTED HOLE
TWICE REGROUTED HOLE

(a)

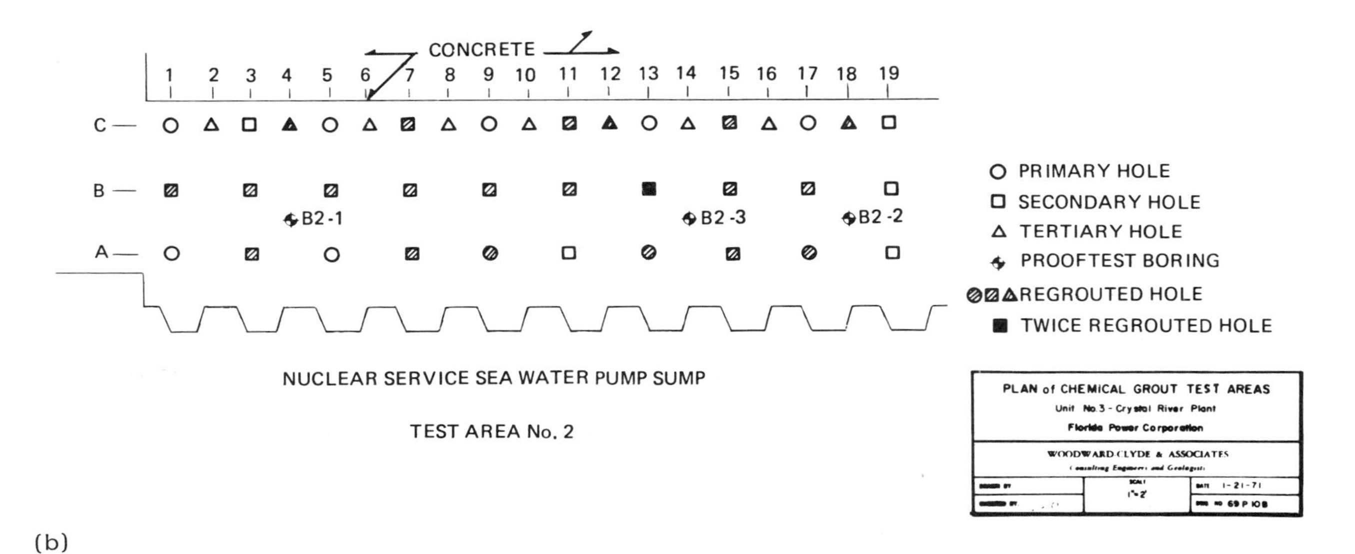

(b)

FIGURE 10.6 Plan of test areas. (Courtesy of Florida Power Corporation, St. Petersberg, Florida.)

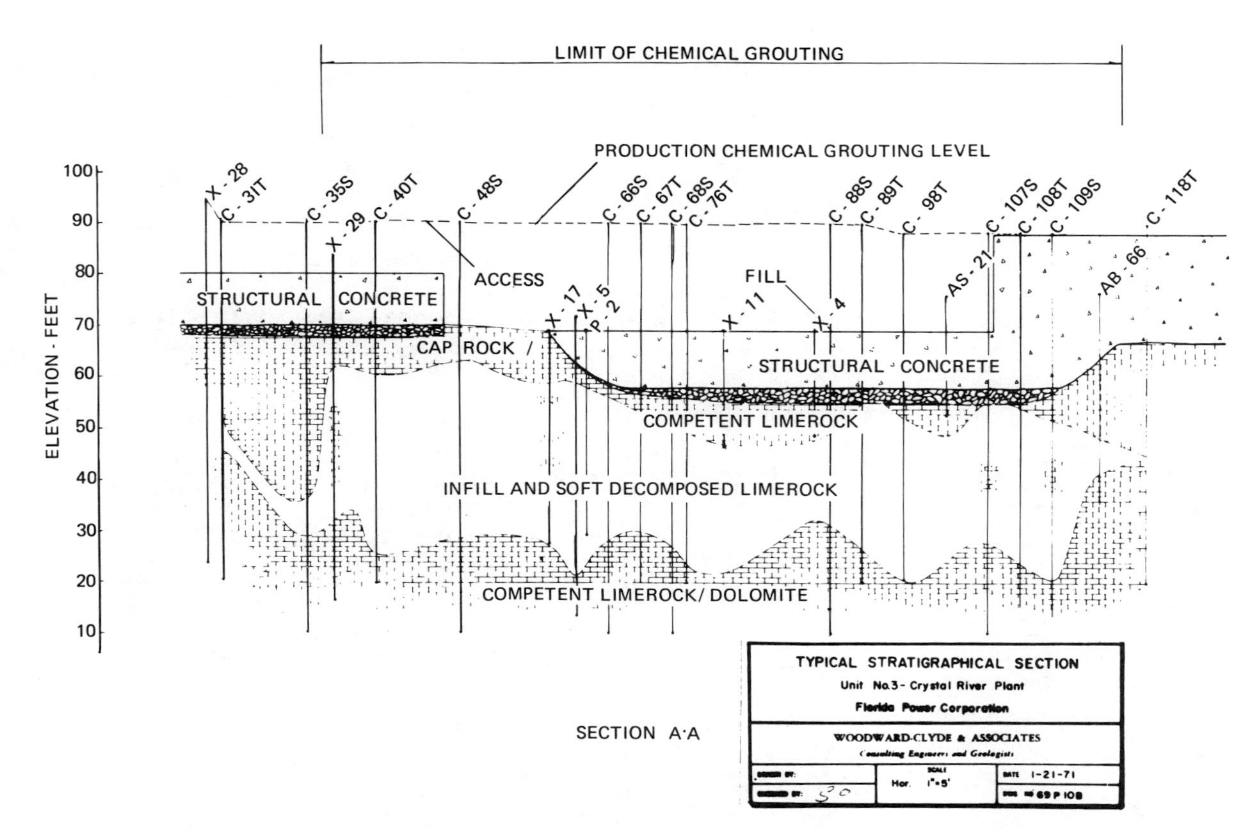

FIGURE 10.7 Stratigraphical section. X, P, AS, AB = prefix designations for various series of exploratory borings. C = prefix designation for cement grout exploration hole. (Courtesy of Florida Power Corporation, St. Petersburg, Florida.)

prevent intrusion of the materials into the tube during drilling. Tests conducted in the field indicate a grout pressure of about 30 psi was required to open the valve. A conceptual sketch of the Stabilator grout injection system is shown in [Fig. 10.8].

A 70 mm solid tubing was used for the bottom (point) injection of the Siroc-Cement grout. Before each injection, the tubing was raised about 14 inches so as to yield a grout injection interval similar to that obtained by the 60 mm valved tubing.

Test grouting was conducted in accordance with predetermined criteria not to exceed an extreme limiting transient grout pressure of 200 psi, and not to exceed a grout pressure of 180 psi under sustained pumping. Based on estimates of the porosity of the subsoils and an assumed radius of grout travel, an injection criterion of 50 gallons per lift was established and maintained unless controlled by the limiting pressure criteria. For production efficiency it was desired to establish a minimum pumping rate of not less than about 5 gpm acknowledging that pumping pressures will usually increase with the rate of pumping.

Early in the test grouting program it was determined that the infill and decomposed limerock materials could be readily permeated by both SIROC and the lower viscosity Terranier grout. The latter appeared to have superior permeation characteristics as was initially anticipated. It was established that pumping rates usually in the order of six to eight gpm were appropriate to satisfy the 60 gallon injection per lift criteria without developing excessive grouting pressures.

Test borings drilled concurrent with the test grouting to evaluate the grouting effectiveness and to establish an acceptance testing criteria, revealed that dilution of the chemical grout was occurring near the base of the grouted zone. This condition was attributed to the significant hydraulic gradients imposed across the test area by dewatering wells located within the near proximity. A reduction in gel time to cause gelation before dilution could occur was generally unsuccessful. The demonstrated ability of the grout to gel above elevation 35 was apparently due to the effectiveness of the upper portion of the peripheral grout curtain. The grout curtain may also have resulted in a confinement of groundwater flow at the base of the test block, accompanied by a significant increase in the gradient across this zone.

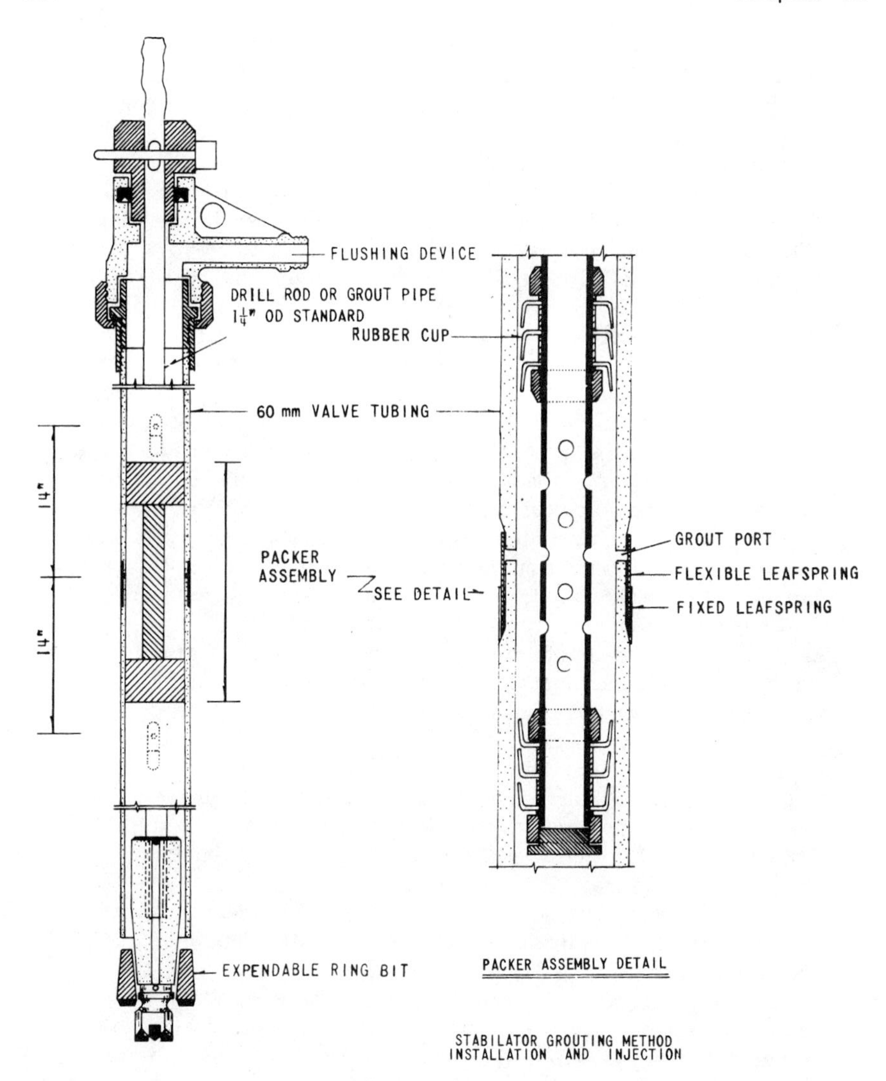

FIGURE 10.8 Stabilator grouting method. (Courtesy of Florida Power Corporation, St. Petersburgh, Florida.)

Grout was injected at 45 locations in Area 1 and at 39 locations in Area 2. In Area 1, six of the curtain holes were grouted with Terranier "C" whereas in Test Area 2, three holes were injected with Terranier "C". (In the absence of the preparatory cement grouting in the test areas, primary holes in both Area 1 and 2 were injected with SIROC-Cement as an expedient method of filling any existing large voids prior to chemical grouting. Initial control of the gel time of the SIROC-Cement proved extremely difficult and it was soon determined that the limerock additive had a detrimental influence on gel set.)

The distribution of grout take in terms of gallons is summarized for Test Area 1 and for Test Area 2 by [Figs. 10.9 and 10.10], respectively.

The average grout take per hole and per linear foot of injection decreased in Test Area 1 with the descending order of the injection hole, as anticipated. Comparison of the grout take profiles with the subsurface conditions observed before chemical grouting indicated the zone of the heavy grout concentration between elevations 58.5 and 22.5 in Test Area 1 and elevation 67.5 and 51.0 in Test Area 2 correspond to the extent of the poorest materials encountered by the pre-grout exploration.

To evaluate the degree of improvement and establish acceptance testing criteria, test borings were conducted using Standard Penetration Resistance (SPR) tests. Samples of the grouted materials were retrieved and examined for grout permeation and the range of the penetration resistance (N) values representative of the grouted materials was established. In addition, a "Dutch Friction Cone" was used to establish cone penetration resistance values (q_c). In situ testing was also conducted using the pressuremeter apparatus.

Comparison of the N and q_c values obtained in the grouted materials with results of similar tests conducted before grouting indicated substantial improvement in the penetration resistance had been obtained by grouting. The average N value of the infill and decomposed limerock increased from 0 to 25 blows per foot. The results indicate that for the loosest infill and softest limerock deposits, an N value of 30 blows per foot should be readily achieved during production grouting.

The before and after grouting comparison of q_c values indicated a similar improvement to that of the N values. A significant increase in q_c between elevations 28 to 30 was evidenced by an average q_c before grout of 39 tsf compared

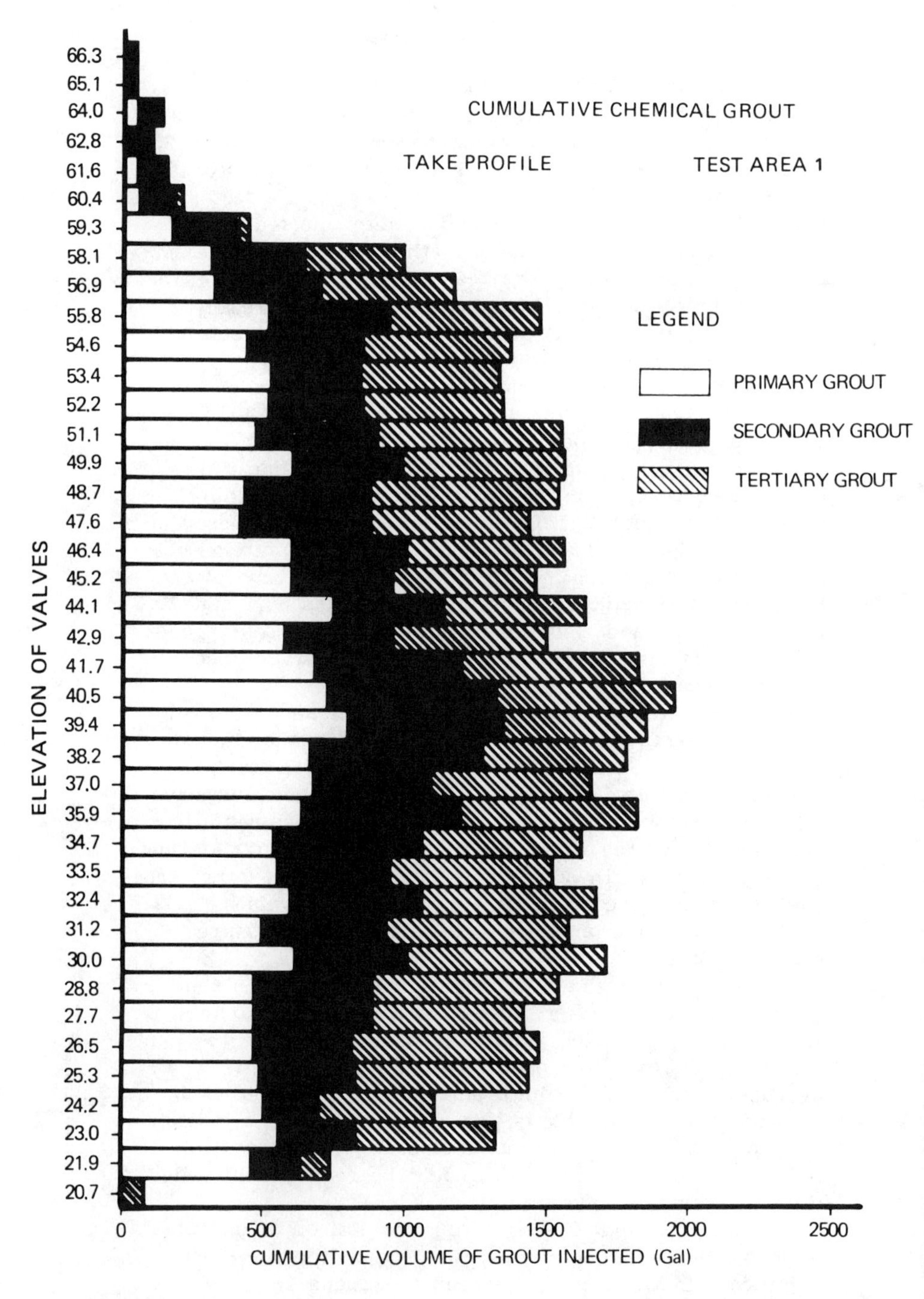

FIGURE 10.9 Grout take profile. (Courtesy of Florida Power Corporation, St. Petersburg, Florida.)

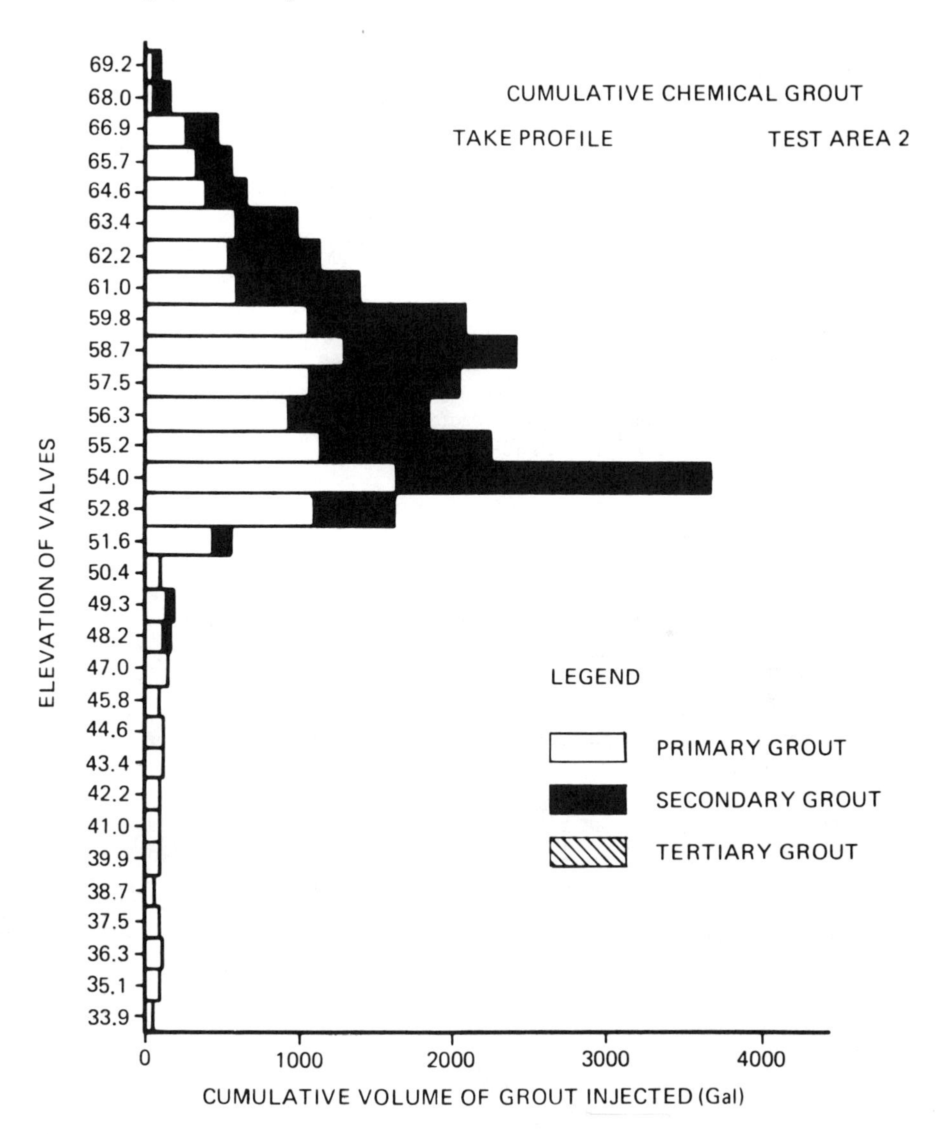

FIGURE 10.10 Grout take profile. (Courtesy of Florida Power Corporation, St. Petersburg, Florida.)

to an average q_c after grout of 120 tsf. (Although the cone resistance test proved to be technically feasible, it was found that advancement of the cone through the frequent hard layers was extremely time consuming and inefficient.)

Two series of pressuremeter tests were conducted to investigate in situ strength and compressibility of the grouted materials. The tests proved very difficult to complete and interpret primarily due to the roughness of the side-walls of the test holes. The calculated in-situ shear strength ranged from 3 to 9 tsf and the corresponding deformation modulus from about 50 to 570 tsf. Because of the irregularity of the side-wall of the test holes, the drilling disturbances and the complex stress conditions implicit in the test, data retrieved from the pressuremeter tests must be considered qualitative and to indicate an "order of magnitude". Thus, the use of the pressuremeter as an acceptance test was considered to be inappropriate.

To evaluate the uniformity of grout permeation into the subsoils, a series of permeability tests were conducted in test holes by pump-in techniques. Comparison of the before and after permeability test values indicates that grouting decreased the before-grout k by a factor of approximately 1000.* The range of after-grout k values can thus be used as a qualitative measure of the average degree of grout permeation achieved.

Based on the grouting test data, production grouting was planned in the 6 to 8 gpm range at about 150 psi. In operation, actual values ranged from 2 to 10 gpm at pressures of 140 to 190 psi. Grouting was done from the bottom up, at 14 in. stages, placing 50 gal per stage or less if 200 psi pumping pressure was reached first. Gel times for the silicates were generally in the 20 to 30 min range, with occasional use at values as low as 5 min. (Stock solutions of silicates not used within 2 h were discarded.) Terranier "C" was used in the 2 to 5 min gel time range. SIROC-cement mixtures, when used, were in the 2 to 10 s gel range. Gel time checks were made at least once every half hour.

During a 4 month period ending in May 1971, over 400,000 gal of grout were placed in about 29,000 linear feet of grout holes,

---

*One very interesting fact that emerged from the test data was that the use of silicate-based grouts resulted in after-grouting permeability of the order of 10^{-5} cm/s. This same value was again attained in the extensive test program of Locks and Dam 26, described in Chap. 14.

about 15% in cutoff walls and the rest in general consolidation of the solutioned zones. Immediate checks could be made on the effectiveness of the cutoff, and they were all positive. No foundation problems have occurred in the grouted zone.

The need for additional formation strength was anticipated in recent underground construction in Brooklyn, New York [3]. A 350 ft long tunnel, 8 ft in diameter, was to be placed under compressed air to connect two existing intercepter sewers. At one point along its path the new tunnel intersected the upper half of an old brick-lined tunnel 12 ft in diameter. At the point of intersection, excess soil movement might damage the old structure, leading to surface problems with structures and traffic. In order to keep soil movements to a minimum, it was decided to treat the zone by grouting.

The soils in the zone to be grouted consist of medium to fine sands with some silt, and organic clay overlying fine sand with some silt. The granular soils could readily run into the tunnel face during excavation. The grading is too fine for penetration by cement, so chemical grout was indicated. Sodium silicate was selected with MC-500 (see Appendix A for details about Microfine Cement MC-500) as catalyst. The required grouted soil strength was set at 2 tsf (28 psi), well within the possible limits for silicates. Limiting MC-500 to 5% or less, would give gel times of 30 min or less, and yield a final product that could be removed by spading.

Grouting was done through tube-à-manchettes, placed vertically from the surface on a 3-ft grid. Grout ports were spaced 18 in. along each pipe. At each port, the planned injection volume was based on the soil voids and the desired overlap between holes. This volume was placed, unless the pressure exceeded 1 psi per ft of depth to the grout port. A total of 72,000 gal of grout was placed.

During the grouting operation, test holes at the center of a 4-hole pattern were used to verify and adjust the grouting process. After completion of the grouting, Standard Penetration Tests (STP) were performed and undisturbed samples were taken to verify the spread of grout and the grouted soil strength. Not all of the strength tests on the samples approached 28 psi, probably due to the difficulty of getting good, solid samples. However, the STP's showed adequate increase in "N" values to indicate overall effectiveness of the grouting. Subsequent construction proceeded with no settlement problems.

Extensive chemical grouting was performed in Baltimore during subway construction, primarily for strength [4]. Slurry walls were to be used for excavation support. However, to place these walls required removal of a portion of existing building footings and/or the loss of footing support. Grouting with sodium silicate was done

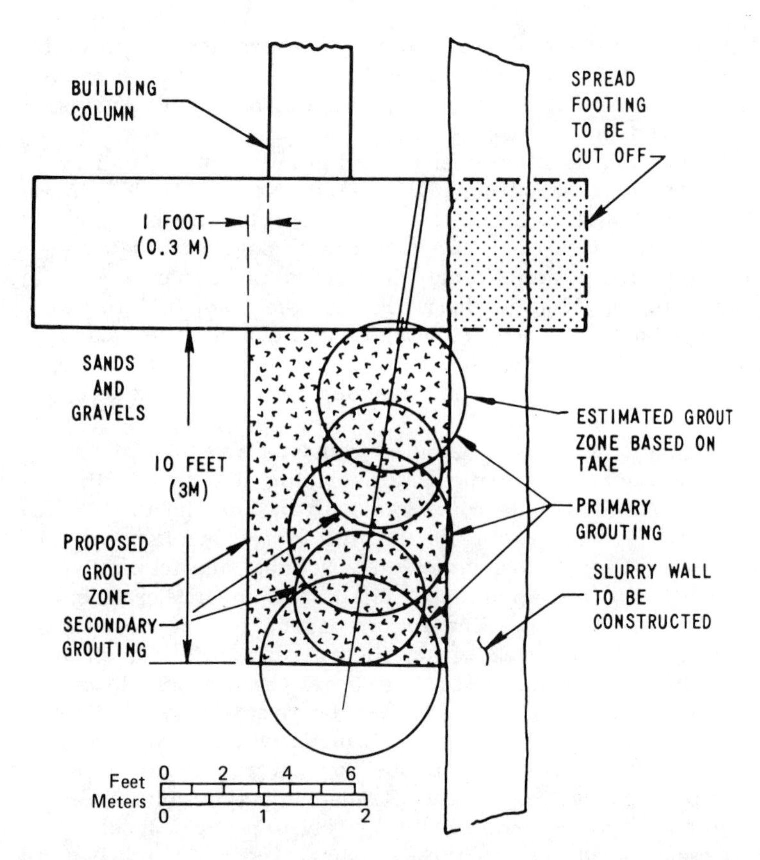

FIGURE 10.11 Cross section of spread footing showing chemical grout zone.

as shown in Fig. 10.11 to add strength and stability to the granular soils underlying the footings.

Grouting was also done in several locations beneath existing underground structures prior to exacvating below them, as shown in Fig. 10.12.

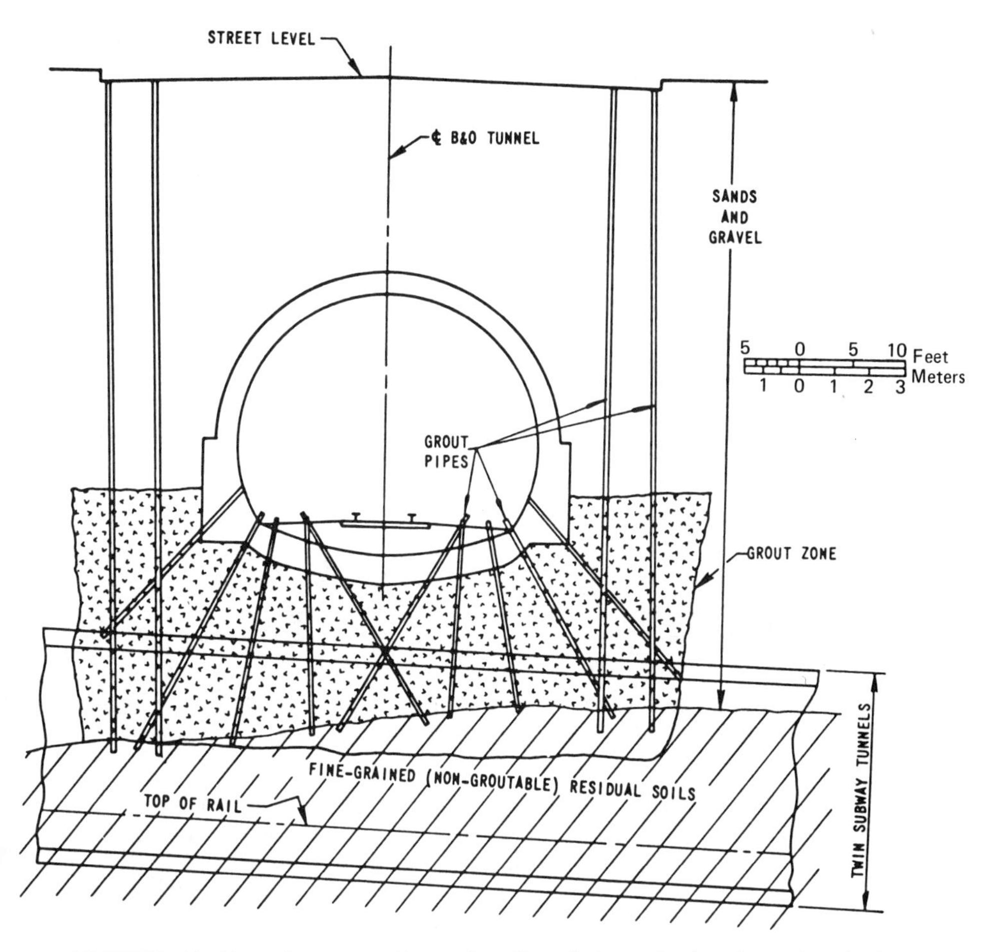

FIGURE 10.12 Cross section of railroad tunnel showing chemical grout zone and grout pipe arrangement.

10.4 SUMMARY

When the primary purpose of grouting is to add strength to a formation, chemical grouts are used when their other properties (sich as low viscosity) are also of advantage. Grouting in granular soils has the effect of adding a cohesion component to the shear strength of a magnitude which varies with the different grouts and grout concentrations. For granular materials near ground surface or close to an open face, the cohesive component added by any grout is a considerable portion of the total shear strength. The strength contribution of grouts is a decreasing portion of the total strength as depths below surface (and lateral pressures) increase.

In contrast to the use of cement for adding formation strength, specifications for chemical grouts almost always specify a required grouted soil strength. The number actually specified is often used to eliminate specific weaker grouts from appearing on bids. Thus, the specified strength may be far beyond what is actually needed, and also beyond what may be attained with any chemical grout, when long term strength (creep strength) is considered. ASTM standards now exist, both for sample preparation and strength testing. These should also be specified, when strength criteria are to be met.

REFERENCES

1. R. H. Karol, *Soils and Soil Engineering*, Prentice-Hall, Englewood Cliffs, New Jersey, 1960.
2. W. C. Parish, W. H. Baker, and R. M. Rubright, Underpinning with chemical grout, *Civil Engineering*, New York, August 1983.
3. A. H. Brand, P. M. Blakita, and W. J. Clarke, Chemical grout solves soft tunnelling problem, *Civil Engineering*, New York, 1988.
4. E. J. Zeigler and J. L. Wirth, Soil stabilization by grouting on Baltimore subway, *Grouting in Geotechnical Engineering*, ASCE, New York, 1982, pp. 576–590.

11

Grouting in Tunnels and Shafts

11.1 INTRODUCTION

In Chaps. 9 and 10, the applications were primarily water shutoff or strengthening a formation. In tunnel and shaft grouting, generally both purposes must be served. In addition, tunnels and shafts (because of their greater depth below grade) often involve the use of much higher pressures than the projects detailed in the previous two chapters (except for mine waterproofing, which may also take place at substantial depths). Further, tunnels and shafts are often very large projects and (like large cutoff walls) are often preceded by extensive soil investigation. This permits prediction of possible water problems and the detailed preplanning of how those problems will be handled. The procedures used in grouting tunnels and shafts are much the same regardless of project site and can be illustrated by case histories of relatively small projects as well as large ones.

11.2 SHALLOW TUNNELS

The tunnel shown in Fig. 1.9 is typical of conditions (unexpectedly encountered in small, shallow tunnels) which require remedial measures. It had been anticipated that mining would take place in clay, using steel supports and lagging. However, a fine, dry sand statum was intercepted for about 300 ft along the tunnel line. It might have been possible to continue mining by using a shield and liner plates. Alternatively, the sand could be stabilized by grouting in order for mining to proceed without danger of loss of ground. Grouting was selected as the more economical procedure. Grout pipes were driven from the surface as shown in the vertical section in rows 6 ft apart. (Grouting in tunnels is often done from the tunnel face. Grouting from the surface, however, permits continuous mining and for this reason may be cost-effective even through longer grout pipes are needed.) A silicate-based grout was used to create an arch about 3 ft thick. The grout volume placed averaged 100 gal per lineal foot of tunnel.

Another project where grouting from the surface proved the most feasible took place in Harrison, New Jersey, where two 12-ft-diameter concrete tunnels were to be constructed under 13 sets of live railroad tracks without interrupting railroad traffic.

The tunnel invert was located 17 ft below track level, and the lower half of the tunnels was in clay. Above the clay was 2 to 4 ft of sand overlain by meadow mat about a foot thick. Mixed cinder and sand fill were above the meadow mat. Two shafts were put down by driving sheet piling. They were 270 ft apart, spanning the tracks. Concrete pipe was to be jacked into place, forming the tunnels. Figure 11.1 shows sketches of the job parameters.

A surface area 38 ft wide by 15 ft long (in the line of the tunnels) adjacent to the shaft was treated from the surface. Drill rods (E size, plugged with a rivet) were to be driven to the top of the clay stratum, and grouting was done by withdrawing in 3 ft stages. After this work, holes were cut in the shaft sheeting, and the jacking operation was begun. After about 17 ft of progress (the limit of the grouted zone) excessive water flow occurred, and quick conditions developed. At this point it was decided to grout the entire tunnel length. The grouting was started with holes and rows spaced 3 ft apart. Spacing of holes in each row was increased to 4 ft after experience proved that this spacing gave adequate stabilization.

The project was done with an acrylamide-based grout. The formation was stabilized to the extent that cave-ins and quick conditions were prevented, and the inflow of water was reduced to the point where it could be readily handled by pumping. The entire grouting operation was completed in 27 working days, using a total of about 34,000 gal of grout.

Pipe jacking was also involved in a 4000 ft pipeline in Alameda, California [1]. Five-foot-diameter concrete pipe was being laid by cut and cover methods. The pipe had to penetrate a 65-ft-high levee, where cut and cover could not be used. After jacking got underway a short distance, sand, gravel, and boulders ran into the pipe, leaving a large open cavity above it. Rather than risk the possibility of the cavity extending itself to the surface of the levee, the contractor chose to stabilize the formation by grouting. A silicate-based grout was selected, since the formation was mainly sands and gravels.

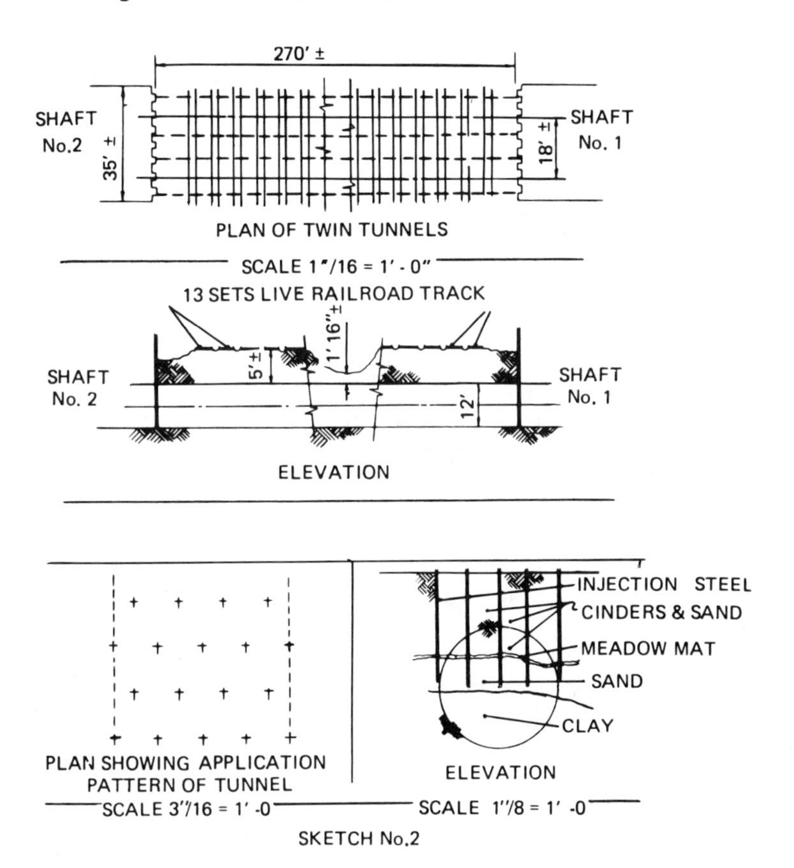

FIGURE 11.1 Plan of grouting for jacked tunnels.

At the heading, 0.5 inch steel pipe was driven by air hammer to a depth of 10 to 11 ft at five locations, arranged in a half-circle pattern flaring out slightly from the upper concrete pipe circumference. Each grout pipe was withdrawn in 1 ft increments, with an average of 10 gal of grout injected every foot. Pumping pressures were generally in the 100 to 120 psi range, and gel times of about 2 h were used. (Much shorter gel times could also have been used effectively.) Grouting was done by a night shift and pipe jacking and mucking during the day.

Similar procedures in tunnel stabilizations are detailed in Refs. 2 through 5. Although these jobs date back 20 to 25 years, the procedures used then are still valid and in use today. The procedures were new at that time and therefore newsworthy. Similar work done today is generally not reported in the technical press.

In more recent work, the grouting patterns for tunnels have become more complex and sophisticated. A case in point is the stabilization of a rock–soil interface detailed in Ref. 6. Other references to tunnel and shaft grouting can be found in Ref. 7.

11.3 EUROPEAN PRACTICE

In Europe, chemical grouting in tunnels is "automatically considered as part of the tunneling plans rather than looked upon as an esoteric tool" (Ref. 8, part 1, p. 9; the several descriptions of tunnel work that follow are taken from the same reference, written in June 1977). As a result, a much larger volume of work and experience exists in Europe, and domestic practice tends to follow the procedures and techniques developed overseas.

A schematic of the subsurface conditions along the axis of a sewer tunnel in Warrington, England, is shown in Fig. 11.2. The tunnel is in sandstone until the surface of the sandstone dips downward; then for about 120 meters tunneling is in mixed face sandstone–sand conditions and subsequently all sand. In the mixed face area and in some of the sand-only tunneling area, the sand was stabilized with a silicate grout. Conventional shield tunneling was performed

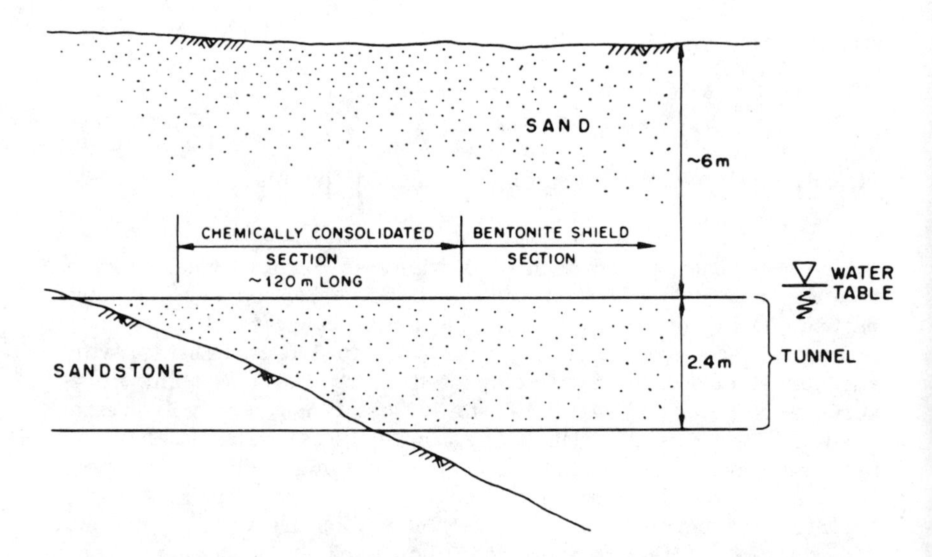

FIGURE 11.2 Tunneling conditions for grouted section of sewer tunnel, Warrington new town development. (Note: This schematic is not to scale.) (From Ref. 8.)

in this region; the bentonite shield operation took over once the tunnel was well clear of the mixed face area. The groundwater table is located near the crown of the tunnel in all areas.

Grouting Scheme

Most of the grouting at Warrington was performed with a silicate-based grout in a one-stage process using the tube à manchette technique. Percentages of the components of the grout varied slightly with the availability of silicate. On average the grout was composed of 44% silicate, 4.5% ethyl acetate, and 51.5% water; gel times were about 50 min. Bentonite–cement grout was used in a few instances to fill large voids before silicate grouting.

The grout was injected into and above the tunnel section as shown in Fig. 11.3a. The grout occupied the zone of soil from 4 m below the surface. The hole pattern is shown in Fig. 11.3b, with one

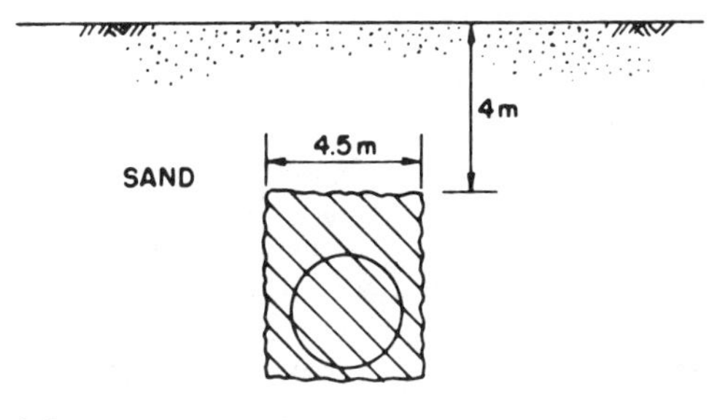

(a)

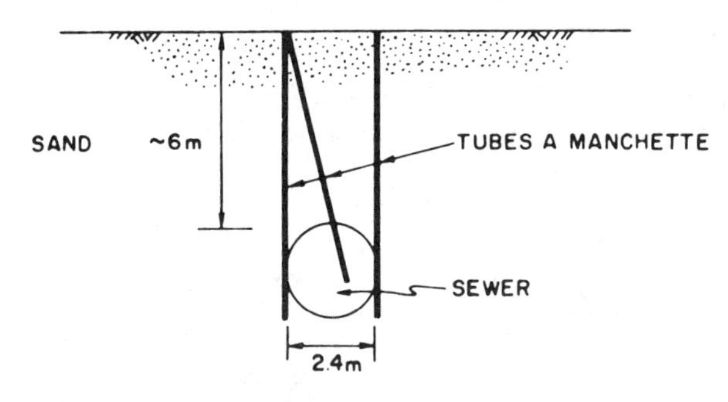

(b)

FIGURE 11.3 Grouting technique and resulting grouted zone. (a) Grouted region and (b) three-hole grouting pattern. (From Ref. 8.)

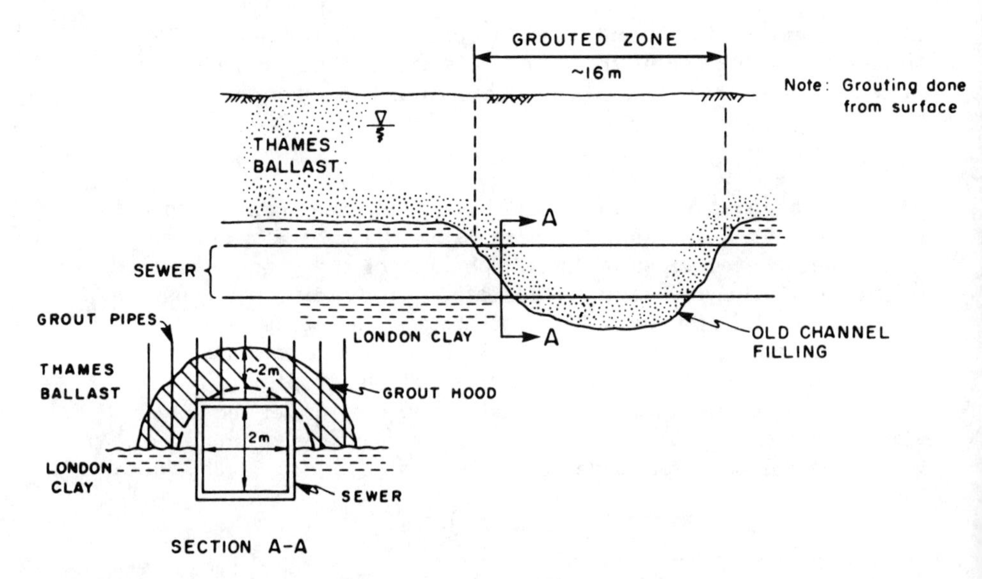

FIGURE 11.4 Use of grouting for sewer relocation. (From Ref. 8.)

vertical hole on either side of the grouted region and one inclined hole through the grouted region. This pattern allowed the grouting to be performed while avoiding utilities which were located immediately above the tunnel. Each set of three holes was spaced 1.5 m apart along the tunnel axis.

Fifty-seven days were required for the drilling of holes for the grouting and 20 weeks for the actual grouting. The total cost of the grouting was $75,000; on a per ft basis this becomes $210 per lineal ft of tunnel.

Another chemically grouted tunnel excavation concerns a relatively small, approximately 2-m-square tunnel, which was being driven through London clay for a sewer relocation in 1974. The geologic setting is shown in Fig. 11.4. Unknown to the designers, the surface of the London clay in one area dipped below the tunnel invert over a 50 ft section because of erosion by an old channel, and the dip was occupied by saturated Thames ballast. Upon encountering the Thames ballast, a run of granular material occurred into the tunnel, which stopped the tunneling completely.

Chemical grouting was chosen to stabilize the Thames ballast in the tunnel area. A semicircular 2-m-radius section of grout was injected above and in the tunnel area as shown in Fig. 11.4. The contractor chose to use a silicate grout with a two-stage process. Grouting was conducted from the surface via a service road. The grout thoroughly solidified the ballast, and tunneling proceeded without further incident.

The total cost of the grouting was approximately $20,000, or about $400 per lineal foot of grouted tunnel. One week of sampling and testing of the soils was needed before grouting, and the grouting required 4 weeks.

In Europe, as opposed to domestic practice, design and quality control of grouting are generally the province of the contractor, and field practice tends to be more precise and detailed than current domestic practice. (This statement does not apply to the Washington Metro work, which was very closely controlled and monitored. See Sec. 11.4 for discussion.) An example of this is shown in Fig. 11.5 (Ref. 8, part 1, pp. 35–36), a typical section design for grouting of a profile with various soil types. A circumferential zone of treatment is defined in heavy lines; within this section are different soil and rock types, and the grout treatment varies accordingly in the percentage of the grout chemicals and the type of grouting procedure. The percentages of the chemicals vary in some cases depending on whether the grout will be in the section to be

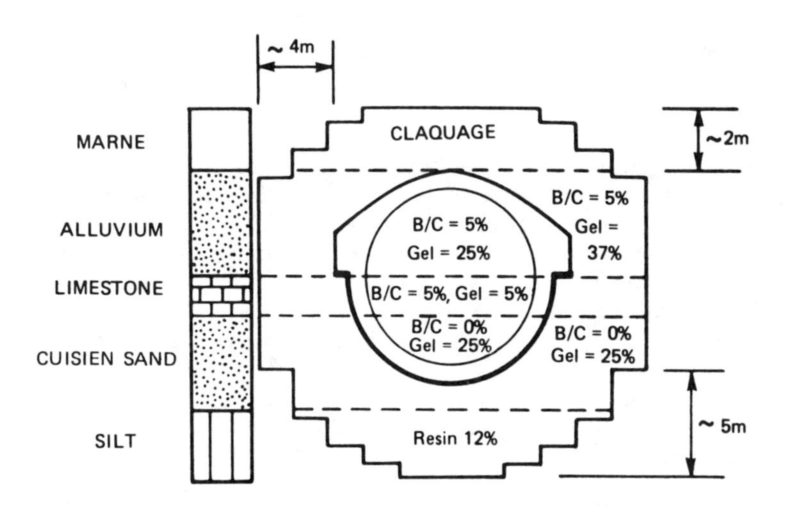

FIGURE 11.5 Typical section design for grout treatment. (From Ref. 8.)

excavated or outside of it. Weaker grouts are injected in the area to be excavated to allow for ease in removal of the soil and to prevent penetration of the hardened grouts in the excavation area. Such a complicated plan could only be carried out using the tube à manchette process where each sleeve can be grouted differently.

In Nuremberg, Germany, extensive chemical grouting has been done for a new subway system. Most of the grouting was done to limit settlements in the existing structures above the tunnel excavation. The sands at Nuremberg are generally coarse, and the authorities specified strength requirements for the grouted sands. Before bidding on the job, each contractor was asked to grout a small test section so that the grouted soil could be cored and tested to determine if the proposed grout formula would meet the required strength. Most major European grouting firms engaged in the bidding; a German firm won the project for a total price of $3 million [8].

The grouting was performed from the surface using the tube à manchette procedure; the grout is silicate based.

Figure 11.6 shows typical grouting zones to be used in Neuremberg above and around tunnel openings and under adjacent foundations. In Fig. 11.6a, the left-hand tunnel passes directly beneath a building foundation (within 5 m of the nearest footing); this opening is covered on top and along the sides by a trapezoidally shaped zone of grouted sand. The opening is, of course, cut after the grout is in place, and the trapezoidal zone should theoretically limit movements of the foundation when the shield passes. The right-hand tunnel in Fig. 11.6a is not directly under a building foundation; the nearest footing in this case is underpinned by a grout column. It should be noted that in the sections for both tunnels some of the grouted zone is shown using crosshatching, while some is dotted. The dotted zones are in the tunnel openings and represent material designed to be weaker than the crosshatched regions. Weak grouts are injected in these areas to provide some cohesion and to prevent penetration of hardened grout into the tunneling area.

A second case of grouting is shown in Fig. 11.6b. Here the left-hand tunnel again passes underneath a foundation. The grouting zone is trapezoidally shaped as in Fig. 11.6a, but in this case the zone extends up to be flush with the bottom of the footings. This provides underpinning to protect against the effects of the opening of the left-hand tunnel and also the right-hand tunnel, which passes nearby.

Grouting of the sands is being done primarily from the surface, as shown in Fig. 11.7. Inclined holes are drilled so as to allow creation of the desired grout zone under the foundation. In plan, each row of inclined holes is spaced 1 m from other rows.

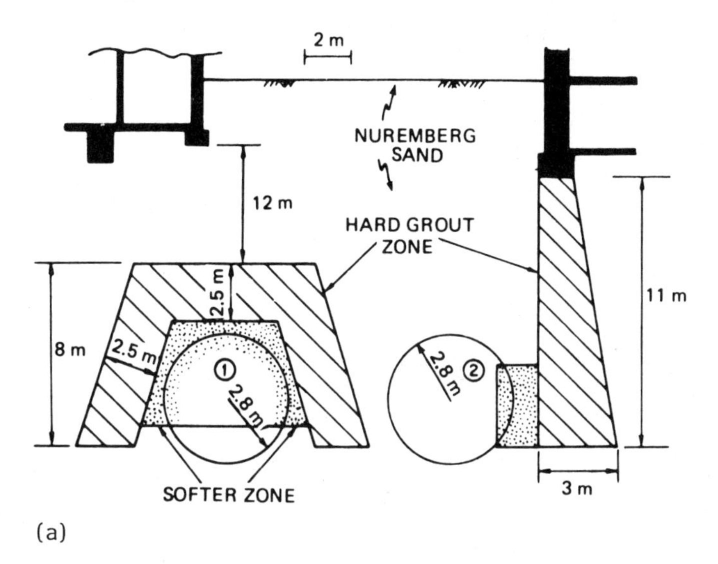

(a)

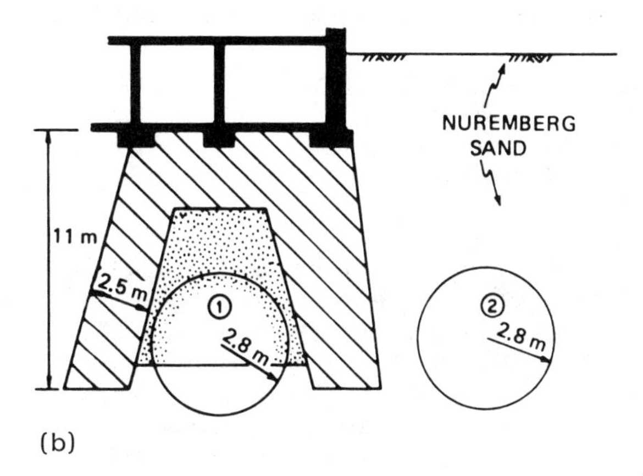

(b)

FIGURE 11.6 Design grouting zones. (From Ref. 8.)

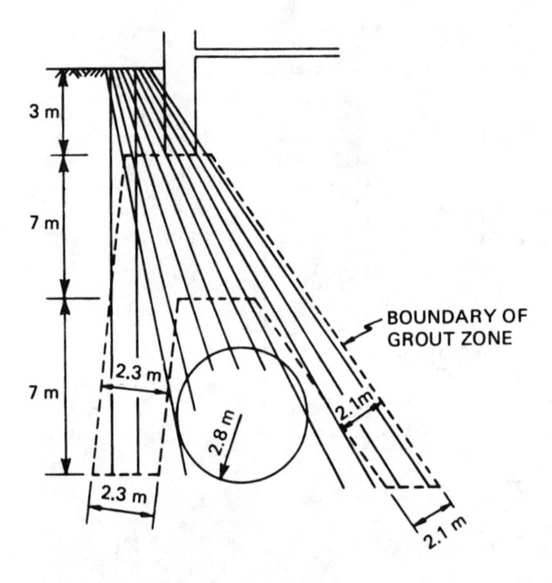

FIGURE 11.7 Grout hole pattern to create underpinning grout arch. (From Ref. 8.)

11.4 RECENT DEVELOPMENTS IN TUNNEL GROUTING PRACTICE

Since the early 1970s, chemical grouting has been used extensively in the Washington, D.C., Metropolitan Area Transit Authority System. Much of this work is available in a series of four reports [8]. The descriptions which follow come from the fourth report. One of the earliest jobs was done in 1972. Large surface settlements (about 11 in.) which occurred during driving of a tunnel were the motivation for the use of chemical grouting on a later adjacent tunnel.

A plan view of the site is shown in Fig. 11.8. Grouting was done from the surface, using 50% sodium silicate with 5% to 10% reactant and water. Hole spacing was 5 ft. Grout pressures varied from 5 to 40 psi, only occasionally reaching as high as 60 psi. Volume was kept at about 7 gpm, and gel times were about 15 min. The results of grouting are shown by the surface settlement data in Fig. 11.9. In addition to reducing settlements, grouting also reduced the risk of running sand flooding the tunnel.

At another location, two tunnels pass beneath an old masonry culvert structure. Grouting was used in lieu of underpinning to support the culvert and prevent sand runs, as shown in Fig. 11.10.

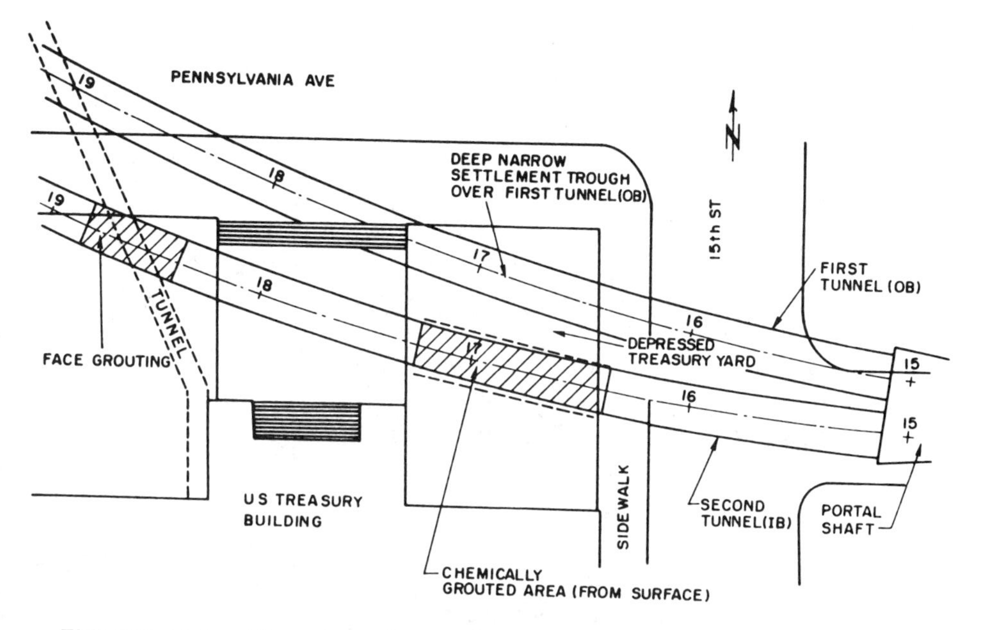

FIGURE 11.8 Plan view of tunnel grouting site. (From Ref. 8.)

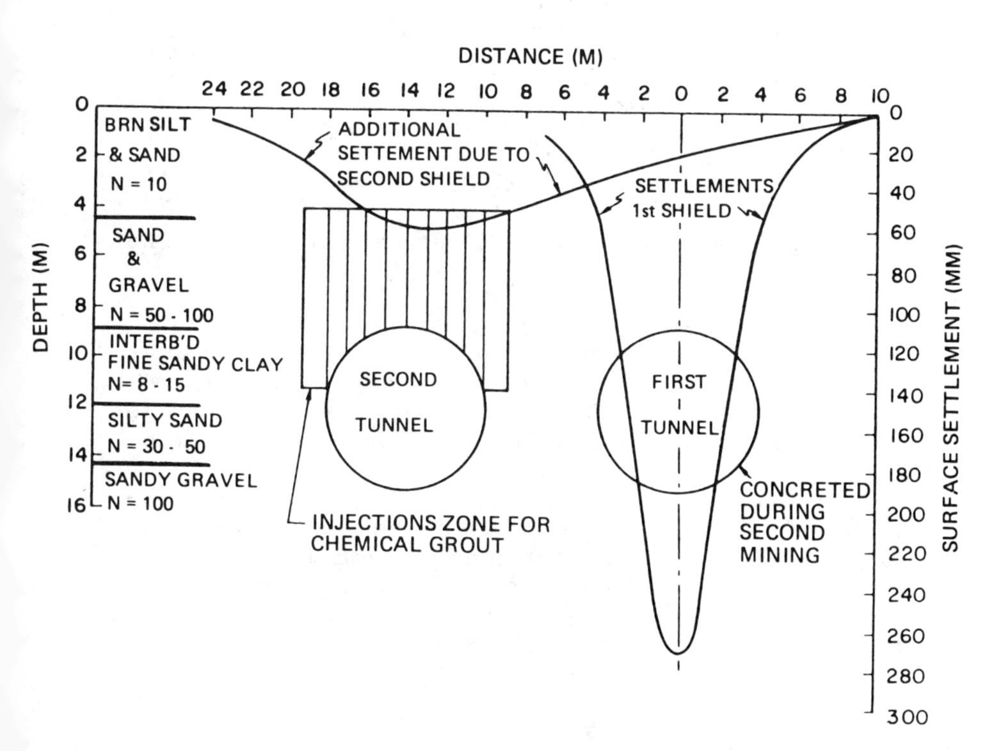

FIGURE 11.9 Surface settlements due to tunneling. (From Ref. 8.)

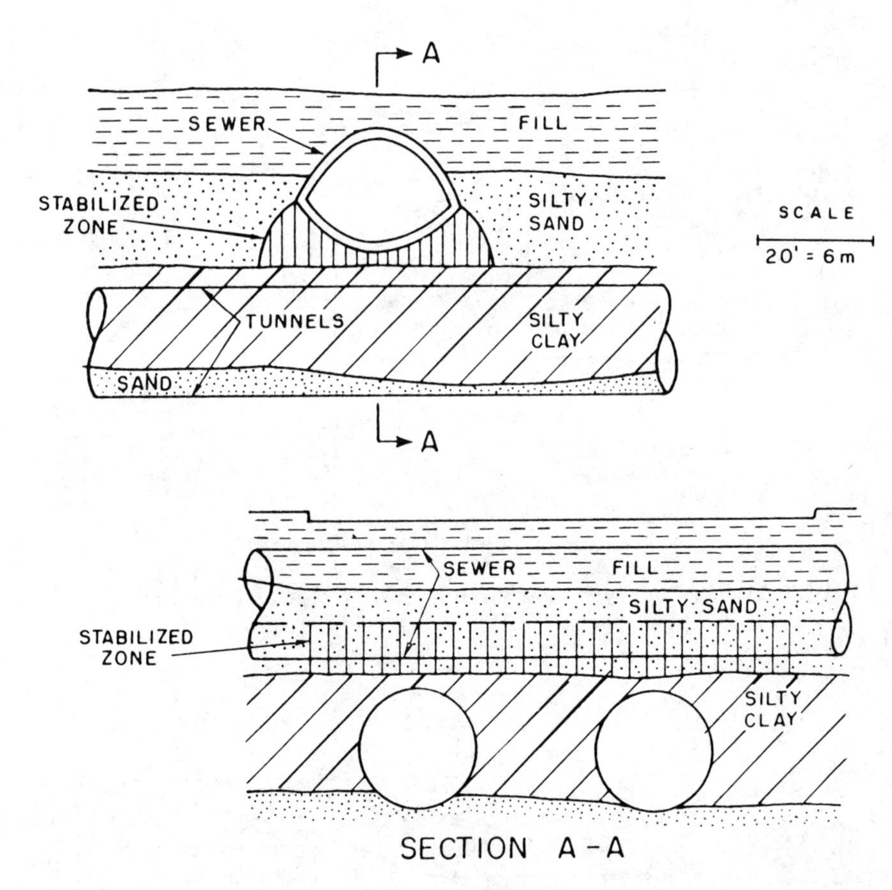

FIGURE 11.10 Elevation through culvert and tunnels. (From Ref. 8.)

The sands in the profile were dense and the clays medium or stiff. Grout injections were made only in upper silty sand, which contained 5% to 50% silt. More than 20% silt makes grout penetrations difficult, so it was doubtful that a continuous grout zone would be formed in the silty sand.

The grouted region extended under the sewer culvert for the distance of 100 ft along the culvert centerline and at a maximum width of about 40 ft at the top of the silty clay layer. Using the tube à manchette system, a three-stage grouting program was conducted. In the first stage, a bentonite–cement mix was injected; in the second and third stages, a sodium silicate was injected.

The sodium silicate solution consisted of 50% type S sodium silicate, 46% water, and 4% ethyl acetate reactant. Two 17-ft-diameter pits 30 ft to either side of the centerline of the sewer culvert were dug to facilitate grouting, as shown in Fig. 11.11.

Grout pipes were placed in holes radiating out from the pits to spread grout over the intended grouting region. The holes were drilled in three sets at different angles from the horizontal. It was planned to place a total grout volume equal to about 35% of the volume of the area to be grouted. The 35% was expected to be 15% bentonite–cement and 20% sodium silicate. During first stage grouting with the bentonite–cement, relatively large heaves were measured in the masonry sewer. When the heave reached 0.5 in., the bentonite–cement treatment was stopped, with only about half of the anticipated volume placed. Second-stage grouting using the sodium silicate was continued until 35% of the ground volume was reached (8% bentonite–cement plus 27% silicate). However, it was found during the second stage that little or no heave of the sewer occurred and that the grout was accepted with relatively low pumping pressures. Thus, after allowing for the first silicate injection to set, a third stage of injection with silicate was carried out to ensure that the voids of the soil were completely filled. In this stage, high pressures were required for grout injection, and small, but visible, heaves of the sewer occurred. This third-stage grouting effort raised the total volume of grout injected to 42% of the soil volume grouted.

The only performance parameter monitored at the site was surface settlement. Movements in the grouted region averaged 1-3/8 in., with a maximum value of slightly under 2 in. Settlements outside the grouted zone were considerably larger, with maximum values approaching 3 in.

The chemical grouting projects for the Washington, D.C., metro system were of major significance in developing effective methods for chemical grouting in tunnels and in expanding our understanding of the potential of a grouting operation. The earlier work was done by Soil Testing Services (a Chicago-based consulting company specializing in soils engineering and one of the first domestic companies to work with the (then) new chemical grouts in the early 1950s; the work referred to is described briefly in this section). Later work was done by Soletanche (a French company with long-term experience in chemical grouting) using procedures common in Europe but relatively new in the United States. Most of the more recent work was done by the Hayward Baker Company (a Maryland-based firm specializing in ground control and chemical grouting), building upon the best of the procedures used domestically and overseas. Much of this work is detailed in Ref. 8.

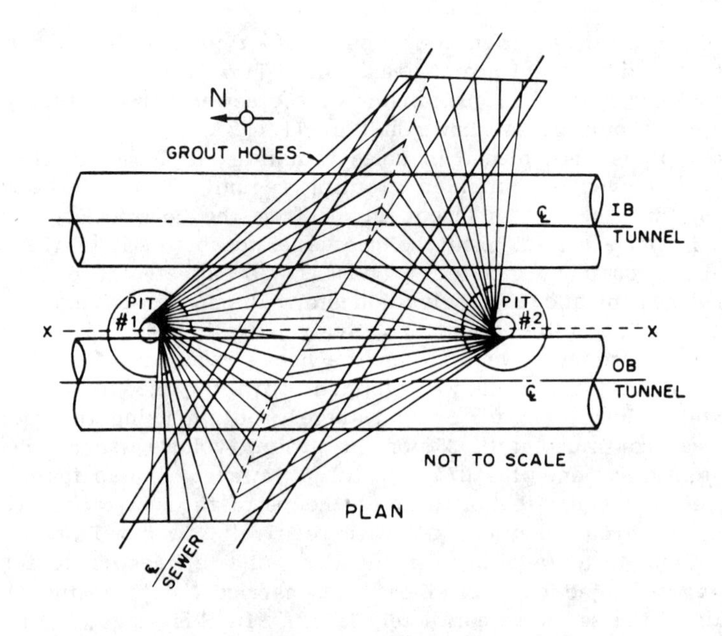

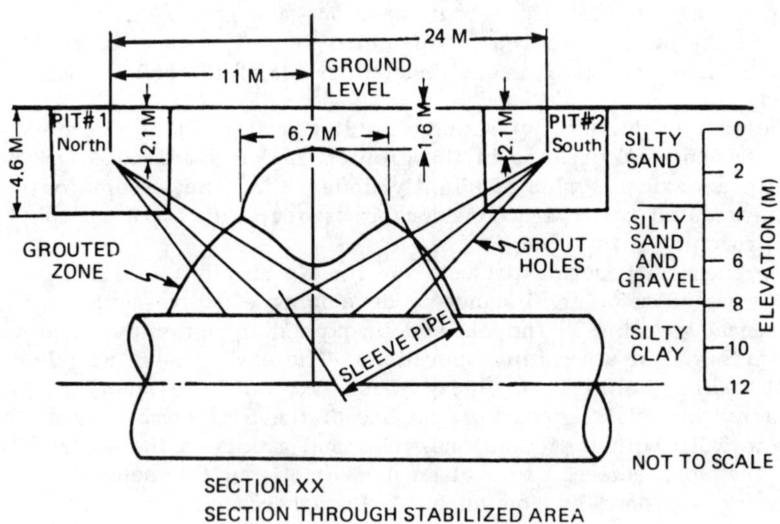

FIGURE 11.11 Grouting plan. (From Ref. 8.)

The instrumentation and field measurements taken during the metro grouting projects have verified the usefulness of chemical grouting in subsurface excavation for the control and reduction of surface settlements (this work also substantiated the results of computer research projects, discussed in Chap. 14). Typical data for similar grouted and ungrouted zones are shown in Figs. 11.12 and 11.13. Such data, combined with soil profile information, led to the following general conclusions:

1. If the soils in the upper half of the tunnel cross-section and above the crown are groutable, very good ground movement control can be achieved by chemical grouting.
2. If the soils in the upper half of the tunnel cross section or above the crown are predominately ungroutable, grouting of intervening sandy layers can still help control ground movements. However, grouting will not be as effective as where the stabilized zone surrounds the tunnel. Adequate ground control can be obtained by grouting only in the tunnel face.

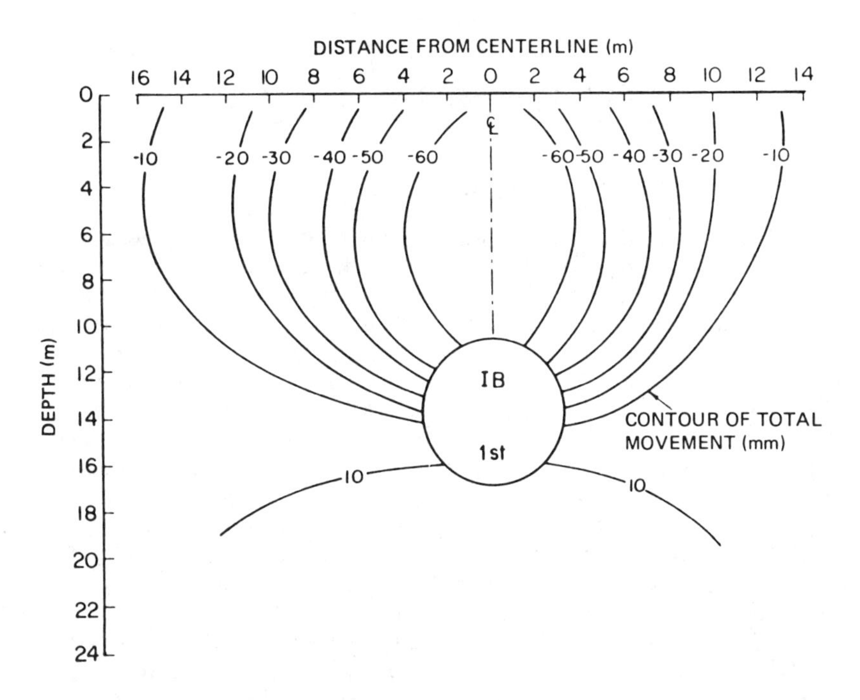

FIGURE 11.12 Vertical settlement contours, grouted zone. (From Ref. 8.)

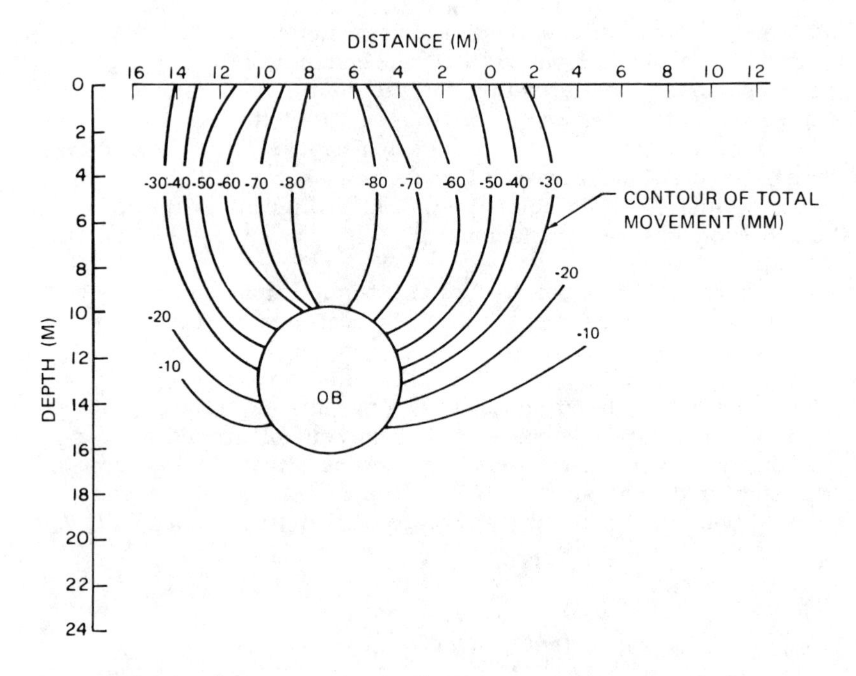

FIGURE 11.13 Vertical settlement contours, ungrouted zone. (From Ref. 8.)

The instrumentation also produced the first published field data verifying the creep phenomenon which lab work had uncovered several decades ago (see Ref. No. 3, Chapt. 4). Data shown in Fig. 11.14 illustrate the time rates of settlements observed in grouted and ungrouted zones. Creep (or relaxation) occurs because the tunneling operation leaves a gap between the tunnel liner and the unexcavated soil. In ungrouted zones, this gap is filled rather quickly by settlement of the soil, which may extend to the surface. In grouted zones, considerable time may elapse before the grouted soil fills the gap. If the gap is filled (with cement grout or other solids) within a short time after its creation, surface settlements can be eliminated or reduced. The data point out the necessity for grouting the gap expeditiously and by inference for supporting grouted tunnel faces which are not being excavated.

The research and development work done on the Washington Metro System was aimed primarily on field procedures for keeping surface settlements to a minimum. Thus, the design procedures developed were for the same purpose, and related specifically to

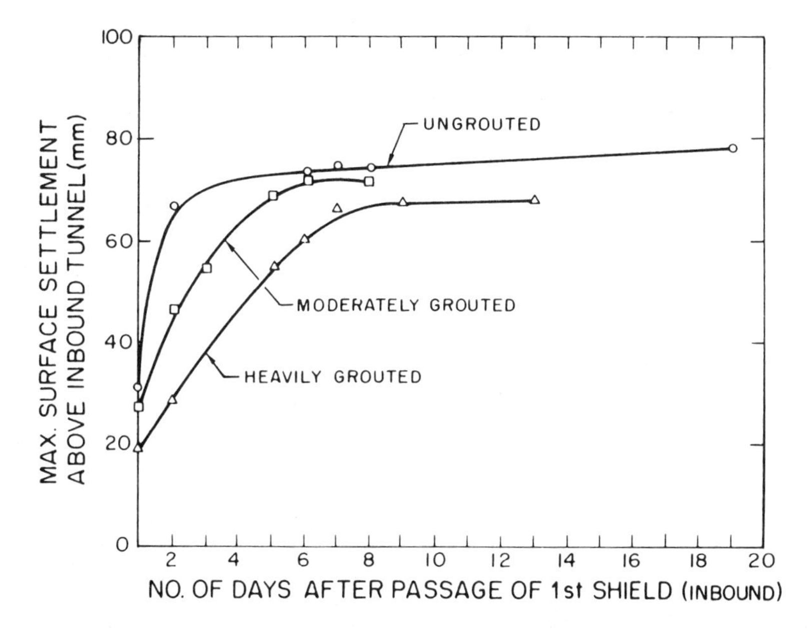

FIGURE 11.14 Creep of grouted soil around tunnel opening (original missing). (From Ref. 8.)

the one-size (7 meter) tunnel used in the metro system [9]. In order to extend the utility of the design procedures, modifications are suggested for other sizes of tunnels.

Due to the limited applicability of the originally suggested design procedures, i.e.,

1. Tunnel cross section is uniform granular deposit
2. Grouted zone is a square, surrounding the cross section uniformly
3. Control criterion is surface settlement

simplification seems to be in order. The following suggested procedure is derived from data presented in [9]:

1. Select a trial value of grouted zone thickness (defined as half the difference between the side of the square grouted zone and the tunnel diameter.
2. From Table 11.1 select the range of average principal stress difference (APSD) for the depth of the tunnel axis, and the trial thickness.

TABLE 11.1 Average Principal Stress Differences

Depth (ft)	Range APSD (psi)			
	t (ft)			
	3	6	9	12
20	26–35	19–23	16–18	14–25
30	40–52	28–34	23–27	20–23
	53–68	37–44	30–36	27–31
	65–85	47–56	39–45	35–39
	80–105	57–67	47–55	42–47
	92–125	67–97	55–63	49–54

TABLE 11.2 Tunnel Diameter Factors

Diameter (ft)	Factor
10	0.23
12	0.31
14	0.41
16	0.52
18	0.63
20	0.77
22	0.92
24	1.08
26	1.25
28	1.44
30	1.63

3. Select a target value for triaxial creep strength of the grouted zone.
4. Use the ratio of APSD divided by grouted strength to get a surface settlement index number. For values less than 0.5, use this number directly in step 5 below. For values between 0.5 and 2.0, multiply this number by 1.8 for loose sands, and by 0.5 for dense sands, prior to use in step 5.
5. Multiply the index number found in step 4 above by the factors given in Table 11.2 for various tunnel diameters, to give approximate surface settlements in inches.
6. If the indicated settlements are excessive, repeat the process using either/and greater grouted thickness or greater target strength.

Caution must be exercised in the use of this data for field design, since the procedure is based on computer study and checked in the field under specific, limited conditions only [9]. The indicated settlements should be considered as indicating a general magnitude of probable settlement, and only then within the limits of the original conditions.

11.5 GROUT PATTERNS

Shallow tunnels are often grouted from the surface, so that grouting and mucking can go on simultaneously. Grout patterns would normally be rows and columns forming squares, with the spacing between holes determined by local conditions (generally under 5 ft). Figure 11.15 (from advertising literature of Soletanche) shows a section through a vertical grouting pattern. Also shown is a grouting pattern for an adjacent tunnel, where drilling footage can be saved by grouting from the completed tunnel excavation.

When conditions preclude working from the surface, and when tunnels are very deep, grouting must be done from the tunnel face. The classical grouting pattern for face grouting is shown in Figure 11.16, for the Seikan Tunnel, in Japan [10], one of the largest tunnel grouting projects ever done.

11.6 SEIKAN TUNNEL

Preliminary studies for a tunnel to connect the islands of Hakkaido and Honshu began in 1946. The major purpose of the tunnel was to eliminate a 4 hour ferry crossing through treacherous seas, before the ferry system became unable to meet growing transportation requirements. Construction began in 1971, and was completed

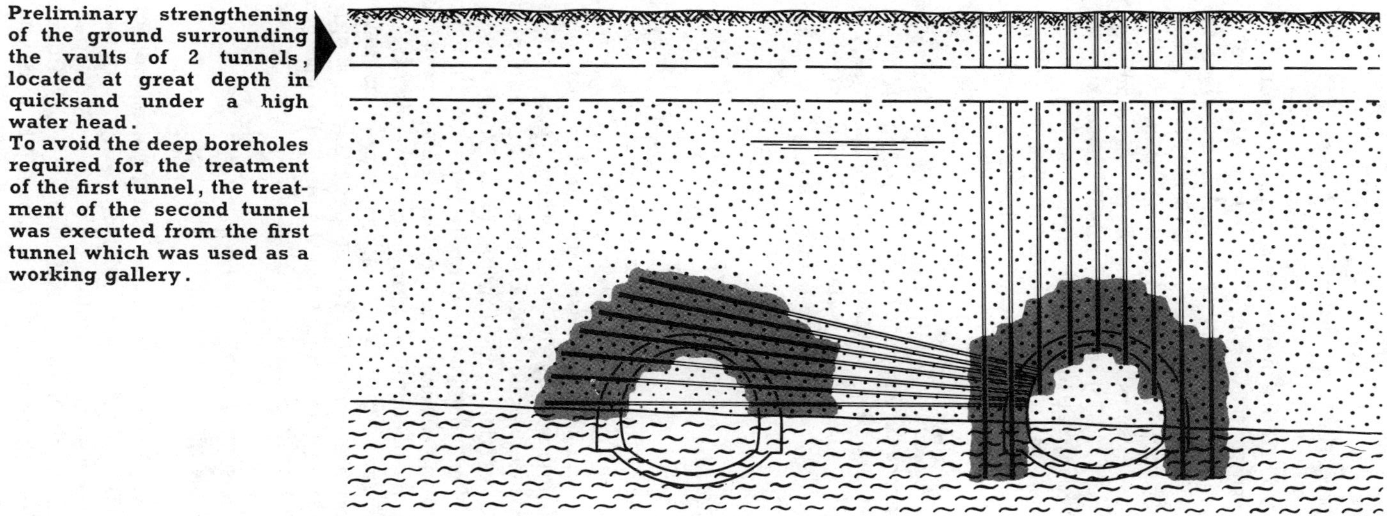

Preliminary strengthening of the ground surrounding the vaults of 2 tunnels, located at great depth in quicksand under a high water head.
To avoid the deep boreholes required for the treatment of the first tunnel, the treatment of the second tunnel was executed from the first tunnel which was used as a working gallery.

FIGURE 11.15 Grout patterns for two adjacent tunnels. (Courtesy of Soletanche advertising literature, Paris.)

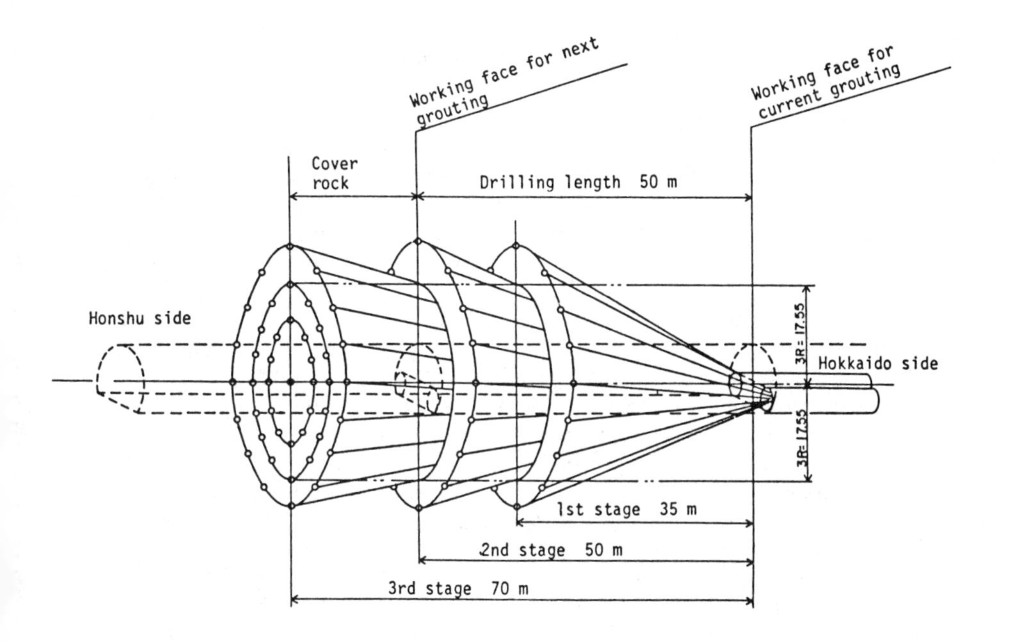

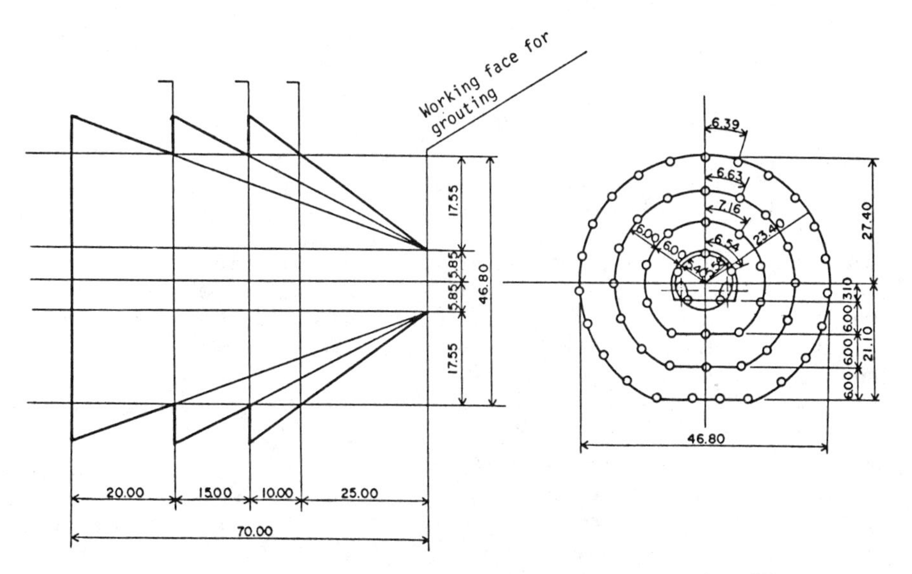

FIGURE 11.16 Grout patterns for the Siekan Tunnel. (From Ref. 10.)

12 years later. Currently, the trip from Tokyo to Sapporo (due to bullet trains) takes under 6 hours instead of the 14-plus hours previously required.

The overall length of the tunnel is 33.5 miles, of which 14.5 miles are under the sea, the longest undersea tunnel ever constructed. Its depth, under the sea bed, is 100 m. Since it was meant to accommodate the bullet train, grade is limited to 12 in 1000, and radius of curvature is limited to a minimum of 6500 m.

The main tunnel has a diameter of about 11 m, and was preceded by two smaller tunnels, a service tunnel 4 to 5 m in diameter, and a pilot tunnel 3.5 to 5 m in diameter, placed first. The pilot tunnel was used to assess the geology to be traversed. All three tunnels were preceded by exploratory drill holes.

Grouting was done along the entire tunnel length, both to provide structural support and reduce water inflow. In the land portions of the tunnel, cement was the main grouting material. Under the sea, sodium silicate-based grout was the primary material. Most of the work was done at short gel times, 10 min or less. Colloidal cement was used with the silicate for catalysis, in part of the tunnel.

The grout pattern is shown in Fig. 11.16. The thickness of the grout ring surrounding the tunnel varied between 3 and 6 tunnel diameters, depending on the upgrouted formation strength.

Despite all the precautions taken, flooding did occur on several occasions. These accidents were controlled, by predetermined procedures, and the completed tunnel is functioning well. Additional details of tunnel design construction and operation can be found in the publications listed in the Bibliography.

11.7 SHAFT GROUTING

Shaft grouting differs from tunnel grouting in one major aspect. In tunnel grouting, to prevent caving, often only a half circle need be grouted to solidify an umbrella above the excavation. In shaft grouting, either for strength or water cutoff, a complete circle must generally be grouted.

One of the earliest uses of the newer chemical grouts was in the lead mines at Viburnum, Missouri, in 1958 [11]. This work verified the performance of low-viscosity grouts in zones of high static groundwater pressure as well as the use of two-row grout patterns. Several years later, similar work was done in the southwestern part of the United States. This job is described by the following excerpts from the field engineer's report (the comments in footnotes have been added by the author):

During the third week in March the shaft sinking operation in Moab [Utah] bottomed out at 1434 feet below grade. A known high pressure, Arkose, sandstone formation existed at 1484. The pressure within this formation is approximately 625 psi. It was estimated that the inflow to the shaft would exceed 3000 gallons per minute if grouting was not performed. During the next eight weeks approximately 44,000 sacks of cement were pumped into the sandstone strata from 1484 feet to 1516, or a total depth of 32 feet. The engineer's personnel felt that most of the cement had penetrated the upper contact zone and refused permission to start sinking.

At this point the writer was contacted by the owner for assistance.

The following points were established during a visit to the job site:

1. Total depth of sandstone—32 feet
2. The upper contact zone made the most water.
3. The formation pressure was approximately 625 psi.
4. The average voids in the sandstone was 20 percent.
5. The formation water was a saturated brine solution with a temperature of 80°F.
6. The mix water also contained some salt.
7. It required a pump pressure of approximately 1500 psi to pump water into the formation at the rate of five gallons per minute.

After a review of the above facts I layed out two grouting patterns. One pattern involved forty holes on three foot centers and the other involved thirty holes on four foot centers. They chose to limit their drilling and use more material and proceed with the thirty hole pattern.*

Consultants retained by the owner recommended that a grout curtain be constructed six and one-half feet outside the limits of the open excavation because of the high pressure within the formation. The O.D. of the shaft was to be twenty-four feet; therefore, the grout curtain was laid out on the basis of thirty-seven feet.†

---

*Based on an economic study of drilling costs versus chemical costs.

†Thus, none of the grouted zone would be excavated during shaft sinking.

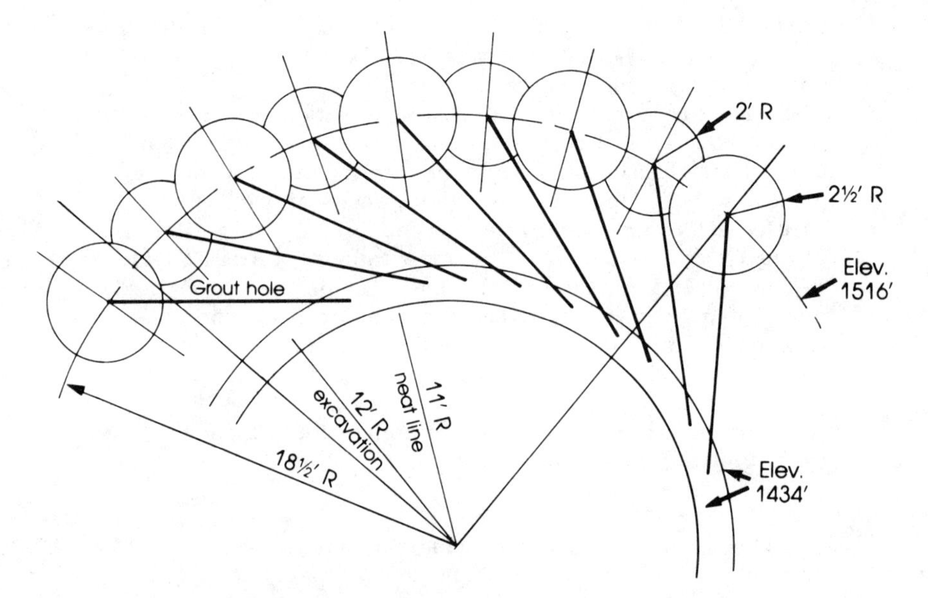

FIGURE 11.17 Shaft grouting pattern.

We laid out a grouting pattern [see Fig. 11.17] with holes on four foot centers with a volume of grout* to provide a five foot stabilized diameter on even numbered holes and a four foot stabilized diameter on odd numbered holes.† Because of excessive water at the contact we decided to split the volume in three parts, one-third at the contact, one-third in each of two sixteen foot stages. These volumes were rounded out to even forty gallon batches because of the physical makeup of the mixing and chemical grouting plant.

Five foot diameter stabilized holes were to receive eight forty gallon batches, or a total of 320 gallons at the contact and each of two stages within the Arkose (960 gallons per fifteen holes = 14,400 gallons).

---

*An acrylamide grout was used.

†Holes were not vertical. They were actually drilled with a radial dip of 1 in 10 and a spin of 20° (dip is measured in a vertical plane plassing through the shaft centerline; spin is measured in a vertical plane tangent to the grout pattern circle at each hole).

Four foot diameter stabilized holes were to receive five forty gallon batches, or a total of 200 gallons at the contact and each of two stages within the Arkose (600 gallons per fifteen holes = 9,000 gallons).

Sum Total for the job = 23,400 gallons.)

Control tests were run in the laboratory with mix and formation water to establish component concentrations.

It is interesting to note that the mix water naturally inhibited the gel formation and the formation water accelerated same. We had a pot life with the mix water of approximately one hundred minutes instead of a graph value of seventy minutes. We expected to pressure off on the holes because of the salt and temperature of the formation water. This condition did not seem to cause us any problems in the field.

The grouting commenced on June 5 and was completed on June 12. All holes in the five foot-four foot pattern were handled separately. The five foot holes were drilled to the contact and grouted followed by the four foot holes to the contact and grouted. This procedure was followed until all six stages were completed, three on the four foot holes and three on the five foot holes. Pumping rates varied between 5 and 7 gpm, with pumping pressures generally in the range of 1750 psi. Several holes pressured off at the pump capacity, 2000 psi. In some of these holes, it wasn't possible to inject the entire planned volume. This was compensated for by injecting more into the next stage, or into adjacent holes.

A report from the job site on June 18 indicated that the sinking operation had penetrated the contact zone with a total flow of less than two gallons per minute.*

A similar problem was solved in a similar fashion at Midlothian, Scotland, for coal mining. Two shafts were to be sunk to 3000 ft. One would be used as a main hoisting shaft with ships, and the other would hose a conventional cage hoist.

The circular shafts would have a finished diameter of 24 ft and would be lined with concrete with a minimum thickness of 12 in. In areas where excessive amounts of groundwater were encountered the thickness was increased up to 3 ft to withstand the hydrostatic pressure.

---

*Later reports showed a total inflow of 6 gpm for the total 32 ft depth of sandstone.

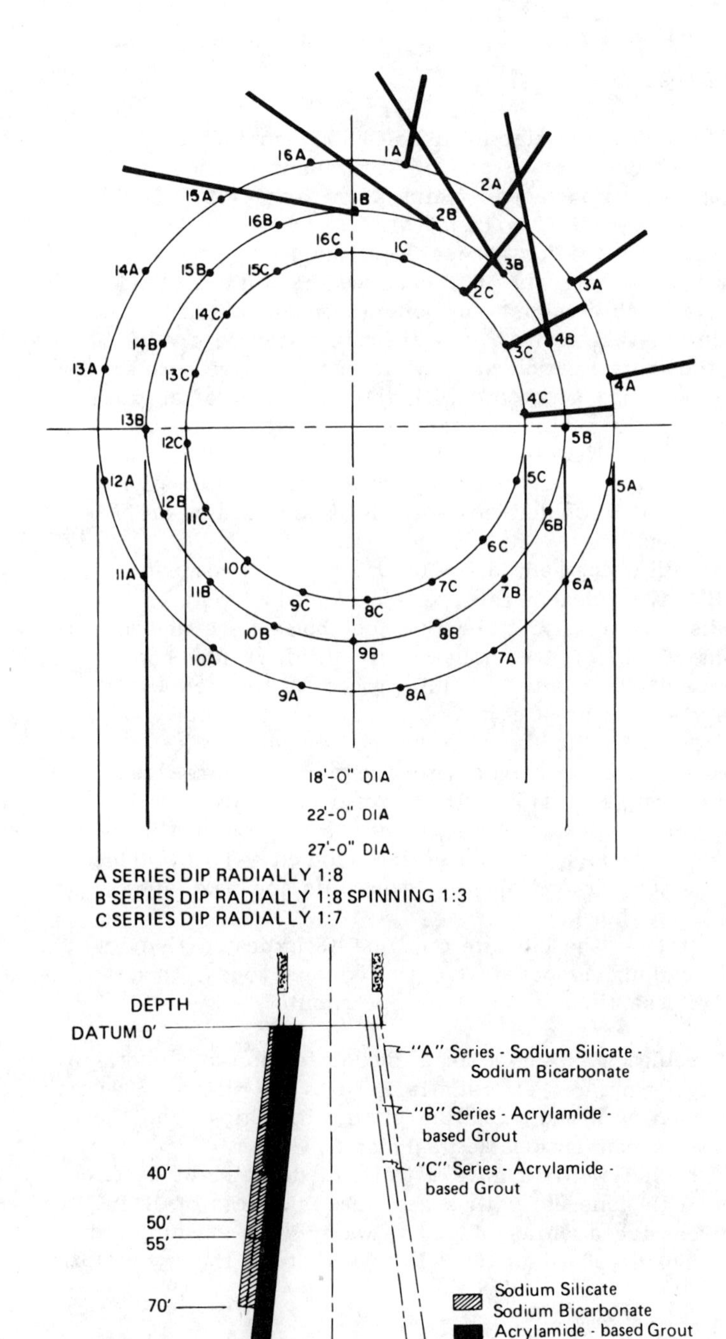

FIGURE 11.18 Shaft grouting pattern.

Slightly below the 2000 ft level porous sandstone was encountered. The inflow of water at this level was great enough to threaten completion of the shafts. The sandstone stratum was about 100 ft thick.

Water flow into the sump from open holes drilled 60 ft below the top of the sandstone layer was measured at 144 gpm. At this point, the need for grouting became apparent. Engineers working on the project determined that conventional grouts, such as cement and cement combined with silicates, were unsatisfactory in some of the sandstone's strata.

Treatment of the sandstone bed to this entire depth was carried out from the bottom of the shaft through injection into a series of holes, as shown in Fig. 11.18.

The first series of 16 holes was drilled on a 27-ft-diameter circle from the bottom of the shaft to 20 ft depth. The dip was 1 in 8. The second series also was drilled with the 1 in 8 dip but was spun at 1 in 3, a diameter of 18 ft.

The first series was drilled to 40 ft after standpipes were inserted and proved. They were then treated with acrylamide chemical grout. The operation was designed to determine the best techniques. A total of 3470 gal of grout was injected into the 16 holes.

The holes were then redrilled to 55 ft and grouted with a sodium silicate–sodium bicarbonate solution. They were then redrilled a second time to 70 ft and injected with the silicate–bicarbonate mixture. This grout was used to close up any large fissures in the sandstone and prevent undue loss of the acrylamide grout.

The second series of holes was then drilled to 40 ft, and each was injected with 100 gal of acrylamide grout. The gel time was set at 0.5 h. The injection rate was 2 to 4 gpm. Pressures at the end of the injection ranged from 1300 to 2500 psi.

A third series of holes was drilled to 20 ft within the other two series. After standpipes had been inserted and proved, the holes were deepened to 50 ft and treated. The volume of grout ranged from nil to 300 gal. The third series was then deepened to 90 ft and again treated with the acrylamide.

When the shaft was sunk through the entire stratum of sandstone, it was found that water inflow from the walls was about 40 gpm, of which at least two-thirds was believed to come from the ground above the sandstone layer or from the upper portion of the sandstone in which the injection pipes had been inserted but not treated with chemical grout. It was felt that the water inflow from the treated area of the shaft had been reduced 95%.

11.8 SUMMARY

Tunnels and shafts whose construction is hindered or halted by water inflow can be treated very effectively by chemical grouting.

For tunnels, grouting should be included in the initial construction and financial planning as a contingency measure to control soft ground movement at the tunnel face and to limit surface settlements above the tunnel alignment. For shafts, if they are deep, it is almost certain that water problems will be encountered. Treatment procedures should be built into the design specifications. Computer studies of the effects of grouting in differing soil profiles, verified by field data, have given good insight into the results that can be expected from grouting.

While tunnel grouting does not require a complete ring of grout, shaft grouting does require a complete curtain around the excavation. Tunnels are often shallow enough so that other construction procedures may be used to control groundwater flow, but deep shafts cannot be controlled by well points, slurry trenches, compressed air, or wells. Freezing is usually the only other alternative to grouting, and grouting must often be resorted to after freezing to correct the porosity resulting from pouring concrete against frozen ground.

REFERENCES

1. Three uses of chemical grout show versatility, *Eng. News Rec.* (May 31, 1962).

2. Chemical grout licks tough quicksand problem, *Railw. Track Struct.* (Feb. 1964).

3. Tough tunnel bows to chemical grouting. *Eng. News Rec.* (March 29, 1962).

4. How chemical grouting keeps sewer tunnel dry, *Constr. Methods Equip.* (May 1962).

5. Chemical grout stops water in dump fill with 70 percent voids, *Civ. Eng.* (Sept. 1962).

6. Tunnel interface crossing solved by specialized consolidation, *Ground Eng.* (Sept. 1972).

7. Bibliography on Grouting, U.S. Army Engineer Waterways Experiment Station, Miscellaneous Paper C-78-8.

8. C. W. Clough, W. H. Baker, and F. Mensah-Dwumah. Development of Design Procedures for Stabilized Soil Support Systems for Soft Ground Tunnelling, Final Report, Oct. 1978, Stanford University, Stanford, California.

9. D. Y. Tan and G. W. Clough, Ground control for shallow tunnels by soil grouting, *J. Geotech. Eng. Div. ASCE* (Sept. 1980).

10. A. Kitamura and Y. Takeuchi, *J. Constr. Eng. Manag. ASCE* (March 1983).

11. J. J. Reed and L. B. Bilheimer, How research advances grouting techniques at St. Joseph lead, *Min. World* (Nov. 1960).

BIBLIOGRAPHY

Pilot bore is test arena for world's longest railway tunnel, *ENR*, May 14, 1970, pp. 30, 31.

Japanese tackle water to drive record tunnel, *ENR*, April 4, 1974, pp. 16, 17.

Flooding sets back work on record Japanese tunnel, *ENR*, May 13, 1976, p. 16.

Tough tunnel job now going smoothly, *ENR*, October 11, 1979, pp. 26, 27.

Record undersea bore in homestretch, *ENR*, June 17, 1982, pp. 78, 79.

Innovations at Seikan, *Tunnels & Tunnelling*, March 1976, pp. 47–51.

Tunnelling in Japan, *Tunnels & Tunnelling*, June 1978, pp. 19–22.

Constructing the undersea section of the Seikan Tunnel, *Tunnels & Tunnelling*, September, 1982, pp. 31–33.

Seikan pilot tunnel opens way for Japan's 23 km undersea rail link, *Tunnels & Tunnelling*, July 1983, pp. 24, 25.

Seikan undersea tunnel, *ASCE Journal*, September 1981, pp. 501–525.

Seikan Tunnel, *ASCE Journal*, March 1983, pp. 25–38.

Japanese tunnel design: Lessons for the U.S., *ASCE Journal*, March 1981, pp. 51–53.

Grouting: More than meets the eye, *Tunnels & Tunnelling*, July 1983, pp. 41–43.

12

Special Applications of Chemical Grouts

12.1 INTRODUCTION

In addition to field jobs, which obviously fall into the categories discussed in previous chapters and whose performance uses techniques and equipment discussed in early chapters, there are uses for chemical grouts that are either not obvious or else use very special procedures and equipment. Jobs in this category include the sealing of piezometers, the sampling of sands, sealing bolt holes in underground corrugated pipe, and eliminating infiltration or exfiltration through underground concrete and vitreous pipe. The last category includes storm and sanitary sewer lines, and this very special work has grown in volume and importance over the past two decades to become the major use of acrylate grout in the United States.

12.2 SEWER LINE REHABILITATION

In many of the older cities, a large percentage of the sewer lines are well over the 40- to 50-year life normally anticipated. Recent studies indicate that in such areas, well over half of the total sewer line flow is from groundwater infiltration [1]. Other studies aimed at cost estimates for joint repair concluded that an excessive infiltration source (one or more joints) totalling 1 to 3 gpm would typically occur every 50 to 100 ft between manholes.

The development of the sewer sealing industry was brought about by the increasing concern over environmental pollution, most specifically the old practice of dumping excess liquid (liquid beyond the capacity of the treatment plant to handle) untreated into streams and rivers. In many cases the high volume of liquid flowing into the treatment plant was due in part to infiltration of groundwater through leaky joints and fractured pipe. (See Ref. 2 for a discussion of inflow evaluation.) If the location of the infiltration zones could be determined, economic analysis could be used to determine whether the line should be repaired or the treatment facility enlarged.

Repairing of sewer lines in the 1940s meant digging up and replacing the pipe—an expensive and inconvenient procedure, especially in cities. With the development of the new chemical grouts in the 1950s, a quicker and much less expensive alternative became available. Grout pipes driven from the surface could be used to impermeabilize the zone surrounding the leak, thus preventing infiltration. Of course, successful work of this kind depended on knowing the exact location of the leak. This necessity was a spur to refinement of sewer survey techniques and contributed to the rapid growth of video equipment for sewer inspection. It soon became obvious that the same mechanical equipment which guided a TV camera through a sewer line could also guide a packer through the line, thus permitting grouting from within the pipe.

The equipment necessary to do internal sewer grouting efficiently is sophisticated and expensive. Private companies made the major contributions to the developing technology, all of which was originally based on the use of acrylamide grout. Two major contributors were Penetryn Systems, now a division of Carborundum,* and National Power Rodding Company through its subsidiary Video Pipe Grouting. Both companies are still active in contracting and in continued technological development.

A full set of sewer grouting equipment may cost $50,000 to $100,000. It has been estimated that as many as 600 such equipment sets are in operation in the United States. The large scope of the industry is apparent from that estimate. Equipment, usually housed in a large van, consists of the video control panel and pumping plant, a (small, self-illuminated TV camera, and a double pneumatic packer together with the cables to pull the camera and packer between manholes, and air and liquid hoses to operate the packer and pump the grout. Hoses, in 500 ft or 1000 ft lengths, are stored in the van on reels. Chemical grout, when pumped, always goes through the full length of the hose. This explains the necessity for a very low-viscosity grout: to minimize pumping effort.

Pumping of grout is generally done with air pressure (closed tanks are used) using valves and flow meters to control the grout and catalyst ratio as well as the gel time and volume. (Pumps can of course also be used. However, the industry developed around

---

*Since manuscript preparation, Penetryn has changed owners, and currently operates under the name CUES.

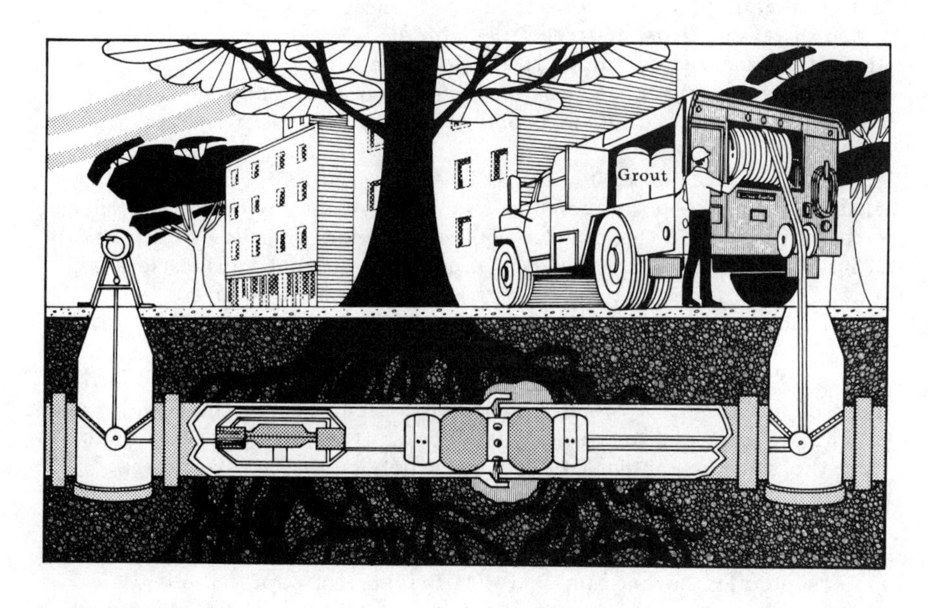

(a)

FIGURE 12.1 Illustration of grout application (a) and diagram of grouting equipment (b). TA_1 = 50-gal, stainless steel mixing tank for chemical grout; TA_2 = 50-gal stainless steel mixing tank for catalyst; SG = sight gauges; P_1,P_2 = 5–20-gpm 40–180-psi stainless steel piston-type position-displacement pumps; P_3 = 2 1/2–7-gpm 200–900 psi stainless steel, piston-type position-displacement pumps; A = air or electric agitator; C = 180-psi 30-cfm compressor; V_1 = quick opening valves; G = diaphragm pressure gauges; H_1 = 1/2-in. 3000-psi pressure hose; H_2 = 1/2-in. 1000-psi air hose; V_2 = spring-loaded check valves; Q = quick disconnectors; R = rotary connections; W = water tank; F = 1/16-in.-mesh inline filter.

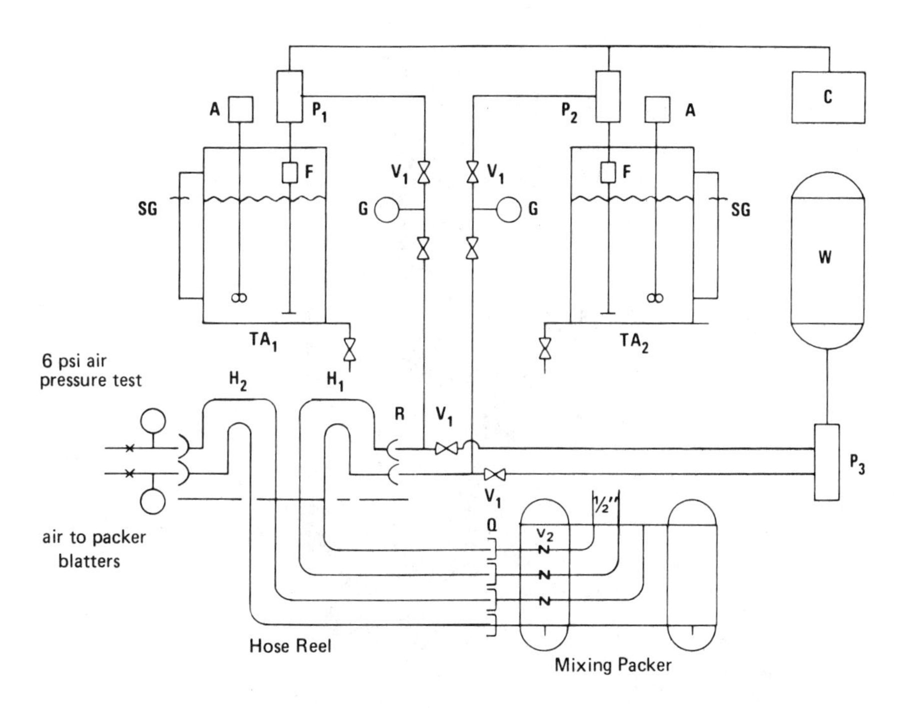

(b)

FIGURE 12.1 (Continued)

pneumatic systems.) Controls can range from very simple hand-operated valves to pneumatic or electronic controllers, preset to give desired results. The total system in operation is shown by the drawing in Fig. 12.1a. Figure 12.1b shows the complexity of the injection control equipment. The TV camera inspects the pipe and joints for cracks and infiltration. Typical of the problems which occur are photos of the TV monitor, Figs. 12.2 and 12.3, showing infiltration at a joint (the most common problem) and a fracture in the pipe. When infiltration is seen, the packer behind the camera is pulled up and located so the two inflatable sleeves straddle the leak. The sleeves are then inflated to contact with the pipe, isolating an annular space at the leak. Grout is then pumped into the annular space, through the leak itself, and into the soil surrounding the pipe. When the grout solidifies, the stabilized soil mass prevents further infiltration.

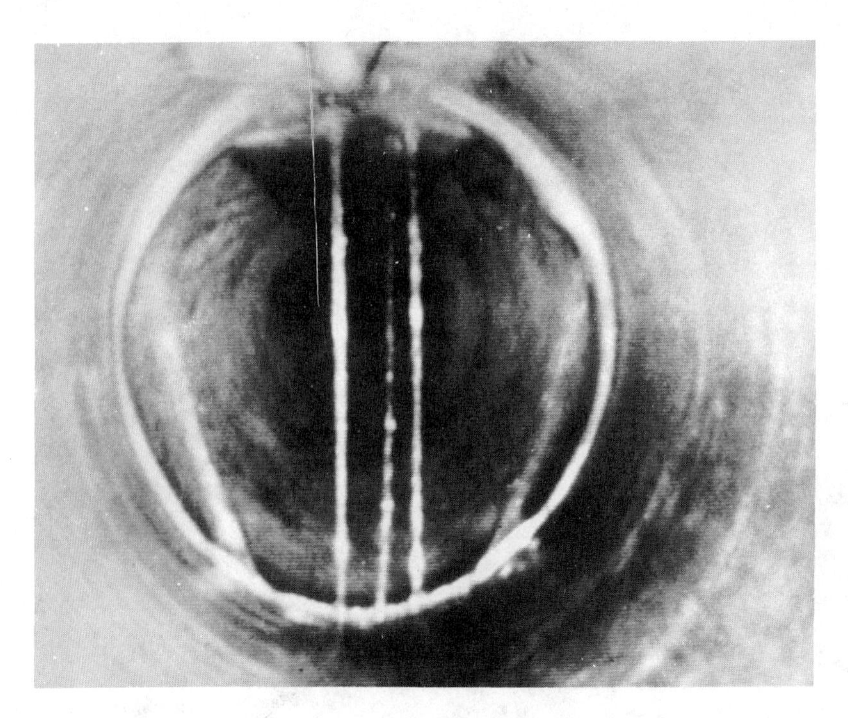

FIGURE 12.2 Video view of leaking sewer line joint. (Courtesy of Penetryn System, Inc., Knoxville, Tennessee.)

Contactors generally have packers to fit all pipe sizes between the 6 and 24 in. sizes. Special equipment has been built for 4 in. pipe as well as for large pipe. In pipe over 30 in. in diameter, however, people can work inside the pipe, and video equipment and packers are not needed.

On the smaller sizes of sewer pipe, of which there are literally thousands of miles and millions of joints in the United States, very small grout quantities (2 to 5 gal) are generally sufficient to seal a leaking joint. Gel times normally used are very fast, 10 to 30 sec, so that in a minute or two after pumping has stopped, the packer can be deflated and the treated joint examined with the TV camera to ensure that the grouting was successful.

The initial requirements for a low-viscosity grout, and one which could also be gelled very quickly with excellent gel time control, had precluded the use of any but the acrylamide-based grouts for sewer sealing. The consternation that reigned in the industry following the withdrawal of AM-9 from the marketplace was short-

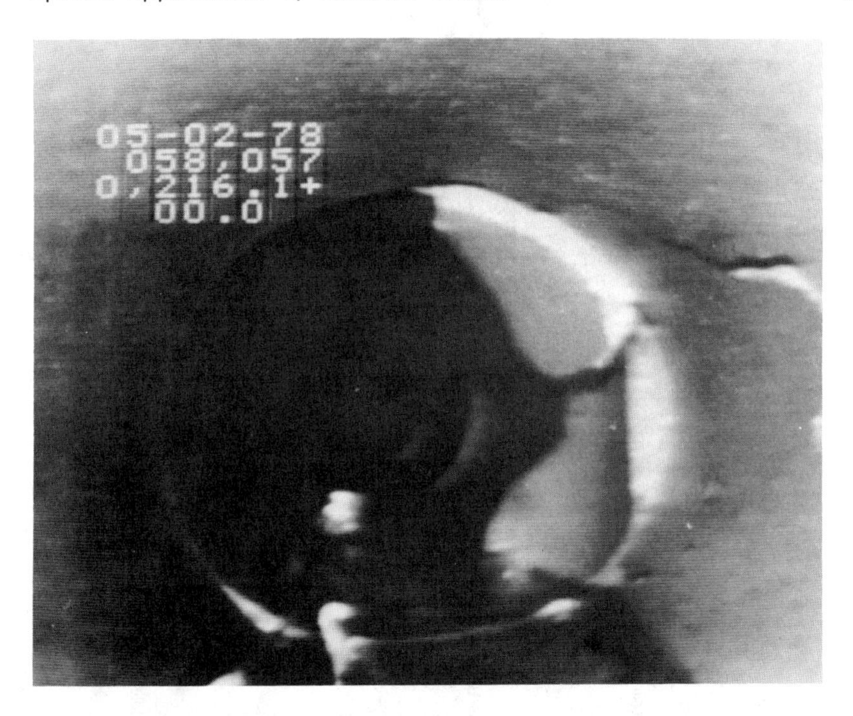

FIGURE 12.3 Video view of cracked sewer pipe. (Courtesy of Penetryn Systems, Inc., Knoxville, Tennessee.)

lived, as substitutes were quickly found. All these are acrylamide-based and present essentially the same environmental hazards as AM-9. There has been (in the United States) no official mandate banning the use of acrylamide grouts nor any public or private pressure in that direction. Efforts are underway to control the use of these materials by training and inspection procedures so as to reduce the risk to personnel of acrylamide intoxication to acceptable levels. At present, acceptable levels are considered to include application procedures that minimize exposure, regular medical inspection to detect the earliest symptoms of intoxication, and of course, termination of exposure if such symptoms are found. (At this stage, intoxication effects are reversible.) The risk of environmental pollution from the sewer sealing process is negligible because of the nature of the process: limited grout quantities, short gel times, and immediate inspection of results.

Since its introduction in the early 1980s, AC-400 (acrylate) has gradually been replacing acrylamide in sewer-sealing applications.

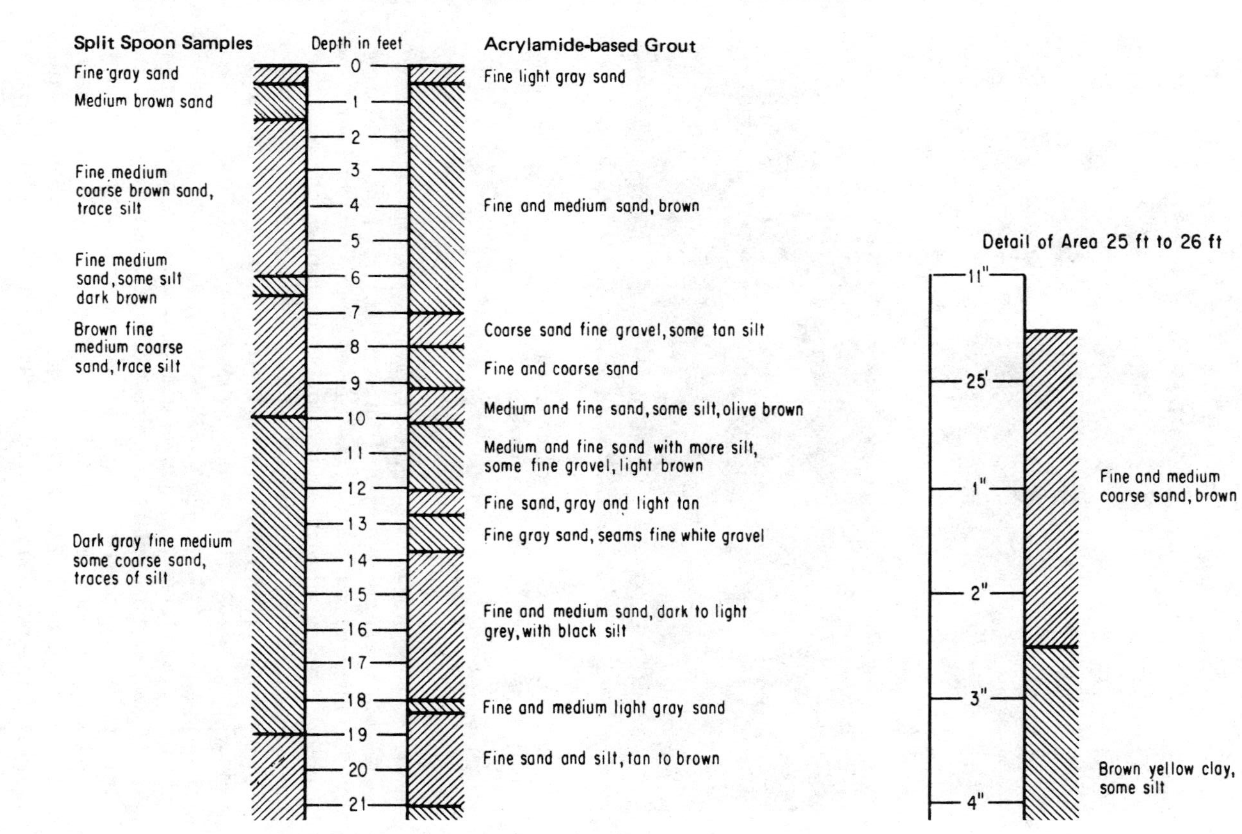
Chemical Grout Method Vs Split Spoon Samples
Split Spoon Samples
Depth in feet
Acrylamide-based Grout
Fine gray sand
Medium brown sand
Fine medium coarse brown sand, trace silt
Fine medium sand, some silt dark brown
Brown fine medium coarse sand, trace silt
Dark gray fine medium some coarse sand, traces of silt
0
1
2
3
4
5
6
7
8
9
10
11
12
13
14
15
16
17
18
19
20
21
Fine light gray sand
Fine and medium sand, brown
Coarse sand fine gravel, some tan silt
Fine and coarse sand
Medium and fine sand, some silt, olive brown
Medium and fine sand with more silt, some fine gravel, light brown
Fine sand, gray and light tan
Fine gray sand, seams fine white gravel
Fine and medium sand, dark to light grey, with black silt
Fine and medium light gray sand
Fine sand and silt, tan to brown
Detail of Area 25 ft to 26 ft
11"
25'
1"
2"
3"
4"
Fine and medium coarse sand, brown
Brown yellow clay, some silt

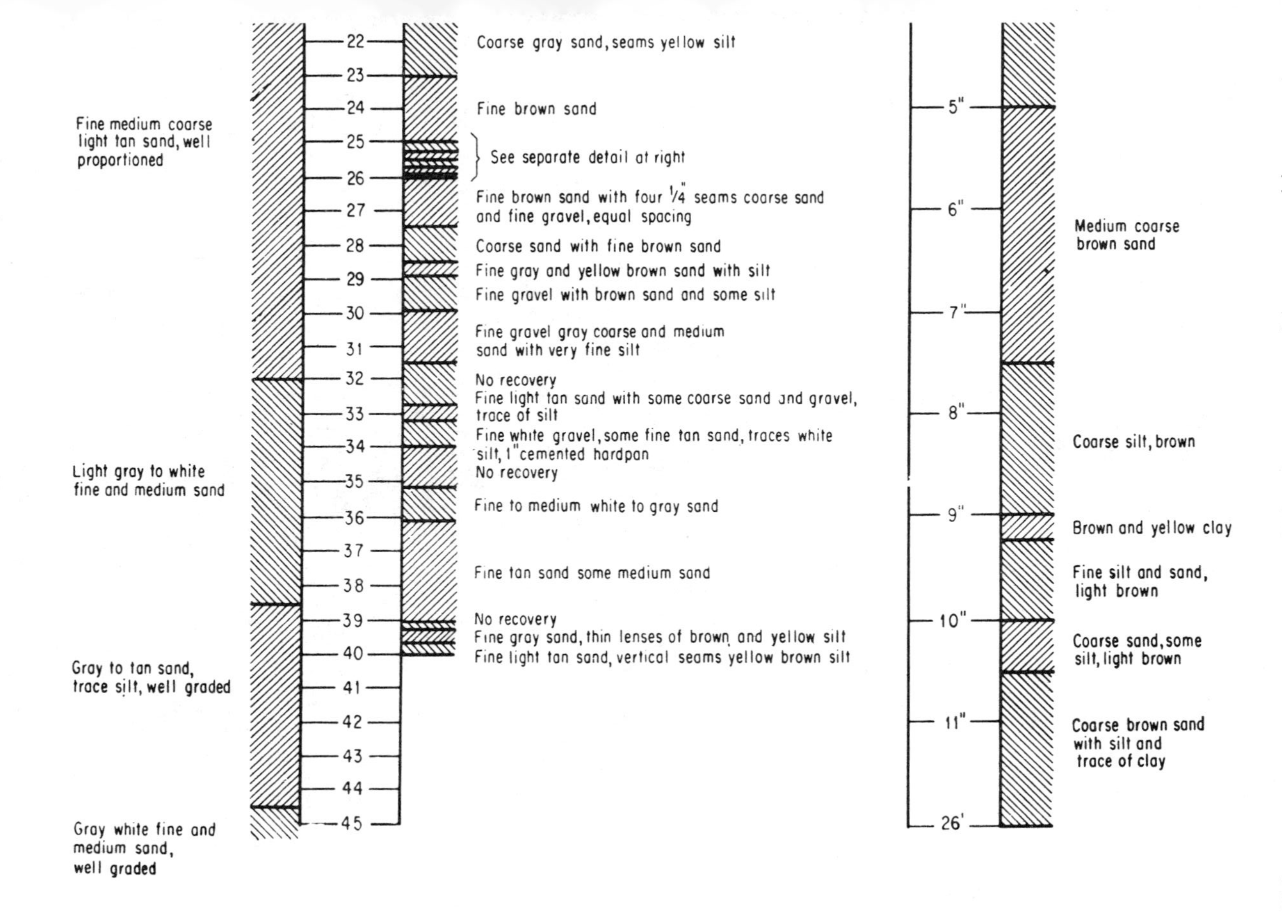

FIGURE 12.4 Boring log comparison.

Other acrylate grouts are also claiming a share of this market, and it is reasonable to assume that acrylamide will gradually be phased out.

Several years before the withdrawal of AM-9 from the market, 3M Company had launched a new product for sewer sealing, using a somewhat different approach. This was a water-reactive polyurethane, whose high viscosity precluded penetration into the soil surrounding the sewer joint. However, the product would fill any open space between the bell and spigot, effectively forming a watertight gasket in place. This process is still in use today, with improved products currently called Scotch Seal Chemical Grouts 5600, 5610, and 5620. These are foaming materials, and despite the high initial viscosity, the mixed foam does have the limited ability to penetrate soil formations.

Further details related to sewer grouting can be found in Refs. 3–6.

12.3 SAMPLING OF SANDS, IN SITU DENSITY

A very specialized use of chemical grouts is for the sampling of granular deposits when it is important to have precise definition of the total profile. In normal split-spoon sampling of materials without cohesion, the very thin strata generally lose their identity, and there is always some degree of mixing and dislocation at the interface between strata. If the soils are chemically grouted prior to sampling, all these problems disappear. The soil must, of course, be grouted ahead of the sampling operation, either in stages at the bottom of the hole, followed by sampling in stages, or continuous sampling may be done a few inches away from a vertically grouted drill hole. The improvement in identification and delineation of soils is shown in Fig. 12.4, a comparison between a boring log taken by standard split-spoon samples and the sampling of stabilized soils. (See ASTM D-1586, Standard Method for Penetration Test and Split Barrel Sampling of Soils, for a description of split barrel.)

The relative density of a granular formation is generally an important design factor when strength and settlement characteristics are involved. The standard penetration test gives data which when coupled with pertinent experience yield reliable estimates of relative density. Chemical grouts may also be used for this purpose.

When unconfined compression tests are performed on grouted granular materials, the short-term UC strength (Chap. 4) is a function of the relative density of the soil. The difference in strength between minimum and maximum densities becomes smaller as the strength of the grout increases and may be negligible for the high-concentration silicate grouts. For the elastic low-strength grouts, the difference is significant and readily measurable. Further, for the acrylic grouts in particular, the relationship between short-term UC strength and density is linear.

To use this relationship to determine in situ relative density, it is necessary to take both stabilized and normal samples of the soil back to the laboratory. Samples are grouted in the lab at minimum and maximum density. These samples are tested and two points plotted on an arithmetic graph of density versus UC strength. The UC value of the stabilized field sample is also determined, and this point is plotted between the other two (it usually falls on or close to the line joining the first two points). The position of the point representing the field sample determines the relative density.

Unfractured field samples of grouted soils are difficult to obtain. Chemical grouts are generally too weak to bind the grains against the disturbance caused by rotating or penetrating sampling tools. In addition, large particles caught under the advancing tool will generally rupture the surrounding grouted soil.

In fine, uniformly graded soils, large thin wall samplers pushed slowly and steadily offer the best chance for success. Close to ground surface, chunk sampling from test pits is most feasible.

In-situ density may also be determined from irregularly shaped field samples, by measuring the total volume by water displacement, then determining the specific gravity and volume of solids by standard test methods. To use this process, the grout in the soil voids must be eliminated. For the acrylics this is readily done by heating to the point where the gel vaporizes.

12.4 SEALING PIEZOMETERS

A very small and specialized use of chemical grouts is for sealing piezometers. These devices, which are placed in drill holes to measure hydrostatic pressures in the formation, must be isolated in the hole to keep them unaffected by pressures in other zones penetrated by the drill hole. Usual procedures make use of bentonite balls or pellets, which well in the presence of water to seal the drill hole with an impervious mass. The swelling takes time, however—as much as 36 h. The use of chemical grouts with a very rapid setting time can create a seal in seconds.

12.5 CONTROLLING CEMENT

Many additives, both active and inert, can be used with chemical grouts. One of these is portland cement, which is compatible with any grout having a pH of 8 or more. Generally, cement will

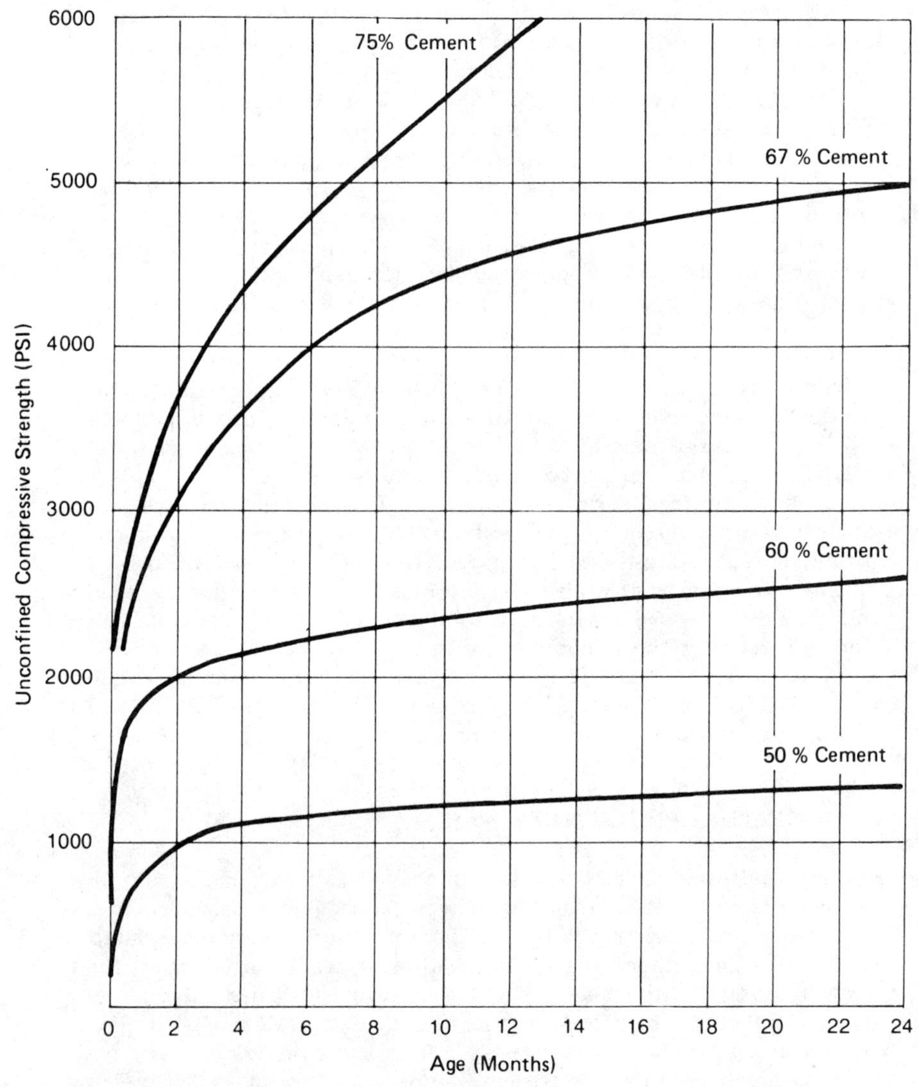

Acrylamide - based grout - 10%
Catalyst DMAPN (dimethylaminopropionitrile)- 0.4%
AP (ammonium presulfate) - 0.5%
KFe (potassium ferricyanide) (to give gel times of 8 -12 minutes)
Water - 89.1%
Bentonite - 1%
Portland cement (as indicated)

FIGURE 12.5 Strength of acrylamide–cement grouts (Bentonite and cement percentages are based on the weight of the grout mixture).

accelerate the gel time of the grout with which it is used, and if added in sufficient quantity, it will also increase the grout strength.

Cement grout is less expensive than chemical grouts and should be the first material considered for a grouting job. If the job requires good gel time control and/or short gel times or if cement will not penetrate the formation, chemical grouts must be used. If cement *will* penetrate the formation and high strength *is* required, chemical grouts can be used as additives to the cement to control setting times.

The silicates and acrylamides are commonly used with cement, and the combinations yield a high-strength grout with excellent gel time control. Unconfined compressive strengths as high as 6000 psi are attainable, as shown for acrylamide–cement combinations in Fig. 12.5. Comparable results can be obtained with acrylates and silicates.

A chemical plant in Virginia was built many years ago on a site underlain by limestone and troubled by sinkholes. When a new plant was built adjacent to the existing one, the foundation area to a depth of 50 ft was grouted with cement to prevent future sinkhole formation. Experience during this grouting program, carried out at 15 psi maximum pumping pressure, indicated that horizontal flow of grout from early injections generally exceeded 20 ft, often exceeded 40 ft, and sometimes exceeded 60 ft. (Holes on a 20 ft rectangular spacing were drilled prior to the start of the grouting and could thus be used to monitor grout spread.)

A railroad spur 1300 ft long was to connect the new structure with the old. A sinkhole area had to be traversed by the spur. It was decided to grout under the spur also to reduce future settlements. However, the unrestricted flow of cement grout would make the grouting very expensive due to the large quantities that would be used. To control the spread of grout, SIROC was added for gel time control. Two pipes were placed in a 6-in.-diameter drilled hole. Cement grout was pumped through one pipe and SIROC through the other. The two materials mixed in the hole as they came out of the separate pipes. Gel times of the order of 5 to 10 min were used. In this fashion, horizontal flow of grout was limited to the loading influence area of the track. In large voids, supporting grout piers were formed instead of filling the entire void.

Cement and sodium silicate act somewhat like mutual accelerators. Each can be used in small quantities to greatly reduce the normal setting time of the other. Of course, the use of Portland Cement adversely affects the penetrability of the silicate. This problem disappears if microfine cement is used (see Appendix A).

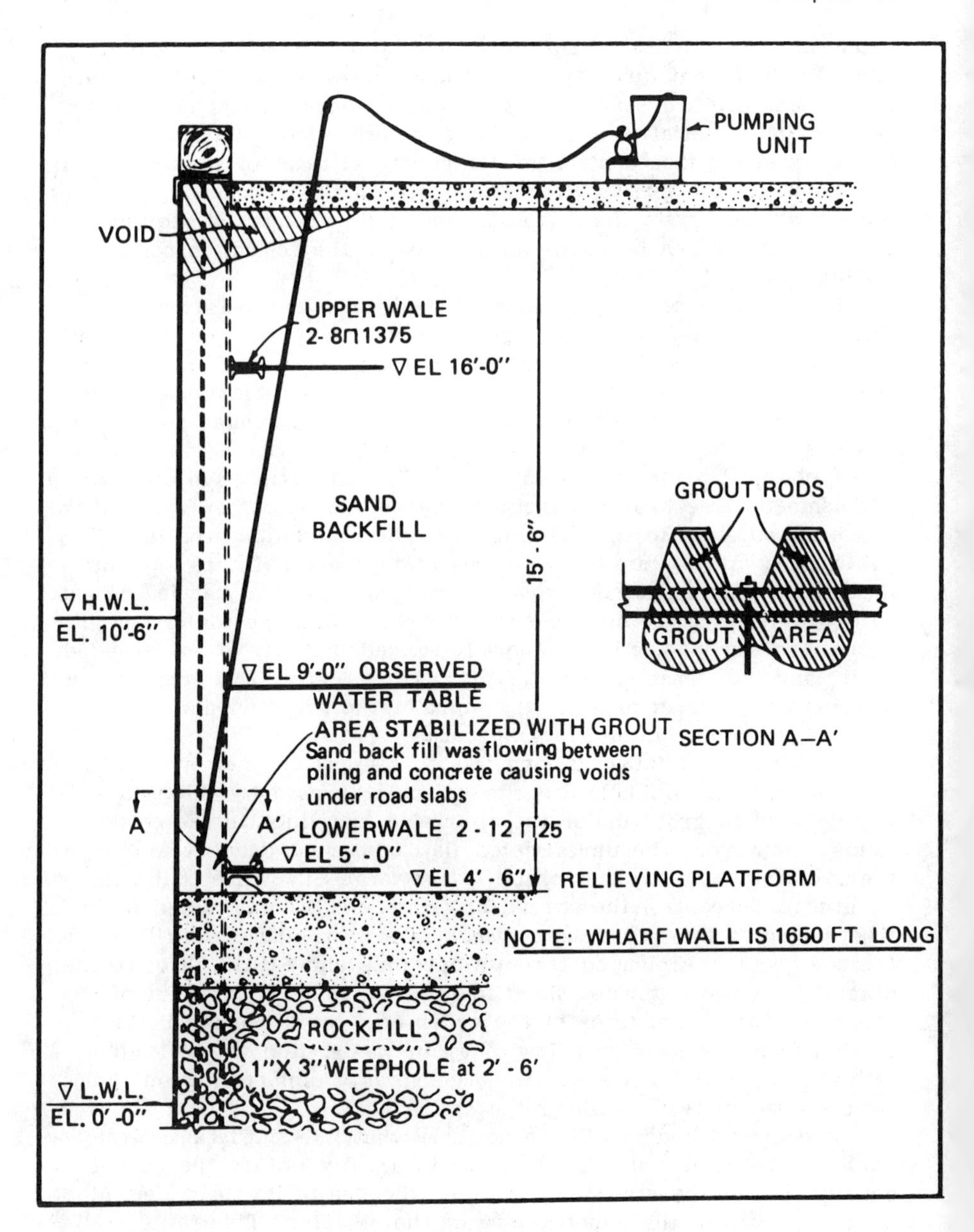

FIGURE 12.6 Cross-sectional view showing grouting location.

12.6 SEALING SHEET-PILE INTERLOCKS

Mating between the male and female ends of sheet piling is not watertight. Water can move quite freely through such joints, although generally in small quantities. Most often, such flow is a minor problem and is handled by removing the inflow rather than stopping the seepage at its source. If the seepage is carrying silt or other fill, however, it may be more desirable to seal the interlocks. This can be done by driving a small bore pipe or tube alongside the interlock and pumping small volumes of grout (1 gal/ft, for example) at very rapid gel times (10 to 15 sec) as the pipe is withdrawn.

When the interlocks between adjacent piles separate, grouting is often the only economical method of closing the resulting gap. Again, very short gel times must be used.

Sheet-piling problems seem to lend themselves to economical solutions by grouting, as a problem on the St. Lawrence River in Sept Iles, Canada, demonstrates. At that location, a loading and mooring dock, 1600 ft long, was showing distress. It had been constructed by driving sheet piling and then placing a heavy rock fill as a base. A concrete relief platform was poured on the rock fill and then covered with 15 ft of dredged sand. A 6 in. deck slab was then placed on the sand. A small gap had gradually opened between the relief pad and the sheet piling. Aided by the tide, sand was seeping through this gap, filtering into the rock base. The voids thus created under the upper slab were causing that slab to cave in under traffic.

Consultants proposed that a wedge of sand at the contact point with the relief pad be grouted to stabilize its action. Holes were drilled through the concrete slab. Pipes were then jetted to the contact between the sheet piling and the relief slab. See Fig. 12.6. Six gallon batches, set to gel at 3 min and 20 sec, were placed at a rate of 2 gpm through each grout hole. Split-spoon samples were taken every 150 ft to check the results of grouting. In this fashion, 15,000 lb of chemicals (approximately 20,000 gal) were placed, during 37 working days, to solve the problem completely, while the facilities were in continuous use. (This work was done with an acrylamide grout.)

12.7 SUMMARY

Specialized applications of chemical grouts include many with very low volume, such as sealing piezometers and the sampling of sands.

A very large industry, however, has grown around the use of grouts for the internal sealing of sewer lines, primarily to control infiltration that overloads treatment plants.

REFERENCES

1. K. Guthrie, Sewer and water: Our decaying infrastructure, *Excavating Contractor*, October 1982, pp. 18–20.
2. Sewer System Evaluation for Infiltration/Inflow, Report prepared for the U.S.E.O.A. Technology Transfer Program by American Consulting Services, Inc., Minneapolis.
3. R. A. Antonio, Chemical grout controls hazardous infiltration, *Water and Waste Technology*, August 1971.
4. H. E. Morgan, How long does sewer grouting last? *Public Works Magazine*, October 1974.
5. T. P. Calhoun, Longevity of sewer grout under severe conditions, *Public Works Magazine*, October 1975.
6. C. G. Sweeney, Grout routs sewer problems, *Water and Waste Engineering*, May 1977.

13

Specifications, Supervision, and Inspection

13.1 INTRODUCTION

There are occasions when for various reasons the owner prefers to have specific problems solved by grouting rather than by other methods. For example, it may be less expensive to replace a shallow leaking tunnel section by cut and cover methods, but a city may decide to use grouting to avoid the inconvenience of a blocked traffic artery. If the owner is a public agency, it then becomes necessary to write specifications for bidding that will ensure the use of materials and techniques satisfactory to the owner. Of course, performance specifications could also be written, but this is not current practice in the United States (it is much more common in Europe). Eventually, it may become necessary for the owner to inspect the job in all its stages and on rare occasions to supervise it.

13.2 SPECIFICATIONS FOR CHEMICAL GROUTING

One of the earliest attempts to compile a general specification for chemical grouts was done by the Grouting Committee of the ASCE [1]. The purpose of those guides was started in the introduction: ". . . intended to make it possible for the specifying engineer to obtain the chemical or chemicals desired, and to effect grouting operations intended with these materials." In the 20 years or so since those specs were written, the chemical grouting industry has expanded greatly, and the exposure of engineers in general to grouting data has increased vastly. Nonetheless, so long as we avoid performance specifications, the 1968 publications remain a good starting point for most chemical grouting jobs.

The major section titles in the guide specs are General (or Scope of Work), Materials, Equipment, Supervisions, Application, and Payment for Work Performed. How these topics may be effectively handled is illustrated in the excerpts from actual job specifications which follow.

The opening paragraphs of a set of specifications should broadly define each phase of the work to be done, so that contractors' extras are kept to a minimum. The first three paragraphs of a Corps of Engineers specification under the heading "Advancing and Casing Holes in Overburden and Grouting" read as follows:

1-1 GENERAL

1-1.1 Scope

This section covers advancing, casing and washing grout holes; making grout connections; furnishing, transporting, mixing and injecting the grout materials; care and disposal of drill cuttings and excavated overburden, waste water and waste grout; cleanup of the areas upon completion of the work and all such other operations as are incidental to the drilling and the grouting.

1-1.2 Program

The work contemplated consists of grout stabilization of loose soil areas and voids, the approximate locations, limits and details of which are indicated on Drawings Nos. 0-PHG-84/2 and 3. Graphic logs of overburden materials are included as Drawing No. 0-PHG-84/6. Typical void grouting details are included as Drawing No. 0-PGH-84/4. The program shown on the drawings and described herein is tentative and is presented for the purpose of canvassing bids. The amount of drilling and grouting will be determined by the Contracting Officer.

1-1.3 Procedures

Grouting mixes, pressures, the pumping rate and the sequence in which the holes are drilled and grouted will be determined in the field and shall be as directed.

The same specification under the heading "Drilling Holes Through Reinforced Concrete Pipe Joints and Chemical Grouting" reads as follows:

2-1 GENERAL

2-1.1 Scope

The work covered by these specifications includes furnishing all labor, materials, supervision and equipment necessary for the chemical grout stabilization of and the cutoff of water flowing through loose soil and voids in the immediate area surrounding leaking joints of the sewer within the limits shown on the drawings. The work includes diversion of surface runoff and wastewater flow; cleaning of the sewer, cleanout of joint annulus; drilling and cleaning of grout holes; furnishing, transporting, mixing and injecting the chemical grout at sewer joint, cleanout and patching of the finished grout holes and sealing of the joints; clean-up of the sewer and such other operations as are incidental to the drilling, grouting, and sealing. The lengths of the various size reinforced concrete pipe sections of the sewer from which stabilization chemical grouting and joint sealing will be conducted are as follows:

Location	Diameter (inches)	Normal joint spacing (feet)	Number of tongue and groove joints to be grouted
INCO Industrial Complex			
M.H. 3 to M.H. 12	66	4*	491
M.H. 1 to M.H. 3	60	4*	50

*The listed normal joint spacing is based on information from visual inspection and is included for the information of the bidder but does not guarantee that additional joints at closer spacing may not be encountered occasionally.

By way of contrast, specifications for grouting under a dam spillway in the Midwest were far less detailed:

ITEM 4—DRILLING AND GROUTING

4.01—Scope of Work

a. *Location and Type of Work.* The work includes but is not limited to the following:

Drilling, casing where necessary, pressure testing as required, and pressure grouting using chemical grout, the stratum of dense sand which exists at about elevation 720 underlying the spillway section of Dam near Indianapolis, Indiana.

b. *General Program.* The program for drilling and for pressure grouting is tentative. The extent of the program will be determined by the conditions developed at the site.

Holes for pressure grouting are designated as "A," "B," or "C" (see Contract Drawing 5A-1). Holes designated as "A" holes will be grouted first. The number and spacing of "B" and "C" holes, and the pressures and gel times to be used, will depend upon the results of water pressure or other tests, and the results of the actual grouting operations conducted for the "A" holes. All pressure grouting will be accomplished prior to any excavation for the spillway sections.

The amount of drilling and grouting that will be required is approximate and the Contractor shall be entitled to no extra compensation above the unit prices bid in the schedule for this specification by reason of increased or decreased quantities of drilling and grouting work required, the time required, or the locations, depth, type or nature of foundation treatment.

Specifications written by large public agencies are generally very detailed, as illustrated by a 1987 specification for sealing leaks in a subway system:

3G1.1 SCOPE

(a) The work shall include the furnishing by the contractor, of all supervision, training, labor, materials, tools and equipment and the performance of all operations necessary for the waterproofing injection grouting work indicated in the Contract Drawings, specified herein, and/or as directed by the Engineer.

(b) The work consists of the injection of liquid chemical grouts into active and inactive leaks through concrete cracks, joints or holes located in roofs, sidewalls, floors, rooms, beams and other locations in the 33rd Street line designated in the document entitled "Leak Locations" incorporated into, and made a part of this contract, and in all other areas designated by the Engineer and located within the limits of the Contract.

(c) The work shall be performed in a skillful and workmanlike manner with special care taken to prevent damage to existing structures, drains and utility lines. Damage caused by improper work procedures or failure to maintain drains, lines, equipment or structures shall be the responsibility of the Contractor.

(d) Documentation of the work shall be performed by the Contractor, including both daily work Reports and color-coded markings of grouting locations marked neatly on the concrete surface adjacent to each leak repaired, at the time of completion of the repair to indicate grouting pass number, grout used, grouting crew identity and date of repair. Contractor shall submit his documentation and coding scheme for approval.

(e) In order to judge performance, all sealed work shall be inspected by the contractor and the engineer within 3 days after a greater than 1/2 inch rainfall in one 24-hour period during the construction phase. The work priorities will be adjusted according to the results of the survey, at the direction of the engineer.

(f) It is estimated that approximately 34,400 lineal feet of cracks and joints may require treatment.

Recent specifications for grouting jobs often include requirements for bidder qualification, as additional assurance of satisfactory low bidder performance. The following sections from a 1987 document show details:

3G1.2 CONTRACTOR QUALIFICATIONS

The actual grouting work specified herein shall be performed by a qualified Contractor with a minimum of five (5) years direct, continuous and recent experience in performing waterproofing chemical grouting work in similar conditions on at least six (6) different projects.

3G1.3 FIELD SUPERVISION QUALIFICATIONS

Field supervision shall be provided by a Grouting Superintendent with at least four (4) years recent experience in waterproofing chemical grouting work and who meets the above project experience requirement, by Journeyman Grouting Foremen with at least two (2) years recent experience with the equipment and chemical grouts specified in applications similar to the proposed project, and by

> Apprentice Foremen who have at least six (6) weeks full-time experience in waterproofing grouting work under the direct supervision of a qualified Journeyman Grouting Foreman.

The same specification also called for submission of qualification data as pre-bidding requirement.

Specifications for grouting materials are generally very detailed, listing generic names, trade names, and manufacturers. Specs written by public agencies must either permit the use of "equal" materials or be prepared to defend legally their decision not to permit bidding on alternates. The spelling out of grout properties in great detail helps ensure those desired and virtually eliminates the offering of alternate materials. This situation is common to all areas where spec writers are requesting specialty products. As knowledge of the products and their use grows, the specifications become more oriented to properties and performance rather than trade names and formulas.

A Corps of Engineers spec starts with a general statement but quickly becomes very specific:

> 1-3.3 Chemical Grout Materials.
>
> 1-3.3.3 General. The void stabilization materials shall be proportioned fly ash with chemical grout and a catalyst system. The chemical grout used shall have a documented service of satisfactory performance in similar usage. All materials shall be delivered to the site in undamaged, unopened containers bearing the manufacturer's original labels. The materials shall be equal to chemical grout AM-9 with recommended catalysts and other materials as manufactured by the American Cyanamid Company and conforming to the specifications described hereinafter.

Succeeding paragraphs cover in detail the basic chemical grout, catalyst, activator, inhibitor, filler, and additives.

A less restrictive paragraph appears in California specs for drilling and grouting test holes at a dam and reservoir:

> (5) Chemical Grout. Chemical Grout shall be a mixture of various chemical compounds which gel or solidify in a given time after mixing. Given gel or solidification time shall range between two minutes and 60 minutes. The viscosity of the grout mix shall be less than 1.5 centipoises before gelling or solidification. The chemicals used shall be so

proportionaed and mixed as to produce a chemical grout that contains no solids, may be pumped without difficulty, and will penetrate and fill the voids in the soil mass and form a gel or solid filling which will be of the required strength, be stable and impermeable.

A far more liberal set of specifications for a tunnel in Asia merely says "chemical grout including acrylamide, monosodium phosphate, sodium silicate, calcium chloride or other suitable approved chemical grout."

A New York City Transit specification listed satisfactory materials in detail:

3G2.0 GROUT MATERIALS

3G2.1 GENERAL

Two different kinds of chemical waterproofing grouts are intended to be used for sealing concrete cracks and joints, to be selected for application at specific locations based on the nature of the crack or joint in relation to the grout's properties. Polyurethane grouts are intended for use in running water conditions or where moderate to large joints and cracks with active water leakage are encountered, and otherwise as directed by the Engineer. Acrylate grouts are intended to be used where inactive leaks are encountered, in fine cracks and otherwise as directed by the Engineer. All grout used shall have a successful history of application for at least four (4) years under conditions similar to the current project.

3G2.2 POLYURETHANE GROUT

Polyurethane grout supplied shall be water-reactive liquid polyurethane base solutions which when reacted expand by foaming to at least seven (7) times the initial liquid volume and when set produce a flexible, closed void solid resistant to degradation by wet and dry cycles and chemicals found in concrete construction. Specific waterproofing grouts meeting these requirements are marketed by the following manufacturers:

3G2.3 ACRYLATE GROUT

Acrylate grout used for waterproof grouting shall be water solutions of acrylate salts which, when properly activated and catalyzed, react in a controlled set time to form a flexible, permanent gel. Specific waterproofing grouts meeting these requirements are marketed by the following manufacturers:

(For each class of grout, two acceptable manufacturers and products were cited.)

Specifications for equipment have also in the past been very detailed, primarily to avoid the problems that would occur if bidders intended to use cement grout pumps for chemical grouting. Most often, the described details cover a page or more, with separate paragraphs dealing with capacity (pressure and volume), materials of construction, number of pumps, drive systems, and control systems including gel time control, pressure and volume control, and systems to prevent overpressuring or inadvertent recycling (see Chap. 7). The following paragraph from a specification on dam foundation grouting is one of the least detailed ways to preclude the use of batch systems:

d. *Grouting Equipment*. All equipment used for mixing and injecting grout shall be furnished by the contractor and shall be maintained in first class operating condition at all times. The equipment will be of proportioning or two solution type, and will include such valves, pressure gages, pressure hose, supply lines, pipes, packers, jacks and small tools as may be necessary to provide a continuous supply of chemical solution at required pressures and volumes. [At this point, "and at required gel times" could have been added.]

A complete detailing of equipment was done in this excerpt from a 1987 specification:

3G3.1 EQUIPMENT

(a) General

(1) The Contractor shall supply all equipment, including pumps, containers, hoses, gages, packers, drills, bits, scaffolds, compressors, generators, vacuums, accessories, and all other items required to perform the work and accomplish the goals outlined in the Specifications.

(2) The equipment shall be of a type, capacity, and mechanical condition suitable for doing the work in an effective and efficient manner. All equipment including all power sources, cables, chemical containers, scaffolds, and

anything used in the performance of the work, shall meet all applicable safety and other requirements of Local, State, and Federal ordinances, laws, regulations, and codes.

(3) All equipment shall be maintained in excellent working condition at all times. Sufficient spare parts and tools shall be maintained on the job to provide for immediate (1 hour) repairs of essential operating items.

(4) Each grout crew shall maintain its own equipment items required herein in order to operate independently of, and separated from other grout crews.

(b) Pumping Units

(1) The Contractor shall supply separate pumping units, including separate chemical containers, hoses, and all other accessories for injection of polyurethane grout and acrylate grout.

(2) Pumps shall be capable of continuous injection of the liquid grout under variable pressures up to a maximum pressure of 2,000 psi and at flow rates of at least 5 fluid ounces per minute at high pressure (2,000 psi) and flow rates of at least 1/4 gallon per minute at pressures of 500 psi and lower, and in accordance with the manufacturer's recommendations and under the direction of the Engineer. Pumps may be electric, air, or hand driven provided that rapid changes in pumping rates and pressures can be obtained by the pump operator without effecting the mixture of the grout being injected and without stopping the pumps.

(3) Pumping Units shall be made of materials compatible with the chemicals being used, and shall be equipped with necessary hoses, chemical containers, gages, fittings, packers and other accessories required to inject the grout properly. Seals and joints shall be such that no grout leakage occurs and no air is aspirated into the injected grout.

(4) Grouting Units shall be so arranged that flushing can be accomplished with grout intake valves closed, flushing fluid supply valves open, and the pump operated at full speeds.

(5) Pumping Units shall be equipped with accurate pressure gages at the pump and near the injection point. Gages shall be accurate to 5% and shall be periodically checked for accuracy against new, undamaged or calibrated gages. Damaged or inaccurate gages shall be replaced immediately. Pumping units shall not be operated without properly operating gages. Replacement gages shall be on hand at all times.

(6) Hoses and fittings shall have maximum safe operating pressure ratings and dimensions as recommended by the manufacturer and under the direction of the Engineer.

(7) Suitable mixing and holding tanks shall be supplied with each grouting unit to permit continuous pumping at maximum pump capacity. Tanks shall have satisfactory covers and shall be stable against tipping under normal usage.

(8) Descriptions of pumping units for both polyurethane grout and acrylate grout shall be submitted for approval by the Engineer as required in these specifications before starting the actual grouting work. Written approval of the pumping units shall be received from the Engineer by the Contractor before actual grouting is started.

(c) Polyurethane Grout Pumps

Grout pumps used for polyurethane grout injection shall be either single or double pump type as recommended by the grout manufacturer. Where double pump types are used, they shall have the same capabilities as required for acrylate pumping units, but shall properly accommodate the more viscous materials used for polyurethane grouts. In no case shall polyurethane grout pumps be used for injection of acrylate materials, or acrylate grout pumps be used for injection of polyurethane grouts, in the same day nor without thorough cleaning, disassembly and appropriate modification nor written notification to and approval by the Engineer. Pumps shall be arranged and operated in a manner consistent with the grouts injected and the grout manufacturer's recommendations.

(d) Acrylate Grout Pumps

Acrylate grout pumping units shall consist of two, parallel high pressure, positive displacement pumps with parallel hoses leading to a mixing chamber or "Y" at the packer. Pumps shall be equipped with check valves to prevent the back-flow of one grout component into the lines of the other component.

(e) Packers

Packers which are specifically designed for the grouting operation shall be supplied and used capable of safety sealing and packing grout holes drilled into concrete and injected at pressures of up to 3,000 psi, and as recommended by the manufacturer of the grout. Packers shall be of the removable type such that the drilled hole can be cleaned and patched to at least 3 inches deep.

(f) Drills

Hand drills capable of drilling small diameter holes of 1/2 to 1 inch in diameter in concrete shall be supplied and operated. The following two types of drills shall be supplied for each grouting crew: (1) Rotary percussion capable of drilling up to 18 inches deep in unreinforced concrete; (2) Rotary flushing type with diamond coring bits capable of drilling up to 24 inches deep in reinforced concrete. Drills shall be supplied with bits of a diameter and length consistent with packer requirements and hole lengths needed for the drilled holes to intersect the target crack or joint as specified. Damaged or worn bits shall not be used. Back-up drills and bits shall be supplied in sufficient numbers so that two drills of either type can be used simultaneously.

Specifications sometimes will have a separate paragraph dealing with supervision, as for example, this paragraph from a dam grouting spec:

(e) Procedures. Supervision of all phases of the contract shall be under the direct control of the Engineer. The Engineer's responsibility includes but is not limited to location of holes, drilling of holes, re-drilling of holes, procedures, methods of grouting, mixing of grout, and maintaining complete records of the grouting operation including cost items.

More often it is inferred by such phrases as "as directed by the Engineer," "in a manner approved by the Engineer," etc.

Sometimes a few words hidden away in other sections remove all engineering decisions from the contractor. In a detailed spec for grouting a tunnel in Hong Kong, the following paragraph appeared under Section 6.11, *Grouting Procedure*: "The grouting methods, mixes, pressures and pumping rates together with the sequence in which the holes are drilled and grouted will be determined by the Engineer."

Sections of specifications dealing with "application" may be very brief if a performance criterion is used; otherwise the scope of the contractor procedural responsibilities must be completely spelled out. These will vary considerably from job to job. A general guide appears in Ref. 1, excerpted here:

APPLICATION

The application of chemicals shall be under the direct supervision of the grouting engineer. Application shall be understood to include:

a. Placement of grout holes—holes may be placed by rotary or percussion drilling, driving or jetting (using water or air), depending on the formation and its response to each method (see Note 18). Casing must be provided for caving formations (the drive pipe or jet pipe may be used for this purpose). Casing must also be provided for formations which will not otherwise permit proper seating of downhole packers. Holes must be placed with sufficient accuracy to insure that planned overlapping of grout from adjacent holes can occur.
b. Grout pattern—this includes the geometric layout of all the holes, the sequence in which each hole is placed and grouted, and the vertical dimension and sequence of grouting the lifts (stages) for each hole. The geometric layout of holes, both in plan or in profile, should be completely planned prior to the start of grouting, and submitted to the owner for approval (or information).
c. Field tests—prior, during and after the completion of the chemical grouting operation, field tests should be performed and records kept to determine the effects of grouting. Such tests should be performed in accordance with generally accepted procedure (see Note 19) subject to approval of the owner or his engineer. The grouting engineer will be responsible for obtaining or constructing adequate test instrumentation and keeping records of field data for testing them during and after grouting.
d. Pumping pressures—maximum value of pumping pressure at the collar of the hole is __________ (see Note 20).
e. Concentration of chemical grout—the concentration of chemicals mixed in the tank (computed as a dry weight percentage of the total solution weight) shall generally be __________ (see Note 21). In no case shall the concentration be less than __________, and it may go up as high as __________.
f. Induction period—Control of the induction period is the responsibility of the grouting engineer. Control should be done mechanically, through the pumping system controls, using the minimum number of stock solution variations (none, if possible). Wherever feasible ground water from the site at the site temperature shall be used

to prepare the stock solutions, to eliminate differences in tank and underground gel times (see Note 22).

g. Gel checks—a sampling cock placed between the Y-fitting and the grout hole or pipe shall be used for checking both induction period and gel strength. Such checks shall be made every time the induction period is changed, or at least once every five minutes during long pumping times, and at least once during every grouting operation of less than five minutes (see Note 23).

Notes:

No. 18 Delete any methods not compatible with local conditions.

No. 19 Pertinent field tests include drop tests, pumping tests and piezometer installations. Methods of performing such tests may be found in "Theory of Aquifer Tests," U.S. Geological Survey Water-Supply Paper 1536-E, 1962, and the U.S. Bureau of Reclamation's "Earth Manual" (Appendix pages 541 to 562) and other technical publications. If desired, specific test methods may be specified in this paragraph.

No. 20 Specify here the maximum allowable pumping pressures, as determined by structural safety considerations. In general, these pressures can be as high as permissible values for cement grouting.

No. 21 Specify here grout concentrations desired and specify also the minimum and maximum percentage limits of each of the materials involved.

No. 22 Mechanical control will not be possible for batch systems.

No. 23 Gel checks for batch system need be made only at the start of the grouting operation, or if chemical concentration in the tanks are changed.

The specifics of grout injections are spelling out very clearly in the following section from a New York City specification:

Step (5) Grout Injection. Injection of the selected grout shall commence immediately after installation of the packer and shall be done using the equipment, materials, and procedures specified elsewhere in this Paragraph 3G3.0. Pumping shall proceed as long as all of the following conditions are fulfilled: (a) grout is entering the crack or joint; (b) the observable loss of grout returning from the crack is estimated to be less than 50% of the volume of acrylate grout or less than 25% of the volume of polyurethane grout being pumped; (c) damage is not being done to the structure; (d) the total volume of grout injected in the current episode in the hole does not exceed five (5) gallons for acrylate grout or two (2) gallons for polyurethane grout; (e) the grout has not extended for more than five (5) feet along the crack or joint away from the grout hole; or (f) the Engineer has not indicated that grouting should stop.

Specification writers often find it necessary or desirable to define in the "application" section terms which are generally used by grouters. The purpose is to avoid ambiguity and misunderstandings of procedures which are done differently in different geographic locations. For example, spec writers for a Hong Kong tunnel found it desirable to define grouting methods:

> 6.8 The following grouting methods, inter alia, shall be adopted as directed by the Engineer's Representative:—
>
> (a) the closure method of grouting, involving grouting in an additional hole located midway between two previously drilled and grouted holes;
> (b) packer grouting, consisting of first drilling a hole to its final depth and then grouting from the bottom upwards in steps defined by a packer set at successive higher elevations;
> (c) stage grouting, consisting of drilling a hole to a limited depth, grouting to that depth, cleaning out the hole, allowing the injected grout to attain its initial set, drilling the hole to a greater depth and then grouting the cumulative length of the hole. The hole is successively drilled and grouted in this manner until the required final depth is reached;
> (d) stage and packer grouting, performed similarly to stage grouting but involving the use of a packer to limit the effect of the second and subsequent stages of grouting to the corresponding lengths of hole.

In general, it is good practice in spec writing to refer to a recognized glossary of grouting terms [2].

Payment for grouting work should be explicitly detailed. Methods will depend on the type of specifications and local practice. Typical of domestic practice is the following excerpt from a basement waterproofing spec for a West Coat power utility:

3.0 Payment

3.1 *Item 1*—A lump sum, which shall include all costs incurred moving Contractor's equipment and personnel to the job site and into position. If partial payments are approved in accordance with Paragraph 6.1 GC the Contractor shall be entitled to 50% for moving onto the job and remaining 50% after moving off the job.

3.2 *Item 2*—A unit price per day for rental of Contractor's equipment from the day equipment is ready to pump grout, up to and including the day Constructor directs that equipment be removed from the job. No payment will be made for Saturday, Sunday or holidays unless Contractor is directed to work on these days. No payment will be made for days on which equipment is broken down or for delays caused by Contractor.

3.3 *Item 3*—A unit price per hour of actual grouting operation which shall include all costs for furnishing labor, tools, and materials, except chemical grout, for operating equipment and handling loading and injecting of grout, connection of hoses, packers, stoppers, etc. Payment shall start at the time the crew is ready to start operating the equipment and shall end when the equipment is shut down for meal time, repairs, maintenance, or cleanup or for Contractor caused delays. Payment will be made for any Company caused delays if Contractor's crew is standing by ready to operate the equipment. No payment will be made for Company caused delays if contractor is given 16 hours or more notice to discontinue the work for one or more succeeding days.

3.4 *Item 4*—Grout material costs which shall include the invoice cost freight and sales tox of all chemical grout material required in excess of the material furnished by Company. Company will pay for all material required to produce a minimum of 4,000 gallons of jel plus any additional material actually injected as directed by Constructor.

An East Coast public agency defined payment for tunnel sealing work as follows:

(b) Payment: Payment shall be made on the following basis:

(1) Mobilization and Demobilization: Payment for mobilization and demobilization will be made at the

Contract Lump Sum price of $100,000.00 as measured and specified in 3G5.0(a)(1) above.

(2) Cracks Treated with Acrylate Grout: Payment will be made for cracks treated successfully with Acrylate grout at the Contract Unit Price per lineal foot as measured and specified in 3G5.0(a)(2) above.

(3) Cracks Treated with Polyurethane Grout: Payment will be made for cracks treated successfully with Polyurethane grout at the Contract Unit Price per Lineal Foot as measured and specified in 3G5.0(a)(3) above.

(4) Acrylate Grout: Payment will be made for "Acrylate Grout" at the Contract Unit Price per gallon based upon the documented actual cost to the Contractor Plus 5% as measured and specified in 3G5.0(a)(4) above.

(5) Polyurethane Grout: Payment will be made for "Polyurethane Grout" at the Contract Unit Price per gallon based upon the documented actual cost to the contractor plus 5% as measured and specified in 3G5.0(a)(5) above.

These brief statements followed very detailed sections on quantity measurement.

The French Association of Underground Construction publishes the document "Recommendation for the Use of Grouting in Underground Construction" (reproduced in translated form as Appendix B of Ref. 3). This document should prove very useful to those who are writing chemical grouting specifications.

13.3 SUPERVISION OF GROUTING

The term *supervision* as used in typical domestic grouting specifications generally refers to all mechanical operations connected with procuring, storing, mixing, catalyzing, and placing grout. Only in performance-type specifications is supervision also responsible for the results of the grouting operation. In either case, however, a good supervisor should have the experience, knowledge, and desire to make his/her own evaluation of the grouting design in terms of its probable effectiveness specifically as related to the field operations to be controlled. Among the many factors to be considered are the following:

1. Is the proposed solidified volume of soil spatially defined? How much grout will be placed in that volume? Is that amount consistent with the percent voids in the formation?
2. Are the grout holes laid out in plan and profile? Do the holes enter the zone to be grouted? Have a sequence and time schedule of hole placement to be selected?
3. Is the method of placing holes consistent with the formation and the depth and accuracy of placement required?
4. Has a grouting schedule been established for each hole, including stage length and grout volume?
5. Have pumping pressure limitations been established? Are these consistent with placement of reasonable volumes while avoiding uplift or fracturing?
6. Has a grout been selected? Are its properties consistent with the ability to penetrate the formation and provide needed strength and gel time control?

If all these questions can be answered affirmatively, while it does not guarantee success, it does at least indicate that the design has been well thought out. Questions that have not been considered or answered prior to the start of the field work will have to be answered while working. This may cause delay and inefficiency, and decisions which should ahve been made in design must sometimes be made in the field by the supervisor in order to keep the job moving. A prestart evaluation by the supervisor may avert this situation.

The factors suggested for consideration are but the major ones, and not necessarily all of those. A separate check list should be made by the supervisor for each job. One final factor which should always be considered is the effects that grouting will have on measurable parameters, such as seepage, leaks, pressure, ground or structure movements, water levels, etc. Whenever it is possible to relate the effectiveness of the grouting operation to such parameters, provisions should be made to record the necessary data. Where no obvious correlations are possible, it is most important that precise records be kept of volumes and pressures of grout for each stage of every treated grout hole.

13.4 INSPECTION OF GROUTING

The job of an inspector is to record field operations in sufficient detail so that a judgment can be made as to whether the project specs are being followed or violated. Normally, this will require the keeping of daily written records, often on work sheets specifically tailored to the project. The responsibility of the inspector may overlap that of the supervisor to the extent that he/she may be required to make decisions as to whether procedures and equipment meet the specifications. The extent of the responsibility of an inspector should be detailed to the satisfactory agreement of both the owner and the contractor prior to the start of field work. A checklist of areas for inspector activity follows.

What to Look for on the Job

Materials

- Is the grout being used that which has been specified?
  - Commercial products
  - Proprietary products
- Are the catalysts those specified by the manufacturer? Are they proper for the grout properties desired?
  - Strength
  - Permanence
- Are site storage conditions adequate?
  - Hazards
  - Aging
  - Moisture
- Are the proper safety precautions used in handling the materials?
  - Protective clothing
  - Container disposal

Equipment

- Is the grout plant capable of controlling flow, pressure, and gel time?
  - Systems—batch
  - Dual pump
  - Proportioning
- Are there readout devices for
  - Volume
    - Direct or indirect
    - Recycling
    - Capacity—accuracy
  - Pressure
    - Range
    - Accuracy
- Are these control devices for
  - Overpressure
  - Changing volumes
  - Accuracy
  - Must pump stop?
- Is there a continuously functioning gel checkpoint?

Grout preparation

- Is the grout being mixed to the specified concentration?
  - Consistent with field conditions
- Is site water used?
  - Possible effects of other water
- Are unspecified additives being used?
  - Effects on
    - Strength
    - Penetrability
    - Gel time

Grout placement

- Are grout holes or pipes located in accordance with plans?
- Are adequate devices being used to keep holes open?
  - Packers
  - Stuffing boxes
  - Drill hole plugs
- Is the sequence of hole treatment following the original plan?
  - Importance of deviation
  - General practice
- Are the grout volumes placed those which were planned?
  - Reasons for deviation
  - Effects on performance
- Are gel times consistent with plans?
  - Does material gel?
  - Importance of deviation
- Are adequate records being kept?
  - Pipe or hole identification
  - Depth treated
  - Starting pressure
  - Final pressure
  - Grout volume placed
  - Gel time
  - Time of treatment

Performance checks

- Changes in water flow
  - Visual
  - Weirs
- Change in permeability
  - Pressure records
  - Pumping tests
- Change in strength
  - Penetration test
  - Previous data needed

13.5 REASONS FOR UNSUCCESSFUL JOBS

Completed chemical grouting projects fall into three categories: (1) those that were successful, as evidenced by some obvious change such as the shutoff of seepage; (2) those that were unsuccessful, as evidenced by the lack of some anticipated change such as the diversion of water; and (3) those that cannot be judged, because (a) there might not have been a failure anyway, (b) there were no methods or attempts made to measure the results of grouting, (c) grouting was used to increase a safety factor or decrease a risk, (d) opinions differed as to the need for grouting in the first place, etc.

It is probable that many jobs which are considered failures are placed in that category because adequate definitions of success or failure were not predetermined. However, there is much that can be learned from those jobs which were obvious failures. Analysis of such work indicates that failure can generally be explained by the obvious statement of "not enough grout in the right place." As trite as the phrase seems, it is still worth looking into the reasons for "not enough grout" and "not in the right place."

To begin with, jobs may fail or be unevaluable because of lack of engineering data. Work is often begun without reasonable knowledge of formation geology, voids, and permeability. Coupled with this shortcoming is often the failure to set up pressure and volume goals that permit the evaluation of the unknown factors. (Grouting each hole to a take refusal is hardly ever an efficient way to use chemical grouts.) Of course, small projects often cannot justify an engineering preevaluation. In such cases, it is vital that mutually satisfactory evaluation procedures (preferably some which relate to ongoing processes rather than end-of-job parameters) be established between the owner and the contractor.

Jobs may be unsuccessful because of the failure to consider the total problem. For example, shutting off several leaks in a porous formation may be readily accomplished but will only result in chasing those leaks elsewhere. Similarly, shutting off large quantities of seepage may raise the water table and create new problems.

Jobs may also be considered unsuccessful because of poor record keeping. If obvious visible changes do not occur, work must be judged on the basis of pumping volumes and pressures as holes were sequentially grouted. If such data were not kept, or were not precise enough, the work cannot be judged.

Even when engineering data are adequate and available, jobs may fail because of poor judgment. The most obvious case of this kind is doing a job for which grouting just is not suitable, for example, trying to permeate organic silts and clays. Less obvious cases of poor judgment include deliberately pumping only enough grout to fill a portion of the voids. (Many successful jobs have been done by the Joosten process on the basis of filling half to two-thirds of the formation voids. Less viscous grouts cannot often be used successfully on the same basis.) With the relatively viscous solutions used in the Joosten process and in the high-concentration single-shot silicates, it has been shown that some of the finer voids (in soils which approach the limits of those solutions' penetrability) remain ungrouted. For such cases, it may be appropriate to design pumping volumes based on filling 80% to 90% of the voids. When using those materials in coarse formations or the grouts whose viscosities approach water, design volumes should be based on filling all the formation voids.

Another case of poor judgment includes wrong assumptions concerning the effectiveness of partial cutoffs. For example, if 75% of a proposed cutoff wall has been grouted, that does not mean that the cutoff is 75% effective. In fact, the long-term effectiveness may be very low, as the previous total volume now flows through a reduced area at a higher velocity and carries away solids to enlarge its channels.

Any of the factors discussed above may result in the placing of insufficient grout—often the direct cause of job failure. Too little grouted volume can also occur for reasons independent of engineering data, design, or judgment. One of these reasons is related to equipment. It is always a mistake to do field work with inadequate equipment. If the equipment used to place grout pipes cannot do so accurately, chances are that there will be gaps in many places where closure is necessary. If the grout pipes are just whatever is handy, rather than specific tools for placing grout, plugged pipes may be misinterpreted as pressure refusal, and grout is not placed where it should be. Using pumps of insufficient pressure capacity will also result in placing less grout than may be planned and necessary. Using batch equipment may preclude the use of gel times short enough to keep the grout flowing away from the zone to be treated or from being excessively diluted. Finally, the lack of accurate measuring devices for the grout volume and pressure may render the grouting records useless and the grouting operation haphazard.

Even with adequate engineering and equipment, stabilized volume may be insufficient because of procedural errors. If the grout does not set up at all, obviously the job will fail. Failure of the grout to set up can be due to equipment malfunction or error in measurement which results in lack of catalysis, to problems with groundwater chemistry or pH (the classic examples are using a grout which sets only under acid conditions in an area previously grouted with cement and using an unbuffered basic grout in coal mines), and to temperature changes in the ground which significantly change the preset gel time.

Even if the grout does set up, it serves no purpose if it does not set up in the right place. Liquid grout can be dispersed by selective flow in unsuspected open channels and by excessively long gel times which permit gravity flow above the water table, displacement by normal groundwater flow, and dilution with groundwater which still further prolongs the gel time. Grout may also be dispersed to ineffectiveness by improper use (or no use) of isolating packers and by an incorrect sequence of grouting holes and stages within a hole.

Grout which sets up too quickly may be just as ineffective as grout with excessively long gel times. (These discussions emphasize the need for regular gel checks at the point where catalyzed solution enters the formation.) Most of the reasons for prolonging gel times (discussed above) also apply to shortening gel times. These include equipment malfunction or measurement errors leading to excess catalysts, improper pumping rates, contamination in tanks and piping, and temperature and groundwater chemistry.

Finally, a job failure (or the consideration of a job as having been ineffective) may be due to lack of knowledge of how to measure job effectiveness or facilities to do so.

13.6 SUMMARY

On large construction projects it is slowly becoming standard practice to include a section dealing with chemical grouting in the jobb specifications. Unless the job is being let on a performance basis, the specifications should be sufficiently detailed to ensure that the work will be done by qualified and experienced personnel.

Guidelines can readily be established for personnel being trained for supervision and inspection of a grouting operation. Only experience, however, can adequately prepare grouters to recognize those occasions when exceptions and deviations from specified procedures are warranted. Good specifications will include provisions for such job-dictated changes.

REFERENCES

1. Guide specifications for chemical grouts, *J. Soil Mech. Found. Div., Proc. ASCE*, 345–352 (March 1968).

2. Preliminary glossary of terms relating to grouting. *J. Geotech. Eng. Div. Proc. ASCE*, 803–815 (July 1980).

3. C. W. Clough, W. H. Baker, and F. Mensah-Dwumah, Development of Design Procedures for Stabilized Soil Support Systems for Soft Ground Tunnelling, Final Report, Oct. 1978, Stanford University, Stanford, California.

14

Grouting Research

14.4 INTRODUCTION

Research related to chemical grouts has been conducted continuously for many decades by individuals and by business organizations. By far the major portion of this research has been devoted to the development of products rather than procedures. Only in the past decade has a reversing trend appeared.

Every active grouting practitioner may be forced to experiment on a field job when the methods he/she normally uses prove ineffective. Such experimentation may lead to new and better procedures and materials that greatly increase the utility and scope of the grouting process. Nonetheless, these efforts are development processes, not research, and the results quite often remain hidden with the developer to increase his/her competitive position in commercial activities.

Two organizations contributed significantly to development work with chemical grouts, a university and a federal government agency. Even prior to the introduction of acrylamide-based grouts, work was being done at M.I.T. to evaluate the potential of various materials as chemical grouts. Some of the early work with acrylates [1] as well as with acrylamide was done at M.I.T. This interest persisted for many years.

Acrylamide grouts were marketed in the early 1950s, and were very successful until their toxic properties prompted their withdrawal by some manufacturers. In the United States, acrylates are steadily replacing acrylamides.

The Corps of Engineers has taken an active interest in chemical grouts concurrent with their development [2]. Reference 3 lists the numerous publications dealing with all phases of grouting, many of which are available, including Ref. 1 and Ref. 3 itself.

Other federal agencies have also occasionally been involved with published data dealing with chemical grouting. One of the earliest is credited to the Highway Research Board [4].

More recently, the Federal Highway Administration and the U.S. Department of Transportation have sponsored research of significant scope, which is discussed in later sections of this chapter.

14.2 GROUTING MATERIALS RESEARCH

A major portion of the research effort to find new grouting materials is done by chemical companies. Cyanamid, Diamond Alkali, Rayonnier, and Borden* have all made significant research efforts which resulted in commercial products. Dow and DuPont have also invested research dollars. In Europe, Rhone-Poulenc markets a wide array of chemical grouts. Large commercial grouting companies such as Halliburton in the United States and Cementation Co. and Soletanche in Europe also do materials research, often terminating in proprietary products. On a more limited scale, smaller chemical companies and individuals also look for new chemical grouts.

With so many organizations doing research, it would seem that we should have a much wider array of materials than are currently available. However, the reverse is true because of the amount of research done by any single organization is tied to the market potential for a possible product. As the markets grow, so does the research effort and the number of commercial products.

14.3 GROUTING APPLICATIONS RESEARCH

The effort spent upon research into the basics of grouting is far less than that spent on materials research because the direct investment return on basic research is nebulous. Most research of this type is carried out by nonprofit organizations, including those in the academic area. It is only in the past few years that universities have shown any interest in grouting research.

Curiously, one of the earliest major research efforts was mounted by a profit-oriented organization. American Cyanamid Co. organized an Engineering Chemicals Research Center in 1956 which devoted 9

---

*Associated with AM-9, SIROC, Terranier, and Geoseal, respectively (AM-9 and Terranier are no longer marketed).

years exclusively to the development of data, equipment, and techniques for field use. These efforts were backed up by basic research into the grouting process itself. The work resulted in over 1000 printed pages of technical resports, almost all of which were freely circulated to the grouting professionals. (These reports are now out of print and no longer available. The data they contained which are generally applicable are included in previous chapters.)

Much of the basic research was applicable to the general run of grouting materials which followed. In 1959, for example, reports on triaxial and unconfined creep tests on stabilized soil samples introduced the concept of a creep endurance limit as a more appropriate design parameter than the unconfined compression test. Three decades later, this concept is finally beginning to take hold, as research at universities corroborate the 1959 findings. Currently, an ASTM standard on creep strength of grouted soils is in the preliminary stages.

Much of Cyanamid's research was aimed at finding out how grout flows through a formation, particularly at gel times shorter than the pumping time. The technical papers summarizing this research [5] had great impact in changing the concepts of how to use chemical grouts, and possibly represent a landmark in the history of grouting.

Not all research leads to positive results, and not all of it makes a usable contribution to field practice. In the early 1960s a large number of laboratory scale grout injections were made to study the field procedures that would lead to most uniformity of penetration in stratified deposits. The variables, in addition to strata permeability, included pipe pulling distance, retraction time for pipe pulling, gel time, and viscosity. Acrylamide was used in all tests, with viscosity controlled through the addition of polyacrylamide. Unfortunately, it was found after the completion of the test program that the polyacrylamide used in the grout is also used commercially as a friction-reducing agent in the pumping of water through pipes. Analysis of the data responded to the friction-reducing effect by indicating an optimum viscosity for uniformity of penetration rather than a trend related to actual viscosity values. Since this did not agree with other and older accepted relationships between flow and viscosity, the results of the experiment were never published. However, if the anomaly is ignored, the report is found to contain other interesting relationships, meriting the reproduction of some excerpts:

TEST PROGRAM

The program consisted of a total of 48 individual injections, made in blocks of 4, into the three 4-inch sand strata.

The upper and lower strata were the same for all 48 injections, the center stratum varied among three different materials.

Washed and graded silica sands were used to obtain the desired stratification. The gradation of these sands is shown in [Fig. 14.1]. The upper and lower strata were always sand "B". The sands designated in the data as coarsest, medium, and finest are sands, S, C, and D, respectively.

The test program extended over a two-year period, due to its time-consuming nature, the number of repetitions required to obtain a usable block of four, and the press of other work. It should be appreciated that it is virtually impossible to maintain all of the variables exactly on their prescribed values. It is felt, however, that the accuracy and control that was exercised in the laboratory testing is far better than could be obtained in field work.

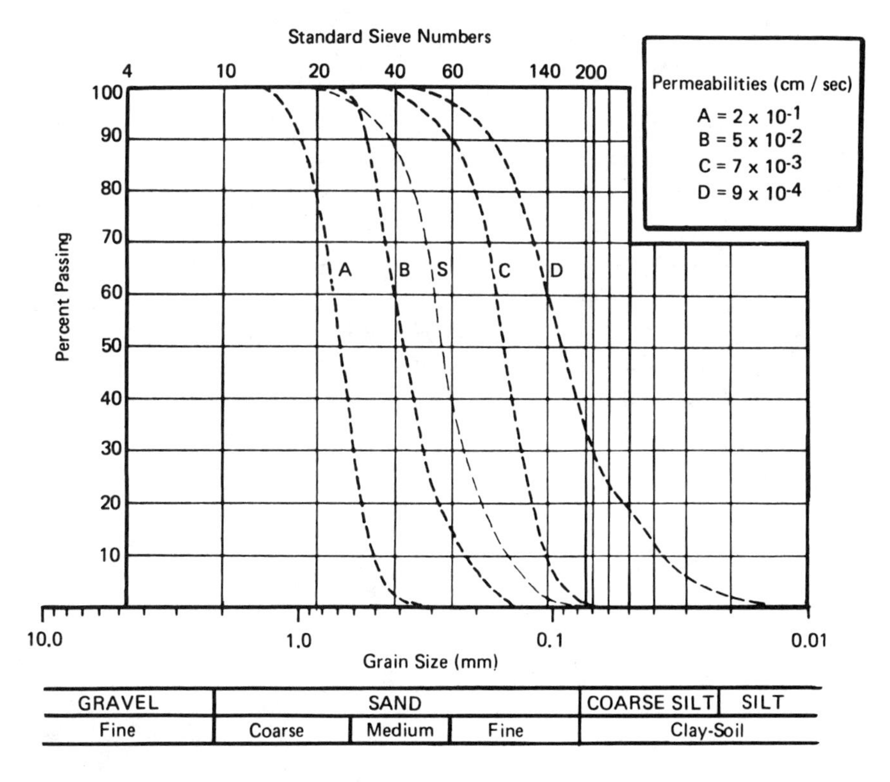

FIGURE 14.1 Gradation of soils used in grouting experiments.

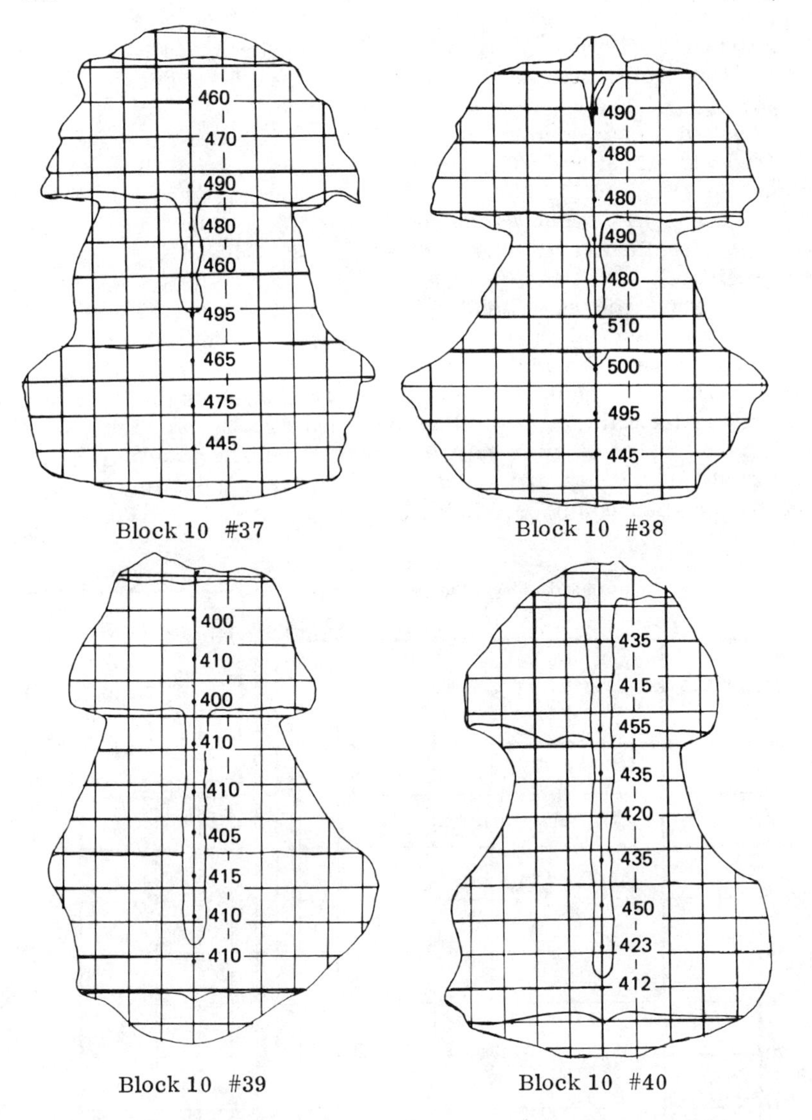

FIGURE 4.2 Size and shape of stabilized soil masses.

At the start of the testing program, data were taken to determine that the methods used resulted in fairly uniform injected volumes at the various lifts. Once this had been established (see block 10 cross-sections), taking of these data was discontinued.

After each block had been completed, the stabilized masses were excavated, weighted and measured, then split vertically through the axis. The cross sections were then sketched or photographed through a one-inch square grid. [See Fig. 14.2.]

TEST RESULTS

The actual levels desired are shown in Table 14.1 for each of the block and injection numbers. Table 14.2 shows the levels obtained, the pumping rate and volume injected, and the data taken from the stabilized masses.

ANALYSIS OF TEST RESULTS

Before any statistical analysis of the data was performed, it was necessary to define a numerical quantity or measure that would adequately represent the injection cross-section. The measure selected, called the penetration index p, is defined as follows:

> The penetration index, p, is equal to the sum of the ratios of the average half-width of the center strata to the average half-width of the outer strata boundaries for the left and right sides of the cross-section.[1]

Under this definition, uniform penetration through the center strata relative to the penetration in the outer strata adjacent to (i.e., within one inch) the center strata is specified by a value of two. This was changed to unity in the graphic representation of test results which are shown in Fig. 14.3.

The values indicated by the graph are the best estimates of the penetration index for a given set of conditions for the independent variables. There is, however, an uncertainty attached to each of these values because of experimental errors and inadequacies of the approximation

---

[1]For practical convenience the penetration index can be defined simply as twice the ratio of the average width of the center strata to the average width of the outer strata measured near the center strata boundaries.

TABLE 14.1 Levels Desired

	No.	X_1, Pulling dist. (in.)	X_2, Retraction time (s)	X_3, Gel time (s)	X_4, X_4, Viscosity (cP)	X_5, Middle sand stratum
Block 1	1	1 3/4	40	30	1.6	Coarsest
	2	3/4	80	30	1.6	Coarsest
	3	3/4	40	30	10.0	Coarsest
	4	1 3/4	80	30	10.0	Coarsest
Block 2	5	3/4	40	90	1.6	Coarsest
	6	1 3/4	80	90	1.6	Coarsest
	7	1 3/4	40	90	10.0	Coarsest
	8	3/4	80	90	10.0	Coarsest
Block 3	9	3/4	40	30	1.6	Finest
	10	1 3/4	80	30	1.6	Finest
	11	1 3/4	40	30	10.0	Finest
	12	3/4	80	30	10.0	Finest
Block 4	13	1 3/4	40	90	1.6	Finest
	14	3/4	80	90	1.6	Finest
	15	3/4	40	90	10.0	Finest
	16	1 3/4	80	90	10.0	Finest
Block 5	17	1 1/4	60	60	4.0	Coarsest
	18	1 1/4	60	60	4.0	Coarsest
	19	1 1/4	60	60	4.0	Coarsest
	20	1 1/4	60	60	4.0	Coarsest
Block 6	21	1 1/4	60	60	4.0	Finest
	22	1 1/4	60	60	4.0	Finest
	23	1 1/4	60	60	4.0	Finest
	24	1 1/4	60	60	4.0	Finest

TABLE 14.1 (Continued)

	No.	X_1, Pulling dist. (in.)	X_2, Retraction time (s)	X_3, Gel time (s)	X_4, Viscosity (cP)	X_5, Middle sand stratum
Block 7	25	1 1/4	60	60	10.0	Medium
	26	1 1/4	60	60	10.0	Medium
	27	1 1/4	60	60	1.6	Medium
	28	1 1/4	60	60	1.6	Medium
Block 8	29	2	60	60	4.0	Medium
	30	1 1/4	90	60	4.0	Medium
	31	1 1/4	60	110	4.0	Medium
	32	1 1/4	60	110	4.0	Medium
Block 9	33	1 1/4	60	10	4.0	Medium
	34	1 1/4	60	10	4.0	Medium
	35	1 1/4	90	60	4.0	Medium
	36	1 1/4	60	60	4.0	Medium
Block 10	37	1 1/4	60	60	1.6	Medium
	38	1 1/4	60	60	1.6	Medium
	39	1 1/4	60	60	10.0	Medium
	40	1 1/4	60	60	10.0	Medium
Block 11	41	2	60	60	4.0	Medium
	42	1/2	60	60	4.0	Medium
	43	1 1/4	30	60	4.0	Medium
	44	1 1/4	60	60	4.0	Medium
Block 12	45	1 1/4	30	60	4.0	Medium
	46	1/2	60	60	4.0	Medium
	47	1 1/4	60	60	4.0	Medium
	48	1 1/4	60	60	4.0	Medium

TABLE 14.2 Levels Obtained

	No.	X_1, Pulling dist. (in.)	X_2, Retraction time (s)	X_3, Gel time (s)	X_4, Viscosity (stormer) (cP)	X_5, Middle sand stratum	Pumping rate (cc/min)	Stabilized wt (kg)	Stabilized vol. (cc)	ABG[a] vol. (cc)	Mass density (glcc)	Volume ratio
Block 1	1	1 3/4	40	29.0	1.6	S	896.0	21,690	11,200	4180	1.94	2.68
	2	3/4	80	29.0	1.6	S	174.0	19,305	10,000	3710	1.93	2.70
	3	3/4	40	29.0	9.5	S	403.0	21,723	11,504	4296	1.89	2.59
	4	1 3/4	80	29.0	9.5	S	418.0	19,720	10,250	3900	1.92	2.63
Block 2	5	3/4	40	88.0	1.6	S	454.0	25,050	12,930	4890	1.94	2.65
	6	1 3/4	80	88.0	1.6	S	434.6	20,610	10,515	4055	1.96	2.60
	7	1 3/4	40	89.8	9.5	S	837.0	20,268	10,300	3900	1.97	2.64
	8	3/4	80	89.8	9.5	S	175.3	18,600	9,750	3740	1.91	2.61
Block 3	9	3/4	40	33.0	1.6	D	397.0	21,580	11,175	4232	1.93	2.64
	10	1 3/4	80	33.0	1.6	D	473.0	19,576	10,198	3780	1.92	2.70
	11	1 3/4	40	33.0	9.0	D	800.0	18,062	9,200	3730	1.96	2.46
	12	3/4	80	33.0	9.0	D	186.8	17,915	9,500	3995	1.89	2.40
Block 4	13	1 3/4	40	88.0	1.6	D	803.6	17,735	9,278	3745	1.91	2.48
	14	3/4	80	88.0	1.6	D	224.1	23,855	12,305	4780	1.94	2.58
	15	3/4	80	90.0	10.75	D	403.3	20,524	10,750	4300	1.91	2.50
	16	1 3/4	80	88.0	10.75	D	414.8	19,025	9,975	3870	1.91	2.58
Block 5	17	1 1/4	60	52.6	3.0	S	436.0	20,224	10,820	4790	1.87	2.36
	18	1 1/4	60	62.5	3.0	S	437.0	19,293	10,590	3930	1.82	2.70
	19	1 1/4	60	62.5	3.0	S	433.0	19,411	10,360	3900	1.87	2.66
	20	1 1/4	60	62.5	3.0	S	436.0	19,921	10,170	3920	1.96	2.60
Block 6	21	1 1/4	60	61.0	4.2	D	413.0	18,086	9,785	3720	1.85	2.63
	22	1 1/4	60	61.0	4.2	D	442.0	18,740	10,020	3980	1.87	2.52
	23	1 1/4	60	61.0	4.2	D	464.0	19,545	10,510	4175	1.86	2.52
	24	1 1/4	60	61.0	4.2	D	463.0	18,402	9,940	4170	1.85	2.38

Block 7	25	1 1/4	60	60.0	9.1	C	370.6	16,280	8,698	3335	1.87	2.61
	26	1 1/4	60	60.0	9.1	C	437.8	19,325	10,350	3940	1.86	2.63
	27	1 1/4	60	60.3	1.6	C	431.0	18,965	10,200	3880	1.86	2.63
	28	1 1/4	60	60.3	1.6	C	445.0	20,180	10,726	4005	1.88	2.68
Block 8	29	2	60	60.0	3.8	C	544.2	17,282	9,010	3265	1.92	2.67
	30	1 1/4	90	50.0	3.8	C	194.7	15,590	8,250	2920	1.89	2.82
	31	1 1/4	60	110.0	3.8	C	369.5	19,409	10,402	3695	1.87	2.82
	32	1 1/4	60	110.0	3.8	C	390.0	20,120	10,610	3900	1.90	2.72
Block 9	33	1 1/4	60	11.5	4.0	C	269.0	12,394	6,520	2690	1.90	2.42
	34	1 1/4	60	11.8	4.0	C	384.0	18,705	9,800	3840	1.91	2.55
	35	1 1/4	90	59.0	4.0	C	235.7	16,745	8,750	3535	1.91	2.48
	36	1 1/4	60	59.0	4.0	C	334.0	15,809	8,330	3340	1.90	2.49
Block 10	37	1 1/4	60	62.5	1.6	C	424.0	20,509	11,152	4240	1.84	2.63
	38	1 1/4	60	62.5	1.6	C	437.0	21,805	11,680	4370	1.84	2.67
	39	1 1/4	60	60.0	10.0	C	367.0	16,910	9,612	3670	1.76	2.62
	40	1 1/4	60	60.0	10.0	C	388.0	18,745	10,306	3880	1.82	2.66
Block 11	41	2	60	59.0	3.3	C	742.0	19,017	9,910	3710	1.92	2.67
	42	1/2	60	64.0	3.3	C	145.0	15,516	8,164	3486	1.90	2.34
	43	1 1/4	30	60.2	3.3	C	814.0	19,068	9,840	3660	1.94	2.69
	44	1 1/4	60	64.0	3.3	C	325.0	18,717	9,760	3654	1.92	2.68
Block 12	45	1 1/4	30	54.0	3.6	C	800.0	19,306	9,950	4000	1.94	2.49
	46	1/2	60	62.0	3.6	C	170.7	20,000	10,390	4095	1.92	2.54
	47	1 1/4	60	59.5	3.6	C	385.0	19,170	9,960	3805	1.92	2.63
	48	1 1/4	60	54.0	3.6	C	410.0	20,843	10,770	4100	1.93	2.63

[a] ABG = acrylamide-based grout.

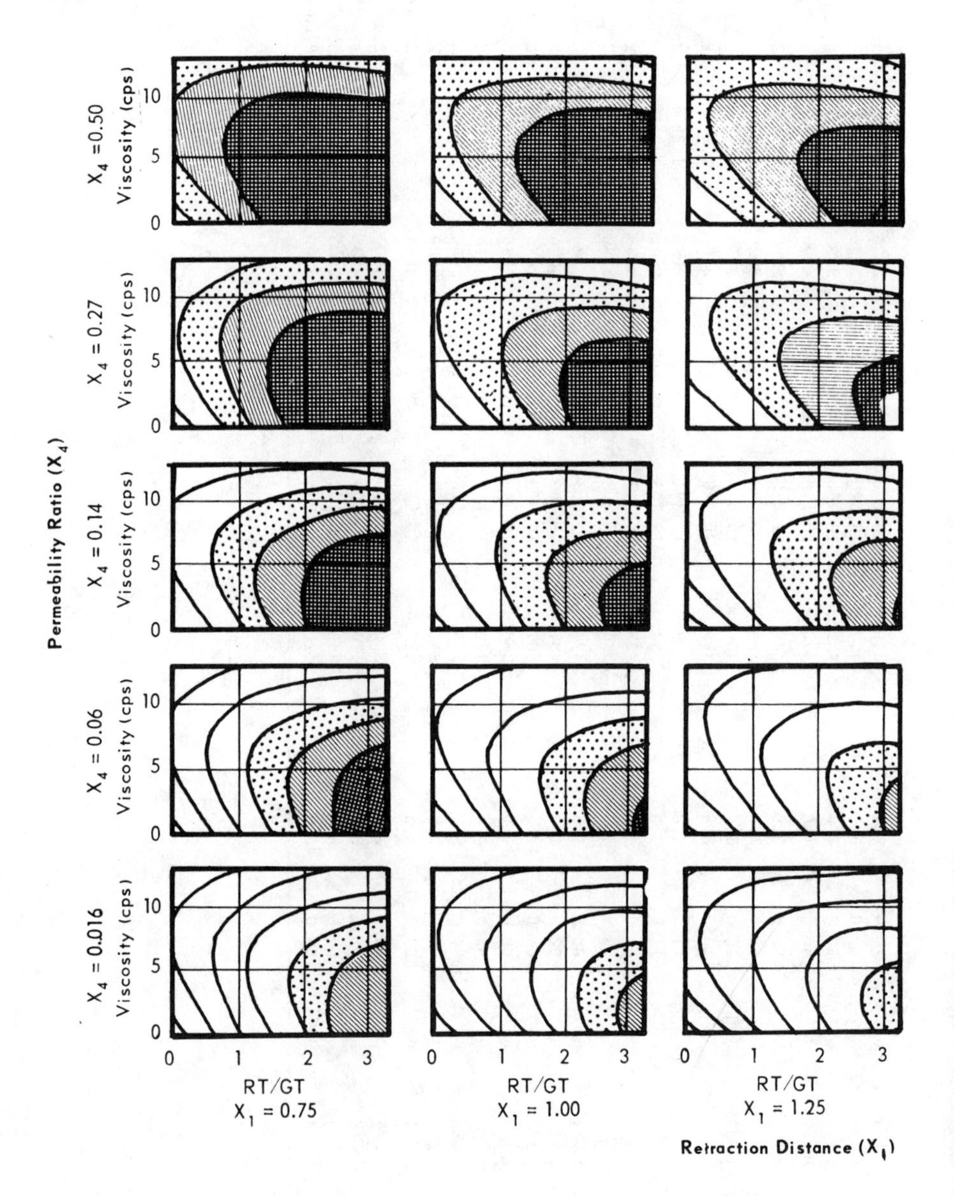

FIGURE 14.3 Relationship of test variables to uniformity of penetration.

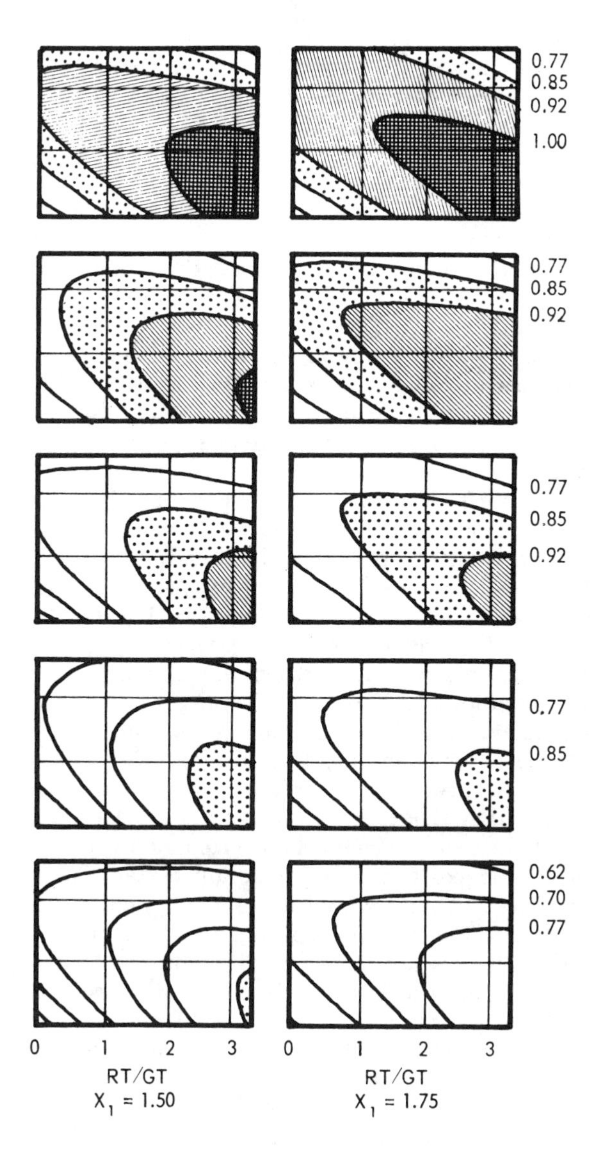

FIGURE 14.3 (Continued)

function. Confidence limits are limits which indicate the range within which there is 0.95 probability that the true value of the response falls. These limits are of the magnitude of 15 to 25 percent of the penetration index.

CONCLUSIONS

Uniform penetration is highly desirable when injecting into stratified deposits. The following conclusions are visually evident from the chart in Fig. 14.3.

1. More uniform penetration is obtained as the pipe retraction distance decreases. The effects are more acute when permeability differences are large.

The theoretically ideal condition in regard to pipe retraction distance would be an infinite number of small retractions, or in effect, continuous retraction. From a practical point of view, however, it would be quite difficult in the field to maintain a constant pulling rate that would correlate exactly with the desired volume of grout per vertical unit distance. Pipe pulling should therefore be accomplished by limiting each lift to the smallest practical distance (the exception to this would be when working in relatively impervious formations, where the pumping rate is so low that long vertical distance of hole must be worked to satisfy the minimum mechanical pumping volumes).

2. Decreasing the gel time (which for any given conditions increases the retraction time-gel ratio) results in more uniform penetration. This trend is adversely affected by a combination of increasing viscosities and retraction distances.

Decreasing the gel time will, in general, tend to give more uniform penetration, as well as decreasing the loss of grout due to dilution and migration. The relationship between gel time and retraction time is of greater significance than the absolute value of gel time. The rate at which beneficial effects accrue increases as the retraction time-gel time ratio increases.

The effects of variations in viscosity should be assessed in the area of other conditions already determined as optimum for uniform penetration, and in the area where uniform penetration is most difficult to obtain. This refers to the use of small retraction distances and short gel times where differences in permeability are greatest. Under these conditions, decreasing the viscosity results in increasing the uniformity of penetration.

Other research sponsored by Cyanamid was done at Lehigh University by R. L. Schiffman and C. R. Wilson, resulting in a 1958 ASTM publication, "The Mechanical Behavior of Chemically Treated Granular Soils."

After a decade of concentrated effort, the grouting research started in the mid-1950s was phased out (in the mid-1960s). Subsequent marketers of grouts (Diamond Alkali, Rayonnier, Borden, etc.) found it unnecessary to do grouting research, since most of the original data was applicable to all grouts. Interest in the type of research detailed above is just now being rekindled. An obvious piece of needed work is the extension of the data presented to other grouts, particularly the silicates. Graduate studies at Northwestern University have started for this purpose.

14.4 GOVERNMENT-SPONSORED RESEARCH

Federal Highway Administration, 1975

In 1975, the Federal Highway Administration became interested in chemical grouting. This interest was stimulated by the highway tunnel construction in Washington, D.C., and the need to keep surface settlements to a minimum. Chemical grout was one of the available construction procedures.

The Federal Highway Administration started from square one and awarded a contract to Halliburton Services for a two-phase report: *Grouting in Soils*, Vol. 1, a State-of-the-Art Report, and *Grouting in Soils*, Vol. 2, Design and Operation Manual, were completed in June 1976 [available through the National Technical Information Service (NTIS), Springfield, Virginia 22161]. Shortly thereafter, a contract was awarded to Soletanche and Rodio to do a (literature) search and evaluation of chemical grouts which are not petroleum based: *Chemical Grouts for Soils*, Vol. I, Available Materials, and *Chemical Grouts for Soils*, Vol. II, Engineering Evaluation of Available Materials, were completed in June 1977. The last two volumes contain a wealth of information about grouting materials, and data drawn from them have been used in previous chapters [6] (also available through NTIS; request Report Nos. FHWA-RD-77-50 and 51).

U.S. Department of Transportation

In 1975, the U.S. Department of Transportation contracted with Stanford University for research to improve the practice of soft ground tunneling. Most of the effort was spent with chemical grouts, and excerpts from the final report [7] are quoted in Chap. 11 and elsewhere. Part of the work consisted of mathematical model analyses, and the general conclusions were later verified by field data and experience. (Reference 7 is in four parts. The last report has been quoted previously. The model study is detailed in Part III, dated June 1978.) A summary of the model studies, as included in Ref. 7, follows:

> Ideally for chemical grouting the soil would be granular and uniform. Then, grout could be injected to form a continuous zone of stabilized soil surrounding the axis of the future tunnel opening. In [Fig. 14.4a] such a hypothetical case is shown. The opening of a tunnel in the stabilized zone creates stresses in the materials around the tunnel. Assuming the stabilized zone to be stiff and strong relative

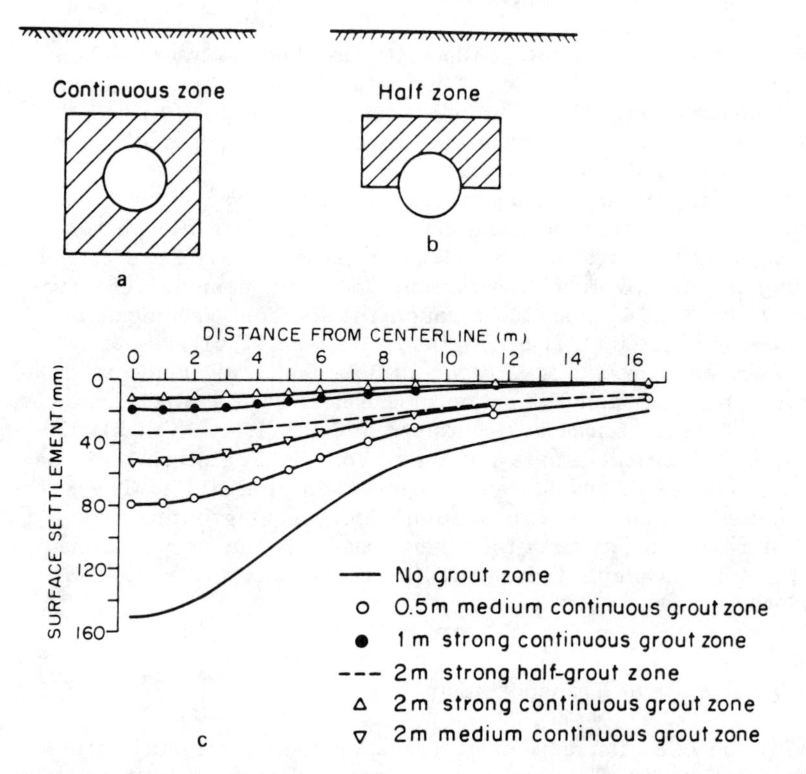

FIGURE 14.4 Surface settlement profiles for two grout zone characteristics. (From Ref. 7.)

to the surrounding untreated soil, the stress changes created by the tunnel opening will be concentrated in the stabilized zone. In finite element studies of this situation by Tan and Clough,* it is shown that a structurally competent stabilized soil zone around a tunnel acts as a compression ring, and prevents stress changes from being felt in the soft soil beyond it. This, in turn, reduces compression effects in the untreated soil and limits settlements at the ground surface above the tunnel. Also, the stabilization of the soil creates a "stand-up"† time for the soil which can be utilized to control movements in the face and crown areas of the tunnels which further reduces detrimental movements. These results show that in order to optimize the benefits of stabilization technology, the stabilized zone should be able to function as a structural support system around the opening.

One of the important design decisions to be made when employing stabilization techniques is how strong and how large the stabilized zone should be in order to control surface movements. An insight into this problem can be obtained by examining the results of a series of finite elements where ground movements were calculated using various sizes and strengths for the stabilized zone around a tunnel in homogeneous, medium, dense sand. The tunnel considered is 7 m in diameter and had a crown depth of 10 m and the zone is assumed to surround the entire tunnel. The thickness, the smallest dimension of the grout zone left after tunneling [see Fig. 14.4], is varied from 0.5 m to 2.0 m. The strengths assigned to the stabilized areas are representative of a weak, moderate, strong and very strong grouted sand, with unconfined compressive strengths of 125, 320, 640 and 1025 kM/m^2 (18, 45, 95 and 150 psi) respectively. The grouted soil is assumed to have a time-independent response. [See the last paragraph in this section.]

Surface settlement profiles for selected cases are plotted in Fig. 14.4 and show several expected trends, i.e., as the stabilized zone gets larger and stronger, surface movements are reduced. In fact, the largest and strongest of

---

*Volume III of Ref. 7.

†See Fig. 11.14 and pertinent discussions.

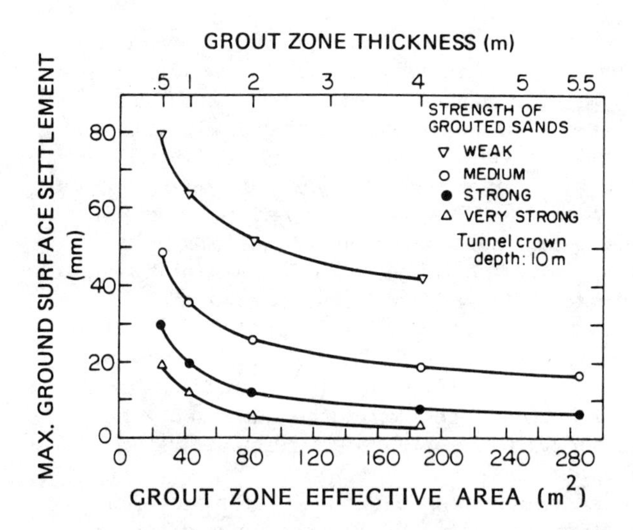

FIGURE 14.5 Effect of grout zone thickness and strength on surface settlement. (From Ref. 7.)

the zones limits surface movements to less than 2 cm.* The nature of the relationship of settlement to zone properties can be better understood by examining [Fig. 14.5], where maximum surface settlements are plotted versus thickness of the grouted zone for several zone strengths. In this diagram it can be seen that the decrease in the surface settlements achieved by increases in zone size are most prominent up to a thickness of about 2 m. Use of larger thicknesses produces only small reductions in settlements, regardless of zone strength. Thus, the 2 m thickness might be viewed as an optimal value; use of larger thicknesses are primarily useful only in providing a factor of safety to account for the fact that the placement of grout in the field is difficult to carry out with a high degree of accuracy.

Figure 14.5 also illustrates that increases in stabilized soil strength decrease settlements, but at a diminishing rate. As with size, there appears to be an optimal value of strength, and the use of higher values than this leads to

*The specific values given refer to the original selected dimensions. They can, of course, be reduced to percentages.

little settlement reduction. The optimal value of unconfined compressive strength appears to be about 525 kM/ m^2 (75 psi).

(Author's Note: At this point the report suggests that laboratory strength testing for grouted samples related to tunnel work be carried out at strain rates that correlate with the unsupported exposure time of grouted soils in the field. Since these times can vary considerably, the creep strength is a more realistic value for design purposes.)

The problems discussed to this point in this report are confined to a 7 m diameter tunnel with a crown depth of 10 m in medium, dense sand. Other cases have been considered in Vol. III [of Ref. 7] and a methodology is presented in that report which allows selection of optimal design values for a range of potential tunneling conditions. [See Appendix D for details.]

In the discussion thus far, it has been assumed that the stabilized soil zone completely surrounds the tunnel opening. This situation is not possible where ungroutable soil horizons are located in the tunnel area. The effects of such conditions are reviewed in the next section of this report. However, it is useful to consider the possibility that a discontinuous stabilized zone might be used even in a homogeneous soil mass. Figure 14.4b shows a "half zone" where only soil above the tunnel is stabilized. Finite element analyses show that such a treatment scheme is ill-advised. Predicted ground movements for a "half zone" are found to be 75 percent larger than those of a continuous zone of smaller thickness but using the same area extent of grouted soil. Thus, for a given volume of grout, a continuous zone is more effective in controlling movements than a "half zone."

In many instances of tunneling, the soils which can be treated by grouting include strata of clay and silt. Under such conditions, a continuous stabilized zone around the tunnel cannot be formed. The degree of influence of ungroutable zones on the effectiveness of the grouting treatment is a function of a number of factors, such as the strength and coherence of the ungroutable layers and their location and thickness. Obviously, if the ungroutable layer is coherent and strong, it can only have a positive influence on the performance of the stabilized zone. On the other hand, if it is weak or non-coherent it will work to the detriment of tunnel performance.

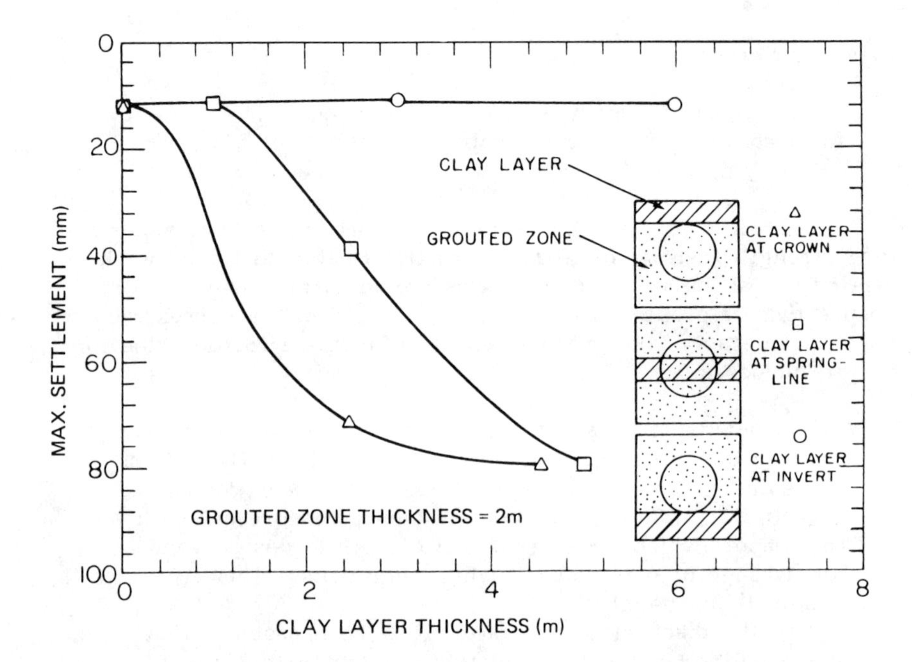

FIGURE 14.6 Effect of ungrouted horizon on surface settlements. (From Ref. 7.)

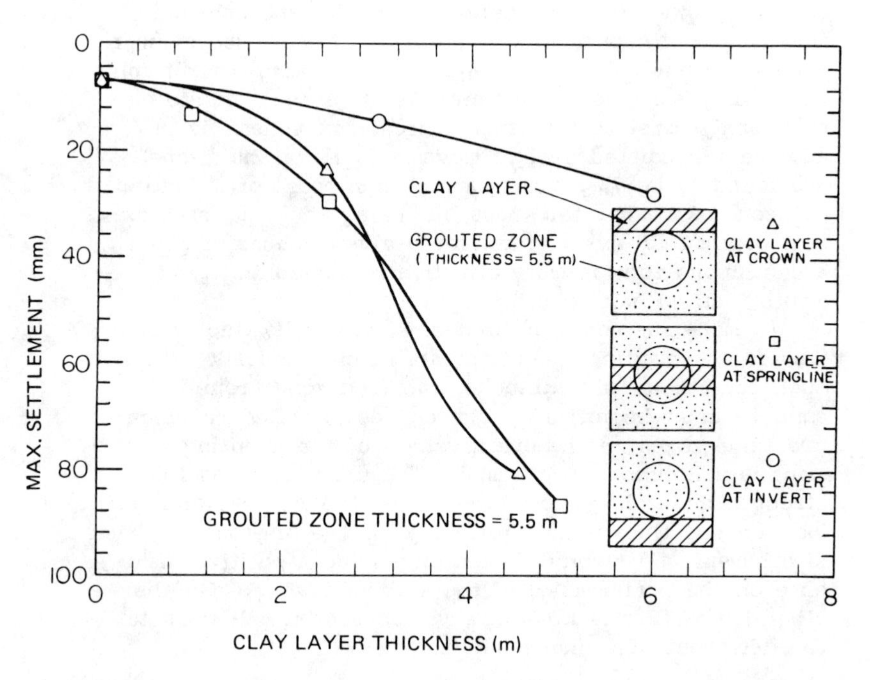

FIGURE 14.7 Effect of ungrouted horizon on surface settlements. (From Ref. 7.)

To illustrate this effect, a series of finite element studies were performed, where a medium, dense sand deposit contains one horizontal layer of a soft clay with an unconfined compressive strength of 48 kN/m^2 (7 psi), which is located in such a position as to disrupt the continuity of the stabilized zone. The positions assumed for the clay layer were varied from the crown, to the springline, and finally at the invert (see Figs. 14.6 and 14.7); its thickness was varied from 0 to 6 m. Analyses were conducted assuming the stabilized zone to have an unconfined compressive strength of 634 kN/m^2 (90 psi) and a thickness of 2 and 5.5 m. The tunnel was given the same dimensions as in the work described previously: diameter = 7 m and crown depth = 10 m.

The predicted ground surface settlement profiles are quantified in terms of the maximum surface settlements at the centerline of the tunnel. In Fig. 14.6 this value is plotted versus clay layer thickness for the 2 m thick grout zone and the differing clay layer positions. It is apparent that the presence of the clay layer at the crown or the springline can lead to larger settlements relative to the homogeneous case, while a clay layer at the invert has relatively small effect. A similar finding can be drawn from Fig. 14.7, a plot of settlements vs. clay layer thickness for a 5.5 m grout zone thickness. A comparison of Figs. 14.6 and 14.7 leads to the following conclusions:

(1) The location of the clay layer at the invert, regardless of thickness, has only a minor effect on maximum settlement.
(2) The location of the clay layer at the crown or the spring line can increase settlements relative to the homogeneous case, the degree of the effect depending upon its thickness relative to that of the stabilized zone.
(3) With the clay layer at the springline or the crown and with a thickness of the clay layer that exceeds 4 m, the size of the stabilized zone (within the range of 2 to 5.5 m thickness) has little influence on settlements.

These results show clearly that the presence of ungroutable horizons in the area to be stabilized can have an important influence. If the ungroutable layer is at the

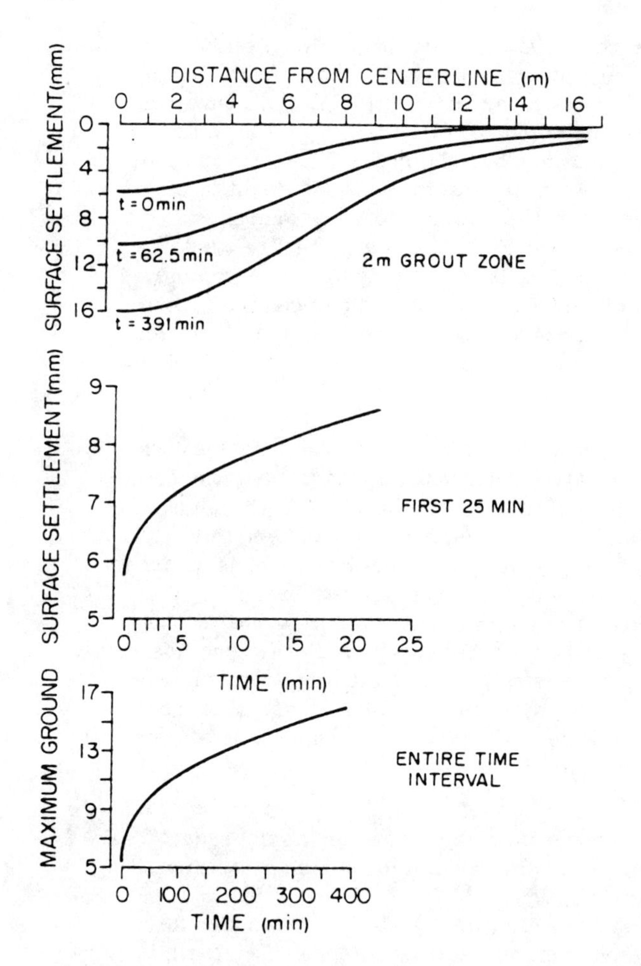

FIGURE 14.8 Creep phenomenon in grouted soils. (From Ref. 7.)

invert or below, it apparently does little to reduce the effectiveness of the stabilization treatment. However, if the layer is at the springline or above, it can work to the detriment of tunnel performance. In these cases, the structural action of the stabilized zone is disrupted. Once the clay layer thickness becomes large relative to the tunnel diameter, increasing the thickness of the stabilized zone above certain minimum values has little influence on behavior. Often calling for larger grouting zones to compensate for the presence of ungroutable layers will have little positive effect. In such cases, the primary benefit to be derived from grouting is control against runs at the face and local reductions of soil compressibility.

Under constant load, chemically stabilized soils typically creep and yield time-dependent deformations. [Field research verifies the creep phenomenon. See Fig. 11.14 and the Locks and Dam 26 Project in Ref. 8.] If unrestrained, creep in the stabilized zone around a tunnel will lead to increased ground movements. Figure 14.8 shows a plot of movements versus time calculated via finite element procedures with allowances for creep in the grouted soil, for several points above a tunnel where no liner is installed. Time is clearly an important variable where grouted soils are unsupported. In tunnel work, the faster a liner or temporary structural support system is placed to restrain the ground, the smaller the movements will be. Thus chemical stabilization can not be used as a substitute for good construction procedures.

Federal Highway Administration, 1977

In 1977, the Federal Highway Administration invited research proposals to develop better methods for field grouting. Hayward Baker Co. was subsequently awarded a contract for "Improved Design and Control of Grouting in Soils." The research objectives were expressed (in part) as follows: "(a) improve design concepts and criteria directed toward enhancing the use of grouting as a more attractive alternative in tunnel construction, and to (b) develop construction control measures that can be employed to assure satisfactory grout treatment of soils. . . . emphasis will be directed toward the evaluation of available mapping methodologies . . . and the development of practical criteria to design, control and ascertain the effectiveness of grouting in a given situation."*

The project was envisioned as a three-part effort. The first part would consist of a review of the problem areas and the associated solution technologies, followed by laboratory scale studies to identify the technique most likely to perform satisfactorily in the field.

The second part would involve field scale pilot testing of the methods selected in part 1 to compare measured grout distribution with the actual distribution determined by excavation.

The third part would be an actual full-scale field job, evaluating the recommendations developed in part 2.

The first phase of the project identified electrical conductivity, acoustic and short pulse radar methods as showing sufficient promise for field scale testing. The lab scale injections made for this study also verified a phenomenon for the silicates which had been reported several decades earlier for the acrylamides: that pump pressures much greater than 1 psi per foot of depth could be used without extensive fracturing of the soil. The discussion of this phenomenon (taken from draft copies of the final report) is of interest:

> Almost all hydrofracture experience in the past has been obtained in impervious materials such as rock and clay. Traditionally, one expects a low flow regime wherein pressure is linear with flow, and a high flow, post-fracture regime wherein flow rate increases dramatically with small increases in pressure. The intersection of the two flow regimes indicates hydraulic fracture of the injected medium. For the more permeable soils the transition from the permeation flow to the fracturing flow regimes is very gradual, indicating that the fracturing process itself is gradual. In contrast to the usual view of a fracture propagating indefinitely as soon as the critical pressure is exceeded, it appears that the crack length in a permeable medium is very limited. A fracture of small length exposes a large surface area in comparison with that of the injection sleeve. Continued propagation of the fracture in a porous medium would require increasingly high flow rates in order to

*The work done on this project was based on the use of sodium silicate grouts. It was envisioned that methods would be developed for monitoring the location of grout during and after pumping. Such methods are needed for grouts which add significant strength to the soil, while for those grouts used for water proofing, in situ permeability tests were considered to be adequate and sufficient.

overcome permeation directly into the already exposed fracture surface. It appears that fracture length in porous media is limited by flow rate. For a fracture to extend indefinitely would require ever increasing flow rates, a situation contrary to that found in impermeable materials. The critical parameter for the hydrofracture during grout injection is not the pressure or flow rate, but the extent to which the soil has been made impervious by previous injection. Thus, ground fracturing due to chemical grouting would be expected to occur much more frequently during grouting of secondary holes rather than primary holes. A fracture will probably run through grouted soil until it intersects an ungrouted zone, into which grout would then permeate.

Field tests were performed in a stratum of silty sand bonded above and below by clay strata. A total of 5530 gal of grout were injected into seven pipes, and measurements were taken of electrical resistivity, acoustic velocity, and radar pulses. Both radar and electrical resistivity methods were found to be effective in monitoring the flow of grout during injection. Electrical resistivity has the disadvantage that probes within the grouted zones are required.

Acoustic emission data for one of the holes is shown in Fig. 14.9. The report comments as follows on these data:

> As the pressure is increased, the number of acoustic emission events abruptly increases, indicating fracture. The count rate peaks about two minutes into the test, and then declines although the pressure is still increasing. It is assumed that at this time the fracture is still propagating, but at a much slower rate. The acoustic emissions drop to the background level when the maximum pumping pressure is achieved at 3 1/2 minutes. The fracture is no longer extending even though pumping continues. This is contrary to expectations for an impermeable material, but typical of our results in sand. When pumping is terminated at six (6) minutes, another burst of events is recorded, indicating that the fracture is closing. In these data it is not clear whether the acoustic events are associated with soil failure at the tip of the fracture, or with stress redistribution throughout the soil mass surrounding the fracture zone. It is clear that the acoustic emission system provides a sensitive detector of hydraulic fracture that can operate in the presence of noisy surface equipment.

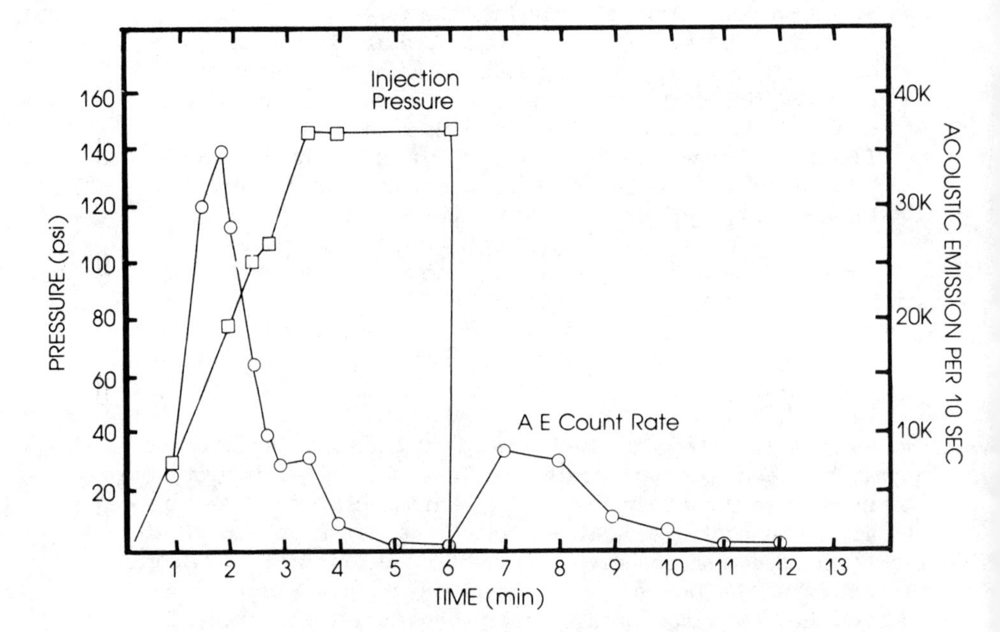

FIGURE 14.9 Acoustic emission data. (Courtesy of Hayward Baker Company, Odenton, Maryland.)

Some use of the developed methods has been made with success on actual grouting projects. This work has not been completed, or the final report written, at the time of this writing. However, it is becoming increasingly obvious that automatic collection and plotting of job-related data can greatly increase the effectiveness of field grouting operations. Development and use of automatic data systems for pressure, flow rate, and acoustic emission are recommended.

Recent Field Research

A very extensive field research program was completed early in 1979, and the final report (prepared for the U.S. Army Corps of Engineers by Woodward-Clyde Consultants) was made available to the public in 1980 [8]. Within the limiting parameters determined by the contract,* the specific purpose of the research† and the grout pumping plant,‡ an extensive amount of data was gathered to evaluate not only the effectiveness of field procedures but the precision of various methods for evaluation of grouting effectiveness.

The direct costs of the project, excluding engineering, earthwork, and dewatering, were over 0.5 million. The total soil volume grouted (almost 40,000 ft^3) was completely excavated and mapped as an irrevocable proof of the grout location. The scope and broad details of the program are defined by the following excerpts from the Phase IV report [8]:

> The purpose of the chemical grouting test program was to assess the feasibility, applicability, and effectivness of injecting silicate-based grouts into Mississippi River alluvial sand. Both low-strength and high-strength grouts were used for the tests. The primary intent of low-strength grouts was to decrease potential displacement and rearrangement of sand grains, and thus increases the stability of the sand, when subjected to vibrations induced by construction activities. The secondary intent was to moderately increase the strength of the sands, which would significantly augment the lateral resistance of piles. The tertiary intent was to increase resistance to erosion and to reduce the permeability of the sand.
>
> The primary intent of high-strength grouts was to increase substantially the bearing capacity and the stability of the sand. The increased bearing capacity must be permanent. The secondary intent was to increase the resistance to erosion and reduce the permeability of the sand.
>
> The objectives of the chemical grouting test program were:

---

*Chemical grouts were limited to those based on sodium silicate in varying concentrations and with different catalyst systems.

†To determine the engineering and economic feasibility of chemical grouting to rehabilitate Locks and Dam No. 26 on the Mississippi River.

‡The grout pumping plant, although well engineered and sophisticated, was designed to be used as a volume control system. A pressure control system would have been more appropriate.

(1) to investigate the technical feasibility of satisfactorily grouting the sand without inducing objectionable heave, lateral movement, and excess pore pressure;
(2) to compare various grouts and provide a basis for selection of chemical grouts that will produce the desired grouted soil properties;
(3) to compare two common grouting methods, the open-bottom pipe and the sleeve-pipe methods;
(4) to establish an optimum grout-hole spacing by comparing the effects of two spacings, 4.2 ft and 6 ft, in achieving the desired grout penetration and uniformity;
(5) to provide bases for establishing criteria for acceptable and optimum grout quantities, grouting pressures, and optimum and maximum grout flow rates; and
(6) to provide cost elements for future estimating purposes.

The intent of the test program was to inject high-strength and low-strength silicate grouts into the upper 20 ft of the recent alluvium deposit underlying the test area between el 400 and el 380. The actual combination of the test variables is shown in Table 14.3.

Open-Bottom Pipe (Method O_1). One grouting method was tested with open-bottom pipes in Subareas 1, 2, and 9; see Fig. 14.10. In this method, an AW steel rod (1.75-in.-od, 1.22-in.-id), fitted with an expendable bottomplug, was driven into the ground to el 376 with a 140-lb or 360-lb drop-hammer. The expendable bottom plug was separated from the grout pipe when the initial grout pressure was applied and the pipe was slightly raised. Grout was injected through the bottom of the grout pipe. The grout pipe was raised 1 ft after each injection step. During each injection step, grout was injected until a predetermined volume of grout, equal to 25 percent of the volume of soil to be grouted, had been pumped or until the grouting pressure reached or exceeded 1 lb/in^2 per foot of soil above the bottom of the grout pipe. Grouting was also discontinued whenever grout leaks out of another pipe or along the grout pipe itself were noticed.

Sleeve-Pipe (Method S_1). Three grouting methods were tested with sleeve-pipes. The sleeve pipes consisted of 2.36-in.-od, 1.77-in.-id PVC pipes. The exterior wall of the pipes were fabricated in 13-in. sections. A number of sections were screwed together to form the required grout pipe lengths. The pipes were provided with two diametrically opposed ports every 13 in. The ports were covered with tight fitting 3-in.-long cylindrical rubber sleeves. The

TABLE 14.3 Proposed Field Variables

<table>
<tr><th></th><th>Grout type</th><th>Grouting method[a]</th><th>Grout hole spacing (ft)</th><th>Maximum grout pressure or rate of pumping</th><th>Test subarea no.</th></tr>
<tr><td rowspan="8">Low-strength</td><td rowspan="5">35% SIROC 142
25% Silicate/aluminate</td><td rowspan="2">O_1</td><td>4.2</td><td rowspan="2">1 psi per foot</td><td>1</td></tr>
<tr><td rowspan="5">6</td><td>2</td></tr>
<tr><td>S_2</td><td>85% of hydraulic fracturing rate of pumping</td><td>4</td></tr>
<tr><td>S_1</td><td>1 psi per foot</td><td>5a</td></tr>
<tr><td rowspan="3">S_3</td><td rowspan="4">85% of hydraulic fracturing rate of pumping</td><td>6</td></tr>
<tr><td rowspan="2">Some cement-bentonite and 25% silicate/aluminate</td><td>7</td></tr>
<tr><td>4.2</td><td>8</td></tr>
<tr><td>28% Silicate/R-600</td><td>S_2</td><td>6</td><td>3</td></tr>
<tr><td rowspan="6">High-strength</td><td>45% SIROC 132</td><td>S_2</td><td>6</td><td>85% of hydraulic fracturing rate of pumping</td><td>5</td></tr>
<tr><td rowspan="3">55% SIROC 142/143</td><td>O_1</td><td>4.2</td><td>1 psi per foot</td><td>9</td></tr>
<tr><td rowspan="2">S_2</td><td>6</td><td rowspan="4">85% of hydraulic fracturing rate of pumping</td><td>10</td></tr>
<tr><td>4.2</td><td>11</td></tr>
<tr><td rowspan="2">46% Silicate/R-600</td><td rowspan="2">S_3</td><td>6</td><td>12</td></tr>
<tr><td>4.2</td><td>13</td></tr>
</table>

[a] O_1 = the open-bottom pipe method with low-pressure criterion and single-stage injection; S_1 = the sleeve-pipe method with low-pressure criterion and single-stage injection; S_2 = the sleeve-pipe method with maximum rate of grout flow criterion and single-stage injection; and S_3 = the sleeve-pipe method with maximum rate of grout flow criterion and multiple-stage injection.

Source: Reference 8.

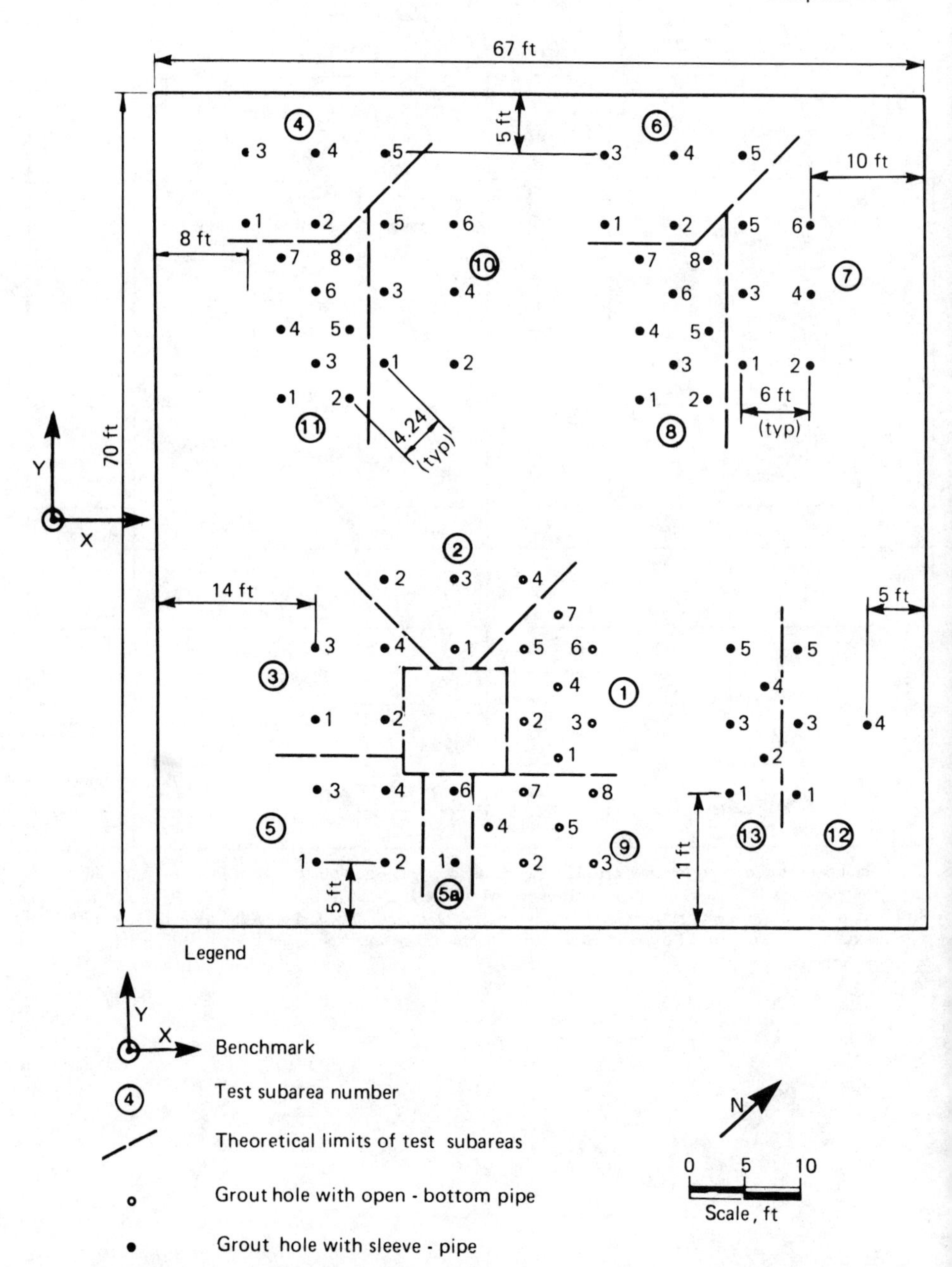

FIGURE 14.10 Layout of grout holes and test subareas. (From Ref. 9.)

sleeve pipes were installed in 3.6-in.-diameter boreholes drilled by rotary method to approximately el 378. Some boreholes were drilled with bentonite drilling fluid; others were drilled with Revert. After the sleeve-pipes were inserted into the borehole, the annular spaces between the pipes and the borehole walls were filled with cement–bentonite grout (sleeve grout), having the following composition:

cement/water (by weight): 0.4
bentonite/water (by weight): 0.03 to 0.04

The purpose of the sleeve-grout was to prevent upward travel of grout along the sleeve-pipe during subsequent injections.

The procedure for injecting grout through a sleeve-pipe involved the use of a double packer connected to a grout pipe extending to the ground surface. The grout pipe was in turn connected to a grouting pump through a rubber hose. Grout injection proceeded from bottom of the sleeve-pipes upward through one or two sleeves at one time.

In grouting Method S, tested in Subarea 5a, chemical grout was injected once at every sleeve level. The grouting pressure was kept below 1 lb/in^2 per foot of soil above the sleeve being injected. This pressure criterion is usually low for sleeve-pipe grouting. Method S was tested on only two holes, mainly for the purpose of demonstrating that the low-pressure criterion is incompatible with sleeve-pipe grouting.

Sleeve-Pipe (Method S_2). This grouting method, tested in Subareas 3, 4, 5, 10 and 11, involved injection of a predetermined volume of chemical grout once at each sleeve level, at a rate of grout flow not exceeding a predetermined maximum allowable value. The volume of grout was determined at 45 percent of the volume of soil to be grouted. The maximum allowable rate of grout flow was established on the basis of the contractor's experience in alluvial grouting and attempts were made to confirm this maximum allowable rate by hydraulic fracturing tests. The intent was to estimate the rate of grout flow inducing hydraulic fracturing during these tests and to establish a maximum allowable rate of grout flow for subsequent injection equal to 85 percent of the fracturing rate. In fact, interpretation of the fracturing tests results was difficult and ambiguous, and the selection of the maximum allowable rate of grout flow generally relied on engineering judgment. Injection was discontinued and restarted at a later time whenever evidence of grout leaking along the sleeve-pipe or through adjacent grout-holes was noted.

Sleeve-Pipe (Method S_3). This grouting method was tested in Subareas 6, 7, 12 and 13. It involved injection of predetermined volumes of chemical grouts at two or more separate times at each sleeve level, at a rate of grout flow not exceeding a predetermined maximum allowable value. The total volume of grout to be injected was determined as 45 percent of the volume of soil to be grouted. Two-thirds of the intended grout volume was injected in a first injection stage. One-third of the intended grout volume was injected in a second injection stage. Generally, the two injection stages were carried out a few days apart. In some grout-holes, where low grouting pressure was recorded during the second stage grouting, a third injection stage was implemented at selected sleeve levels. The volume of grout injected in this third injection stage ranged from 1.2 percent to 13 percent of the volume of soil to be grouted. During any of the grouting stages, injection was discontinued and restarted at a later time whenever evidence of grout leaking along the sleeve-pipe or through adjacent grout-holes was noted. This grouting procedure resulted in actual volume of grout injected greater than 45 percent of the soil volume.

The grouted zone consisted primarily of stratigraphic units E, F, and G with the grain size distribution shown in Fig. 14.11. Properties of these materials before grouting are shown in App. 1. Appendix 2 shows properties of the grouted materials and summarizes the results of all tests.

The first of the objectives listed previously was accomplished. The test results show that the sands in question can be grouted with silicate-based grouts without excessive heave, lateral movement, or pore pressures. In the process of establishing the feasibility a vast amount of data was gathered which may be useful in future projects. Much of these data must be studied in detail to absorb its significance, and perusal of the full report is recommended. Specifically, the results of the before-grouting and after-grouting tests with the pressure meter, the static cone and standard penetration spoon, and the shear wave velocities define ranges of values which can be projected to other sites to gage grouting effectiveness.

One major point of general interest from the test data was that the reduction in formation permeability due to grouting ranged between 2 and 3 orders of magnitude (with acrylamide-grouts, twice the reduction would most probably have been obtained). This would indicate that a sufficient number of voids were left ungrouted so that a limited number of interconnected open passages still existed in the grouted mass. (In zones where the volume of grout injected equaled or exceeded the pore volume, it can be inferred that some lenses and sheets of pure grout must have formed.)

The second objective was also accomplished, leaning more heavily on the laboratory studies than on the field work. All other factors being equal, the rate of grout acceptance by a formation is primarily a function of the grout viscosity. Other factors related to a grout's effectiveness are mainly functions of reliability, strength, and permanence, all of which can be studied efficiently in the laboratory. Of course, the degree of permanence of the silicate-based grouts is well documented in the technical literature and needed no confirmation by field studies. (An exception to this statement might be the grouts which make use of the newer organic catalysts or activators.)

One very important area related to permanence which has had no recognition until very recently for the silicate grouts is that of creep strength. This project is to be commended for its inclusion of both field and laboratory creep studies. These tests verified the conclusions reached from creep studies on other grouting materials that design strength values for grouted soils must be based on creep strength which cannot be determined from typical short-term unconfined compression or triaxial tests. Neither the laboratory nor field tests were continued long enough to give definitive values. The total test time duration was hours or days, whereas it should be weeks and months before a valid conclusion can be drawn that a sample will not fail. Nevertheless, the data are an excellent impetus toward the development of rational design procedures for grouting in areas which are unsupported, such as tunnels and shafts.

The third objective was not met, since the test conditions de facto precluded a comparison of results. Work in open-ended pipes was done with smaller volumes (in fact, the grout volume selected was knowingly only two-thirds that required to fill all the voids, and the pressure selected was 1 psi per foot of overburden depth) and lower pressure than those used with sleeve pipes. Naturally, the penetration was less. On this particular site, the chances are that if the same volumes and pressure had been used, the open-ended pipe would have been just as effective as the sleeve pipe. (Sleeve pipes do have distinct advantages, however, in providing somewhat better control, particularly when manifolding procedures are used, and also make it somewhat easier to regrout a treated zone.)

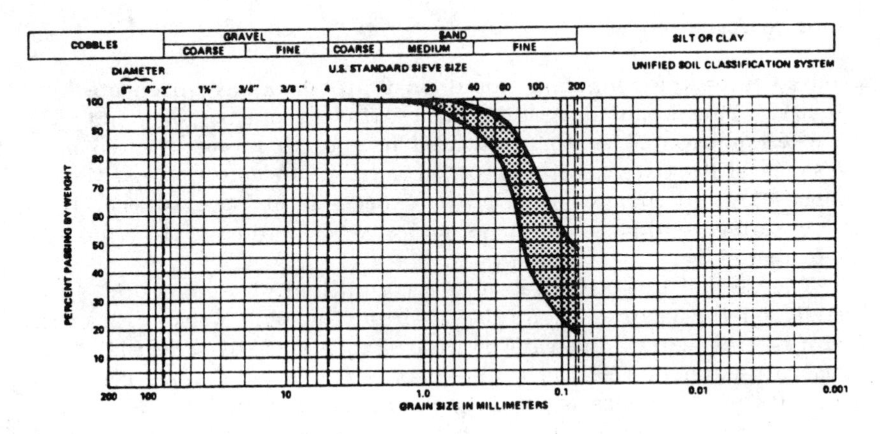

Stratigraphic Unit C

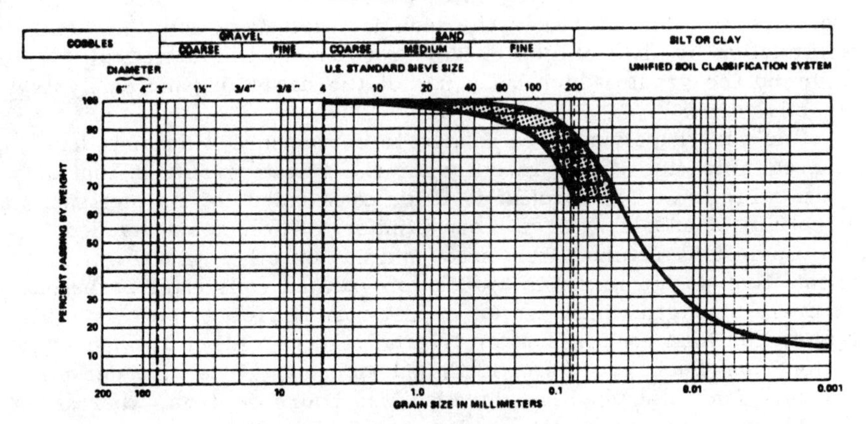

Stratigraphic Unit D

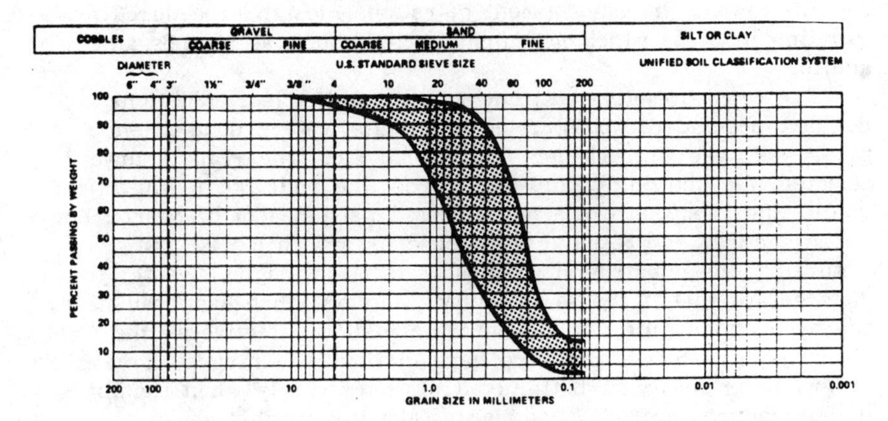

Stratigraphic Unit E

FIGURE 14.11 Grain size distribution curves for soils within the test depth. (From Ref. 9.)

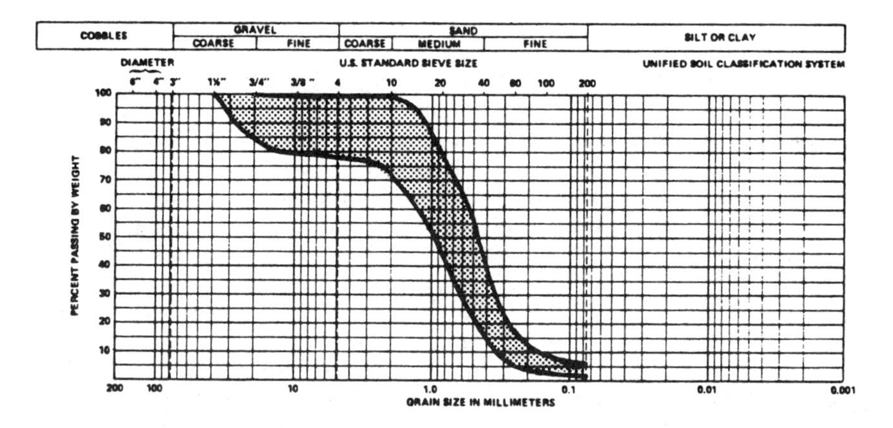

Stratigraphic Unit F

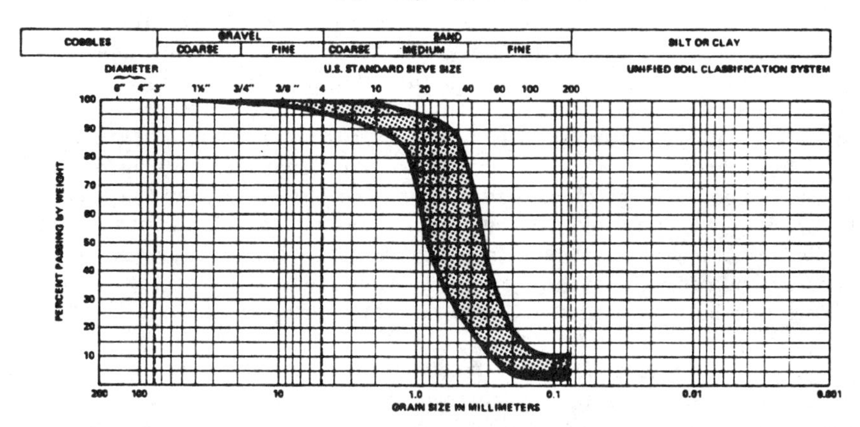

Stratigraphic Unit G

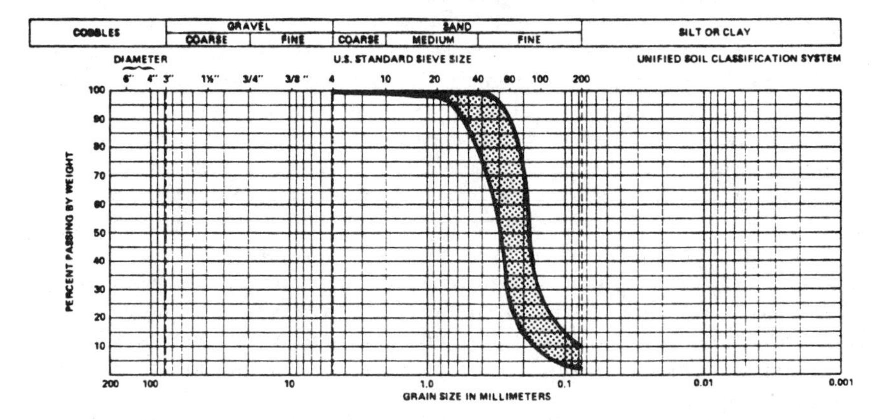

Stratigraphic Unit H

FIGURE 14.11 (Continued)

The common practice of not exceeding 1 psi pressure per foot of overburden is not effective in sleeve-pipe grouting, because a significant portion of the pressure head may be lost in keeping the rubber sleeve expanded against the sleeve grout. Further, if the cracks in the sleeve grout are narrow, there will be additional head loss. To compensate for these factors, sleeve pipe grouting is normally done at pressures well above 1 psi per foot of overburden. On this project, it was desired to permeate the soil without fracturing. Therefore, 85% of the fracturing pressure (as determined by tests on the site) was selected as the maximum grout pressure. However, the grouting equipment did not have pressure control instrumentation. For this reason, the maximum pressure was translated into a corresponding flow rate, and flow control was used as the criterion. This process, of course, did not work well and negated the plan to pump without fracturing. While the pumping rate was closely controlled, the nonuniformity of the formation contributed to continual pressure fluctuations, as shown in Fig. 14.12. Many times the peak of the fluctuations exceeded the fracturing pressure, and fracturing did in fact occur extensively. This was first evidenced by the data which indicated that the volume of grout placed in several zones exceeded the pore volume in those zones. Fracturing was later confirmed by excavation and visual examination, which found fractures in every subarea which accepted grout. Possibly significant is the noticeably reduced degree of fracturing in the two subareas where R-600 activator (Rhone-Poulenc, France) was used.

There is little doubt that fracturing contributes greatly to grout penetration and travel. There is also little doubt that the deliberate formation of chemical grout-filled fissures within a soil mass will decrease the shear strength of the soil mass (unless a very strong grout with strength comparable to cement is used), particularly when the creep phenomenon is taken into account. If strength is not an issue, fracturing will generally give better overall stabilization as well as reduce grouting costs. Of course, fracturing will result in vertical and horizontal movement within and at the periphery of the grouted mass. These movements must be monitored to ensure that they do not become objectionable. Surface heave can be measured by conventional methods. Horizontal movements can be measured with an inclinometer. The report contains full data on both movements, which occurred in measurable amounts. (Subarea 5a was also planned for 25%, but the program was canceled.) There are, of course, no data to determine whether heave and horizontal motion would have occurred if the formation had not been fractured.

The fourth objective was not established, since only two different spacings were used, and both were successful under some

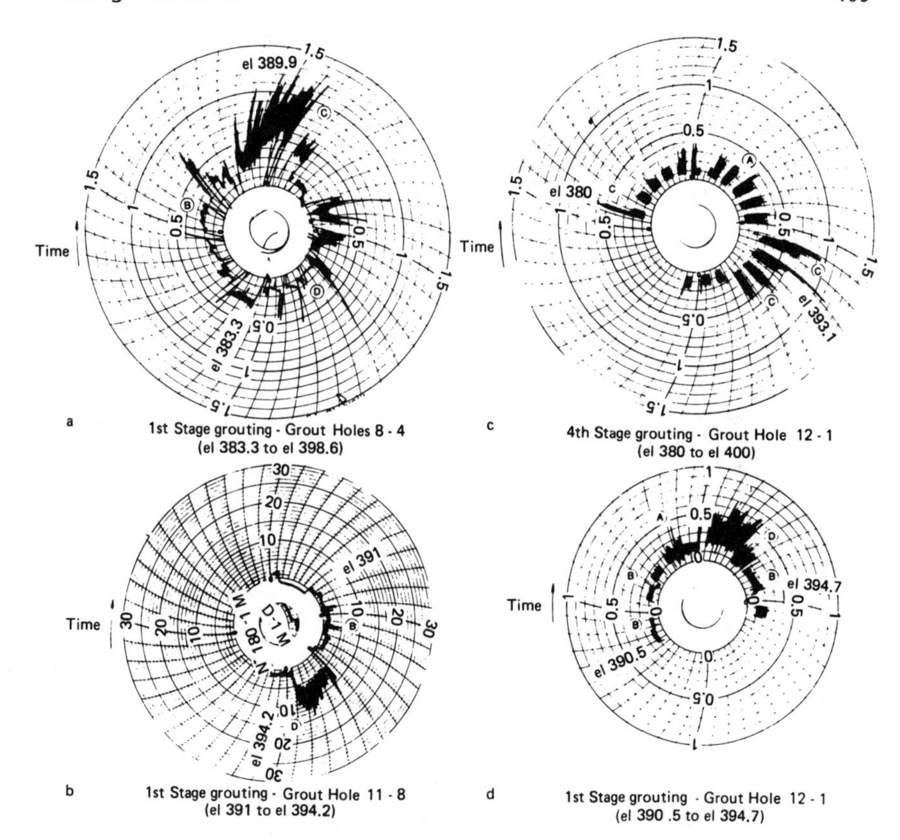

FIGURE 14.12 Actual pumping pressures during grouting. (a) Good grout penetration in fine to medium sand, (b) good grout penetration in medium to coarse sand, (c) difficult grout penetration in fine or already grouted sand, and (d) hydraulic fracturing. (From Ref. 9.)

conditions and unsuccessful under others. There were four separate areas where spacing was compared, as shown in Table 14.4, but in each area at least one other factor varied. Further, the variations in spacing from 6 ft to 4.24 ft were done by adding a grout hole in the center of a square formed by four grout holes 6 ft apart. Thus, the comparison becomes one between two-row and three-row patterns rather than between different spacing. Of course, three-row patterns show up advantageously.

With the exception of test subareas which used open-ended pipes and a grout volume equal to 25% of the soil volume, the

TABLE 14.4 Grout Volume Injected in Each Subarea

Test subarea	Grout hole spacing (ft)	No. of grout holes	Grout type	Grouting method	Theoretical vol. of grout to be injected	Actual[a] vol. of grout injected
1	4.24	7	35% SIROC 142	O_1	4,640 (25)	5,584 (30.1)
2	6	4	35% SIROC 142	O_1	5,330 (25)	5,348 (25.1)
3	6	4	28% Silicate/R-600	S_2	6,710 (45)	6,231 (41.8)
4	6	5	35% SIROC 142	S_2	11,860 (45)	11.836 (44.9)
5	6	4	45% SIROC 132	S_2	9,550 (45)	9,548 (45)
5a[b]	6	2	25% Silicate/Aluminate	S_1	2,045 (25)	303 (3.7)
6	6	5	25% Silicate/Aluminate	S_3	11,930 (45)	12,246 (46.2)
7[c]	6	6	25% Silicate/Aluminate Cement-bentonite[d]	S_3	13,743 (45)	16,049 (52.5)

8[c]	4.24	8	25% Silicate/aluminate Cement-bentonite	S_3	8,900 (45)	11,100 (56.1)
9	4.24	6	55% SIROC 142	O_1	4,000 (25)	3,357 (21)
10	6	6	55% SIROC 132/142	S_2	14,360 (45)	13,750 (43.1)
11	4.24	8	55% SIROC 132/142	S_2	9,550 (45)	9,353 (44.1)
12[e]	6	4	46% Silicate	S_3	9,350 (45)	11,243 (54.1)
13[e]	4.24	5	46% Silicate/R-600 Cement-bentonite	S_3	5,870 (45)	7,574 (58.1)
				Total volume of grout (gal):	118,000 (approx.)	123,522

[a] First figure is volume in gallons; second figure in parentheses is volume expressed as a percent of volume of soil to be treated.
[b] Grout take was practically zero; test was discontinued.
[c] A very small volume of cement-bentonite grout was injected in every grout hole as a first-stage grouting.
[d] Cement-water by weight = 0.25.
[e] A very small volume of cement-bentonite grout was injected in one grout hole as a first stage grouting.
Source: Reference 6.

Source: Ref. 9.

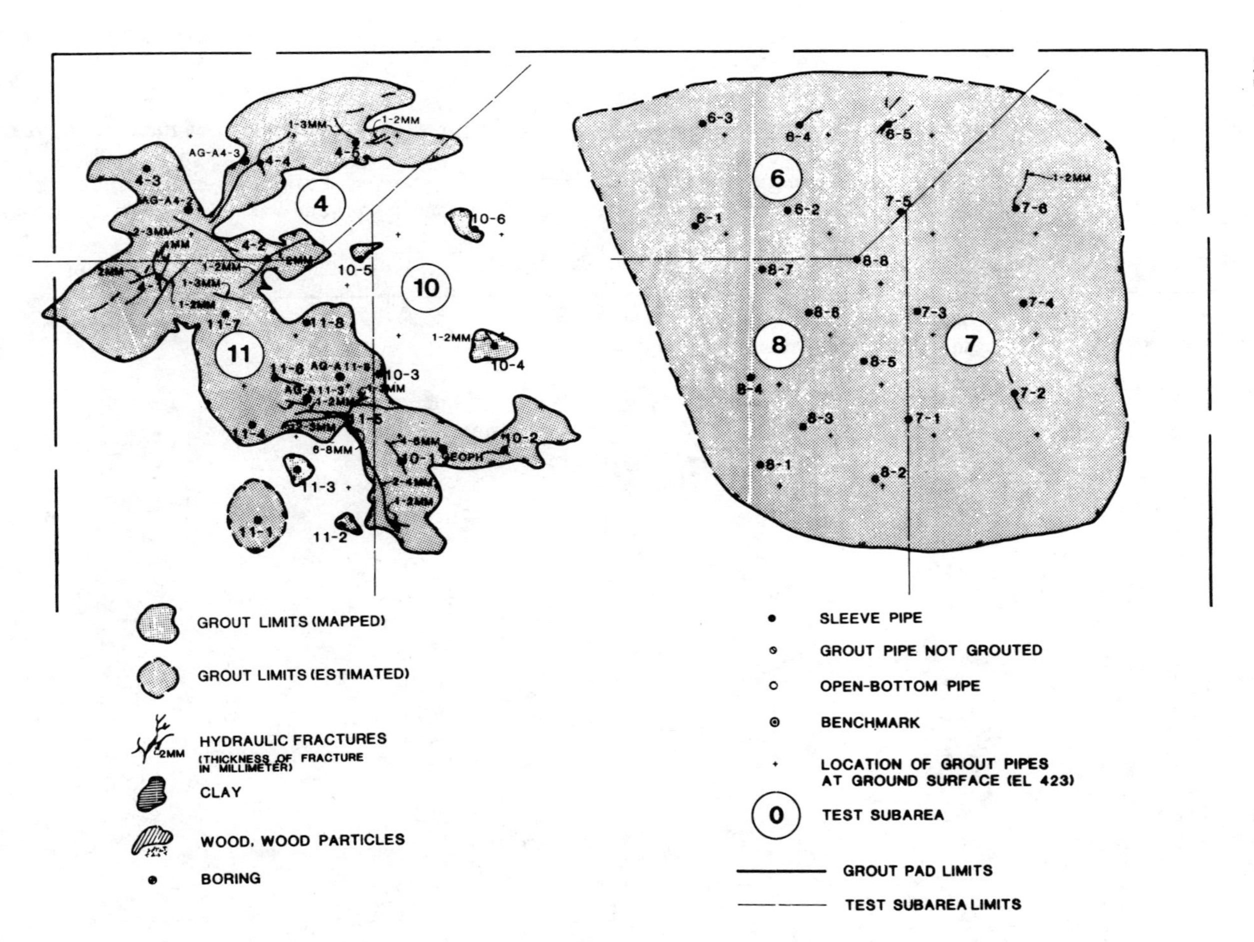

GROUT LIMITS (MAPPED)
GROUT LIMITS (ESTIMATED)
HYDRAULIC FRACTURES (THICKNESS OF FRACTURE IN MILLIMETER)
CLAY
WOOD, WOOD PARTICLES
BORING
SLEEVE PIPE
GROUT PIPE NOT GROUTED
OPEN-BOTTOM PIPE
BENCHMARK
LOCATION OF GROUT PIPES AT GROUND SURFACE (EL 423)
TEST SUBAREA
GROUT PAD LIMITS
TEST SUBAREA LIMITS

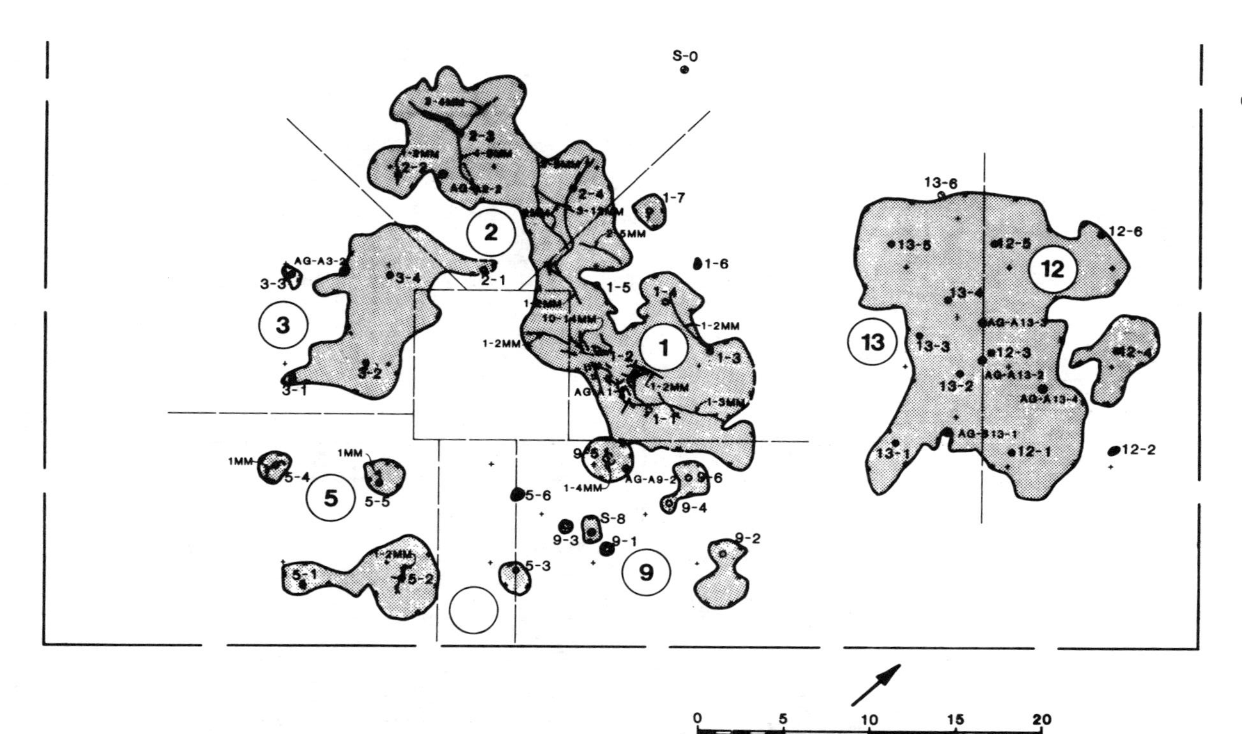

FIGURE 14.13 Horizontal section through test area at elevation 390. (From Ref. 9.)

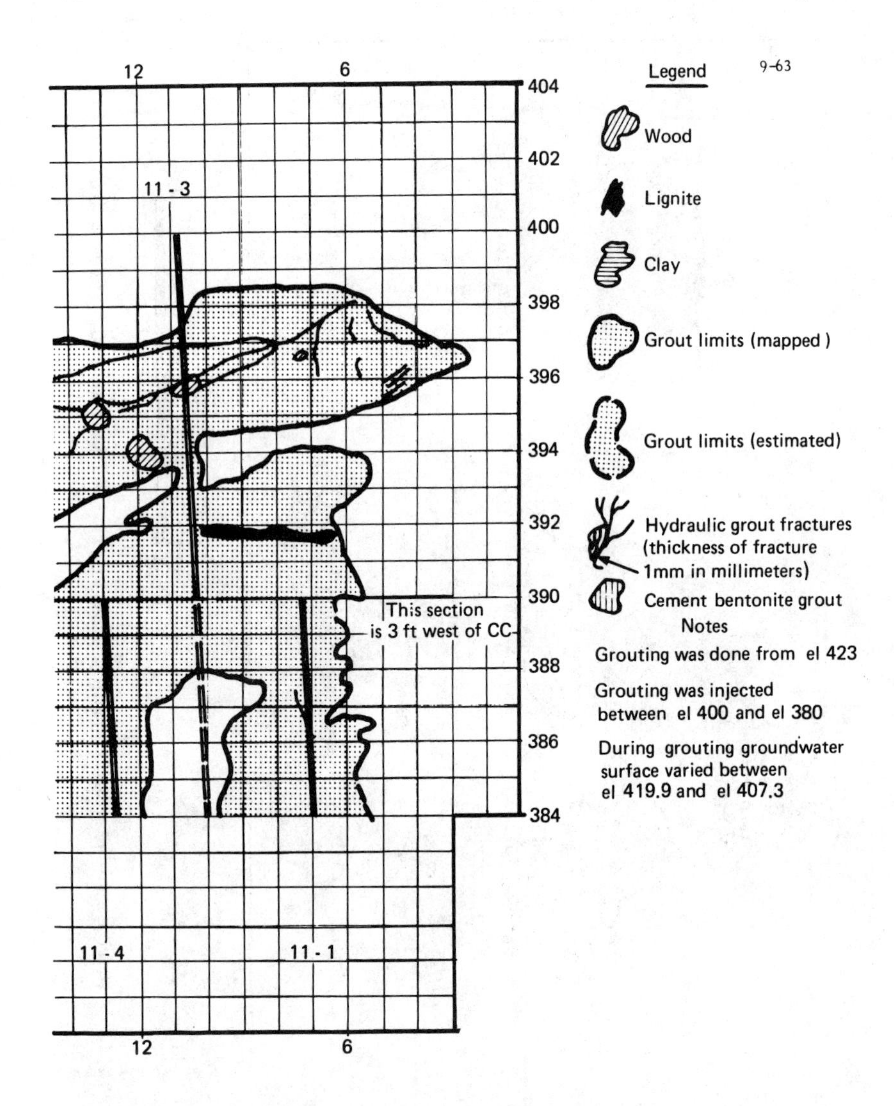

FIGURE 14.14 Vertical section through test area. (From Ref. 9.)

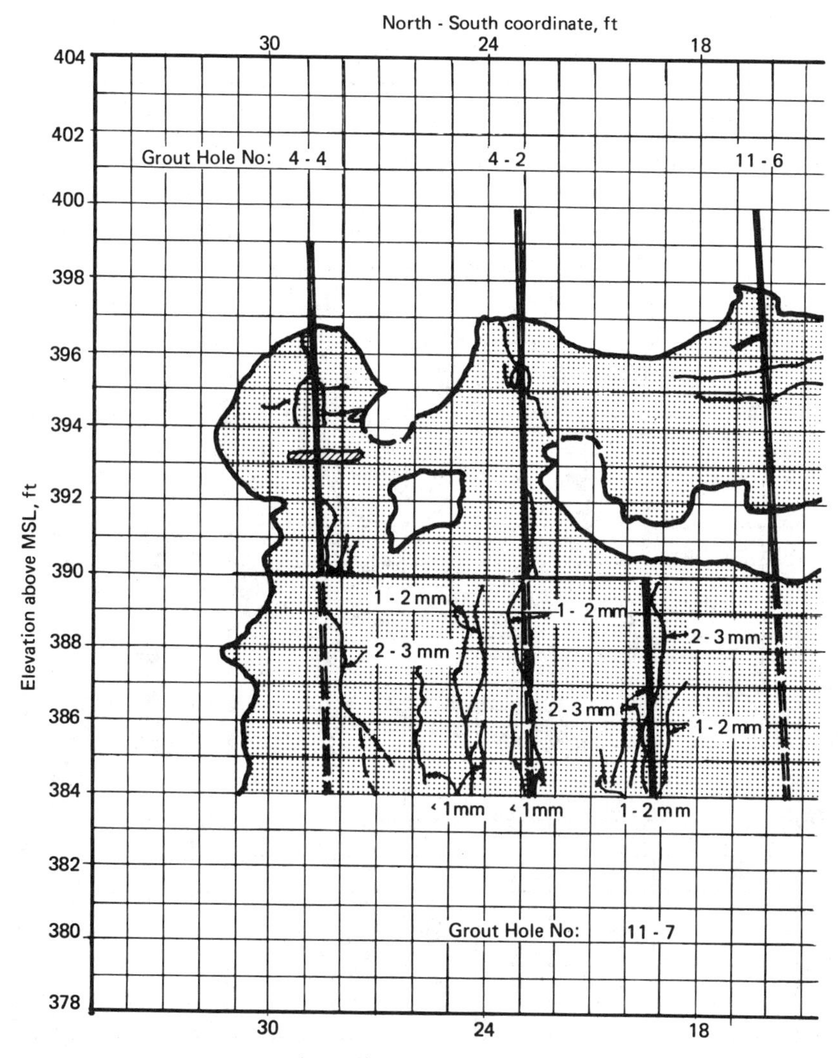

FIGURE 14.14 (Continued)

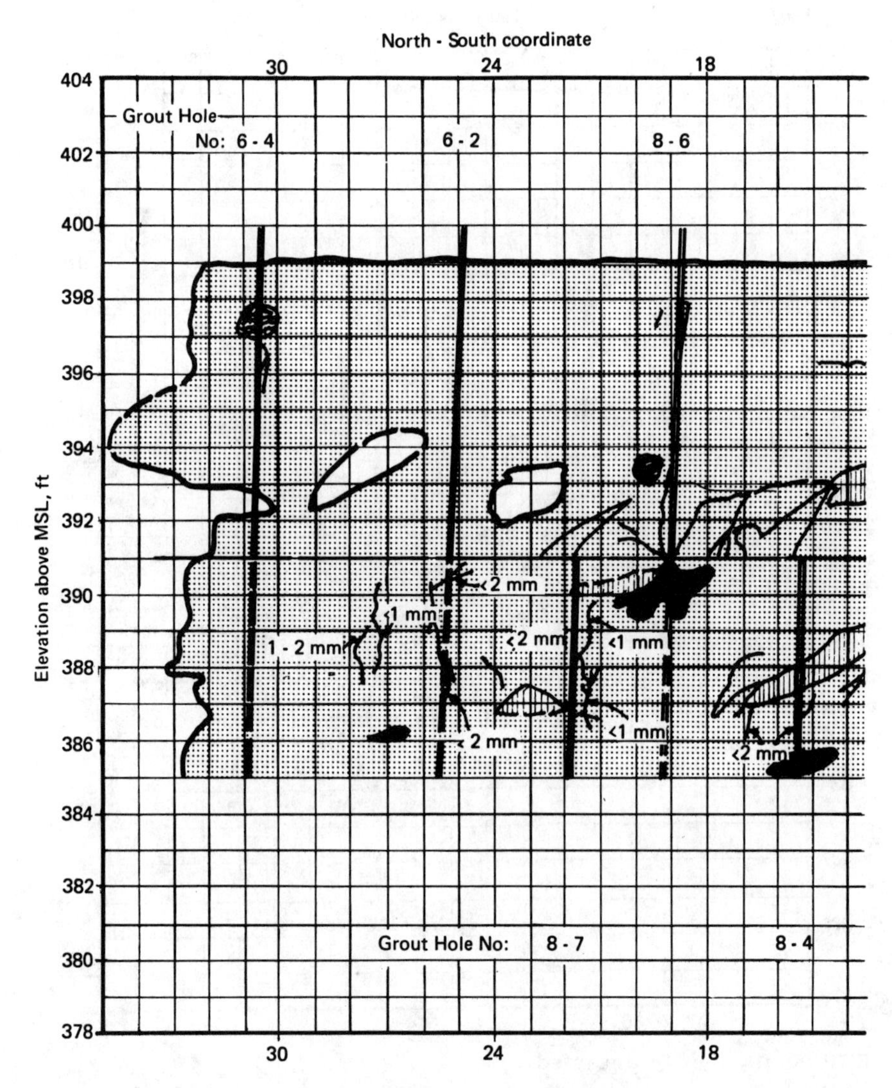

FIGURE 14.15 Vertical section through test area. (From Ref. 9.)

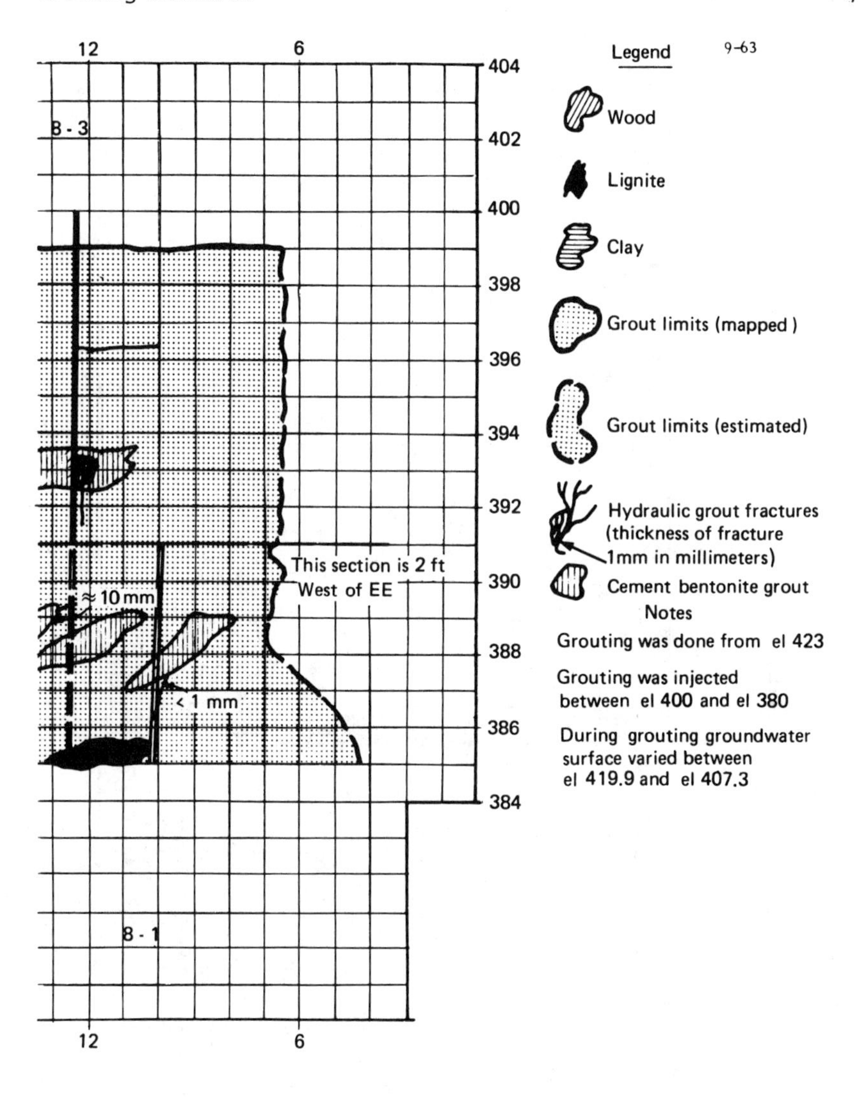
12
6
404
402
400
398
396
394
392
390
388
386
384
B - 3
B - 1
≈ 10 mm
< 1 mm
This section is 2 ft
West of EE
Legend
9–63
Wood
Lignite
Clay
Grout limits (mapped)
Grout limits (estimated)
Hydraulic grout fractures
(thickness of fracture
1mm in millimeters)
Cement bentonite grout
Notes
Grouting was done from el 423
Grouting was injected
between el 400 and el 380
During grouting groundwater
surface varied between
el 419.9 and el 407.3

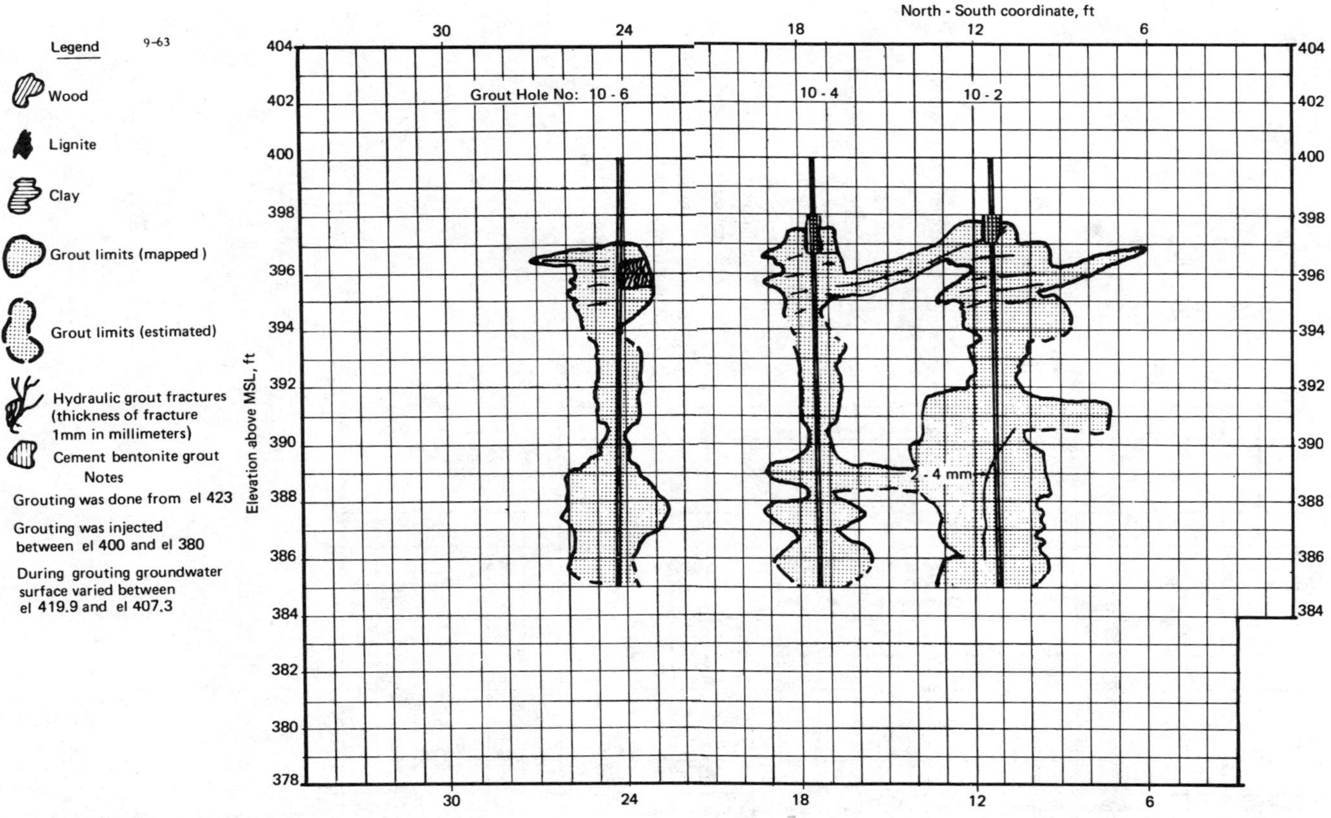

FIGURE 14.16 Vertical section through test area. (From Ref. 9.)

planned grout volume for all other areas was 45% (see Table 14.4 for complete details). The actual volume placed varied from 42% to 58%. The values are much higher than the porosity and indicate greater grout travel and/or fracturing (assuming the grout set up as scheduled). Either condition should result in more complete stabilization of the zones near the grout pipes. The actual horizontal spread of grout at elevation 390, as determined by excavation, is shown in Fig. 14.13. Only in subareas 6, 7, 8, 12, and 13 is stabilization complete around the grout holes. This is consistently true in the other horizontal plots shown in the report and is verified by the vertical sections shown in Figs. 14.14 to 14.17. (The vertical sections can be properly located and oriented by comparing the grout hole numbers with those in Fig. 14.13.)

It is significant that the most complete stabilization occurred in the zones where grouting method S_3 (see page 292) was used. However, both Figs. 14.15 and 14.17 are also in zones where three-row grout patterns were used. Figure 14.14 shows clearly the benefit of a three-row pattern over a two-row pattern. Therefore, the only conclusion that can be drawn is that the best method of all those tested used sleeve pipes with multiplestage grouting in a three-row pattern. This, of course, could have been predicted prior to the experiment.

The value of the field work does not lie in the obvious results which were in reality well known prior to this large project but rather in the collected mass of data which, when finally analyzed and interpreted, may answer the questions posed by anomalies in the data. Although subarea 10 was a two-row pattern, a high-strength grout was placed in large enough volume (see Fig. 14.16) for stabilization to have been much more complete than actually occurred. What happened to the grout? Did it fail to set up, and if so, why? Did it move into other test zones? If so, are the results shown for other zones misleading? Similar questions arise in other areas and await explanation. When provided, these new data may well accomplish the fifth and six objectives of the project.

14.5 AVERAGE SOIL PROPERTIES BEFORE GROUTING

The average soil properties before grouting are shown in Table 14.5.

14.6 SUMMARY OF GROUTED SOIL PROPERTIES

Summary of grouted soil properties is shown in Table 14.6.

TABLE 14.5 Average Soil Properties Before Grouting

			Index properties				Stress	Density					Strength-deformation						Permeability
Unit	w_n (%)	G	Fine content (%)	Gravel content (%)	D_{10} (mm)	D_{60} (mm)	K_o	γ_t (lb/ft^3)	γ_d (lb/ft^3)	$\gamma_{d\,max}$ (lb/ft^3)	$\gamma_{d\,min}$ (lb/ft^3)	D (%)	PMT	PLT	CPT	CSV	$\bar{\phi}$ (°)	P_L (ton/ft^2)	k (cm/s)
A	18	2.65	29	0	—	0.1	—	128	—	—	—	60	—	—	—	—	—	—	—
B	29.7	2.67	—	—	—	—	—	113	—	—	—		—	—	—	—	—	—	6×10^{-5}
C	18	2.65	29.2	0	—	0.18	0.31	128	—	—	—	60	—	—	160	—	32	7	—
D	20	2.65	76	0	—	0.05	1.97	120	—	—	—	50	41.0	—	—	—	—	—	—
E	22	2.65	4	0.4	0.17	0.47	0.64	125	102	108	86.4	65	33.1	—	320	1800	35	8.7	1.7×10^{-3}
F	22	2.65	2.1	5.9	0.28	0.75	0.79	129	106.5	112.5	94.3	70	53.1	197	320	1730	37	20.3	3×10^{-2}
G	22	2.66	2	0.9	0.25	0.53	0.68	128	105.6	114.4	95.5	68	137.8	91	350	1830	39	26.8	5×10^{-3}
H	26	2.66	4.8	0.3	0.14	0.28	0.65	132	104.6	107.4	84.9	90	—	—	600	—	40	49	6×10^{-4}
I	—	—	—	—	—	—	—	—	—	—	—	70	—	—	—	—	—	—	2×10^{-3}
J	—	—	1.9	5.8	0.3	1.0	—	—	—	—	—	70	—	—	—	—	—	—	3×10^{-2}

[a] A = Medium-dense brown fine SAND with some silt to silty fine SAND (SM), FILL. B = Soft to stiff brown silty CLAY to clayey SILT (CH-CL-ML), FLOOD PLAIN DEPOSIT. C = Medium-dense brown silty fine SAND to fine SAND with some SILT (SM), RECENT ALLUVIUM. D = Firm to stiff brown fine sandy SILT to clayey SILT with trace sand (ML), RECENT ALLUVIUM. E = Medium-dense brown to gray fine to medium SAND with a trace of silt (SP-SM/SP), RECENT ALLUVIUM. F = Medium-dense to dense gray fine to coarse SAND with a trace of fine gravel and silt (SP), RECENT ALLUVIUM. G = Medium-dense to dense gray fine to medium SAND with a trace of silt and fine gravel (SP), RECENT ALLUVIUM. H = Dense to very dense gray fine SAND with a trace of silt (SP/SP-SM), RECENT ALLUVIUM. I = Medium-dense to dense gray fine to medium SAND with a trace of silt (SP), RECENT ALLUVIUM. J = Medium-dense gray fine to coarse SAND with a trace of silt and fine gravel, occasionally grading with rock fragments (SP), ALLUVIAL OUTWASH.

Notes:

w_n = Natural water content.
G = Specific gravity.
Fine content = Percent by weight of soil particles passing through a No. 200 U.S. standard sieve (0.074 mm).
Gravel content = Percent by weight of soil particles retained by a No. 4 U.S. standard sieve (4.76 mm).
D_{10} = Effective grain size or grain size for which 10% by weight of the soil particles are finer.
D_{60} = Grain size for which 60% by weight of the soil particles are finer.
K_o = Ratio of in situ effective horizontal stream to effective vertical stress.
t = Total unit weight.
d = Dry unit weight.
d max = Maximum dry unit weight.
d min = Minimum dry unit weight.
D_r = Relative density.
E_s = Elastic deformation modulus.
PMT = Pressure meter test.
PLT = Plate load test.
CPT = Static cone penetration test.
CSV = Shear wave velocity measurements.
= Drained angle of internal friction.
P_L = Pressure meter limit pressure.
k = Coefficient of permeability (measured by falling-head borehole permeability test).

Source: Reference 9.

TABLE 14.6 Summary of Grouted Soil Properties

Subarea number	Grout type	Grouting method[a]	Grout hole spacing (ft)	Standard penetration resistance, N (lb/ft)	Static cone penetration resistance, q_c (tons/ft^2)	Shear wave velocity (intact mass), V_s (ft/s)	Index properties: In situ dry density, γ_d (lb/ft^2)	Index properties: Water content, ω_n (%)	Index properties: Specific gravity, G
U	Ungrouted	—	—	20	125	600	105	22	2.65
6	25% Silicate/ aluminate	S_3	6	—	275	—	106.3	20	—
7	25% Silicate/ aluminate	S_3	6	—	280	—	110	17	—
8	25% Silicate/ aluminate	S_3	4.2	22	280	1100	112.4	17	2.68
5a	25% Silicate/ aluminate	S_1	6	20	—	—	—	—	—
3	28% Silicate/ R-600	S_2	6	60	300	—	103.9	17	—
1	35% SIROC 142	O_1	4.2	45	320	1190	110.3	19	2.68
2	35% SIROC 142	O_1	6	42	250	—	110.2	19	2.68
4	35% SIROC 142	S_2	6	100+	450	—	111.7	17	2.66
5	45% SIROC 142	S_2	6	100+	500+	—	—	—	—
12	46% Silicate/ R-600	S_3	6	100+	500+	—	—	—	—
13	46% Silicate/ R-600	S_3	4.2	100+	500+	2110	105[h]	1.8	—
10	55% SIROC 142/132	S_2	6	100	425	—	—	0	—
11	55% SIROC 142/132	S_2	4.2	100+	500+	3547	104[h]	20	2.72
9	55% SIROC	O_1	4.2	90	375	—	—	—	—

TABLE 14.6 (Continued)

Stress		Stiffness						
		Unconfined compression		CID triaxial compression				
In situ horizontal stress, σ_h (tons/ft^2)	Coefficient of earth pressure at rest, k_0	Initial tangent modulus, E_t (tons/ft^2)	Secant modulus at failure, E_s (tons/ft^2)	Initial tangent modulus, E_t (tons/ft^2)	Secant modulus at failure, E_s (tons/ft^2)	Plate load test modulus, E_{plt} (tons/ft^2)	Pressuremeter modulus, E_{pmt} (tons/ft^2)	Cross hole shear wave velocity modulus, E_{csv} (tons/ft^2)
1.5	0.45	—	—	500	150	144	130	1,800
—	—	—	—	—	—	534	—	—
—	—	—	—	—	—	311	—	—
3.5	2.3	58	39	801	145	246	119	4,000
—	—	—	—	—	—	—	—	—
—	—	90[d]	55[d]	—	—	1,613	—	—
2.2	1.3	1,231	757	902	155	338	151	3,500
—	—	625	471	—	—	1,815	—	—
2.6	1.6	12,000	8000	—	—	1,218	459	—
4.4	3	—	—	—	—	8,233	1576	—
—	—	—	—	—	—	911	—	—
4.7	3.3	3,006	2106	2146	880	962	860	27,000
—	—	—	—	—	—	9,290	—	—
3.8	2.5	5,192	4910	6667	3673	12,098	1327	16,000
—	—	—	—	—	—	5,464	—	—

TABLE 14.6 (Continued)

Subarea number	Grout type	Shear strength: CID triaxial compression: Cohesion, C (tons/ft²)	Friction angle, φ (deg)	Unconfined compressive strength, q_u (tons/ft²)	Ultimate stress plate load test, q_{up} (tons/ft²)	Pressuremeter limit pressure, p_L (tons/ft²)	Coefficient of permeability, K (cm/s)
U	Ungrouted	0	39.5	—	5	25	2×10^{-3}
6	25% Silicate/ aluminate	—	—	—	8	—	—
7	25% Silicate/ aluminate	—	—	—	6.4	—	—
8	25% Silicate/ aluminate	0.34	35	0.32	8	59	1.5×10^{-4}
5a	25% Silicate/ aluminate	—	—	—	—	—	—
3	28% Silicate/ R-600	—	—	0.29[d]	11	—	1.2×10^{-5}[e] 1×10^{-4}[f]
1	35% SIROC 142	0.36 0.70[g]	35.5 39.5[g]	4.0	10	27	2×10^{-5}[e] 2×10^{-4}[f]
2	35% SIROC 142	—	—	1.91	12	—	—
4	35% SIROC 142	—	—	18.38	>14	69	1.2×10^{-4}
5	45% SIROC 142	—	—	—	>17	100+	8×10^{-5}
12	46% Silicate/ R-600	—	—	—	>18	—	—
13	46% Silicate/ R-600	3.5	39.5	11.88	>14.5	100	1×10^{-4}
10	55% SIROC 142/132	—	—	—	>15.8	—	—
11	55% SIROC 142/132	5.82	39	15.60	>14	100+	2×10^{-4}
9	55% SIROC	—	—	—	>16	—	—

TABLE 14.6 (Continued)

Creep		
$\overline{\text{CID}}$ triaxial creep strain rate, E[b] (%/min)	Pressuremeter creep radial strain rate, E[b] (%/min)	Plate load test creep strain rate, E[c] (%/min)
7.8×10^{-5}	—	3×10^{-3}
—	—	—
—	—	—
2.6×10^{-4}	1.9×10^{-2}	1.5×10^{-2}
—	—	—
—	—	5.4×10^{-3}
2.2×10^{-4}	—	2.6×10^{-2}
—	—	6.8×10^{-3}
—	—	3×10^{-3}
—	2.7×10^{-2}	—
—	—	2.6×10^{-2}
5.2×10^{-4}	1.4×10^{-1}	3.2×10^{-2}
—	—	—
1.4×10^{-4}	2.6×10^{-2}[c]	—
—	—	1.5×10^{-2}

Source: Ref. 9.

Table footnotes:

[a] See Table 14.3 for definition of grouting method.
[b] CSR = 30%, time range = 10 to 100 min.
[c] CSR = 50% (CSR = constant stress ratio).
[d] From undisturbed borehole samples.
[e] Laboratory permeability test.
[f] Borehole permeability test.
[g] Reconstituted sample grouted in the laboratory.
[h] From excavation block sample.
General note: Unless otherwise noted, laboratory tests were made on excavation block and core samples.
Source: Reference 9.

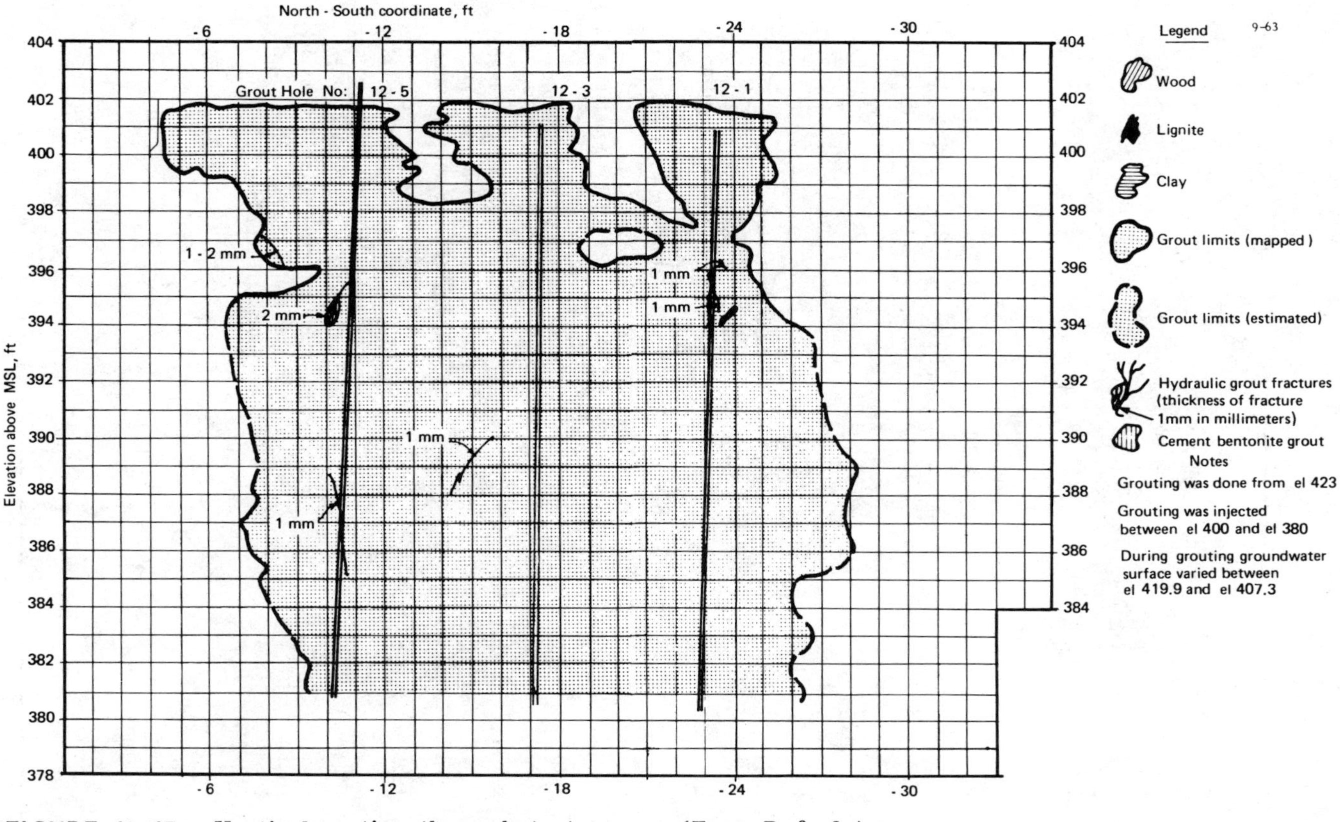

FIGURE 14.17 Vertical section through test area. (From Ref. 9.)

14.7 PRIVATELY SPONSORED RESEARCH

Until very recently, privately sponsored research (except that done by manufacturing companies, whose work has been discussed in Secs. 14.2 and 14.3) was limited to in-house studies done by grouting contractors. Since the major purpose of the research was to provide the researcher with a competitive advantage on grouting jobs, the results of such research were kept in-house as trade secrets. Nonetheless, useful information eventually seeps out and becomes part of the general grouting technology.

In the past several years, at least one university has been doing chemical grouting research with funds provided by private industry (Northwestern University, sponsored by Hayward Baker Company).* This brings to grouting research a much needed point of view divorced from the immediate profit motive. Early results include studies of fluid flow which will add to general knowledge about the design of a grouting operation [10]. As university research grows and progresses beyond the "reinventing-the-wheel" stage, the results of such research will significantly revise and improve chemical grouting technology.

14.8 SUMMARY

Following the acceptance of the Joosten process, research in chemical grouting concentrated for three decades on improving that system. Following the rapid growth of the chemical industry in the 1940s, research by chemical companies quickly produced new and unique materials. As this effort slackened, government agencies took over some of the research effort, concentrating on procedures rather than materials. Finally, in the mid-1970s, universities became interested in chemical grouts, and a limited amount of basic research is underway. By the mid-1980s much of this limited effort had terminated. Currently, research on materials and processes is being carried out mainly in Europe and Japan. As all these efforts coalesce, grouting technology is expanding and improving rapidly.

REFERENCES

1. P. F. Zaccheo, The Stabilization and Impermeabilization of Soils by the Injection of Calcium Acrylate, Master's thesis, Massachusetts Institute of Technology, Cambridge, MA.

---

*Now GKN Hayward Baker.

2. Grouting of Foundation Sands and Gravels, Appendix A, CWI Item No. 506, Technical Memorandum No. 3–408, Jan. 1956, Waterways Experiment Station, Vicksburg, Mississippi.

3. Bibliography on Grouting, U.S. Army Engineer Waterways Experiment Station, Miscellaneous Paper C-78-8.

4. Silicate of Soda as a Soil Stabilizing Agent, Bulletin No. 1, May 1946, Washington, D.C.

5. R. H. Karol, Short gel times with chemical grouts, *Min. Mag.*, 148–152 (Sept. 1967).

6. Chemical Grouts for Soils, Vol. I, Available Materials, Report No. FHWA-RD-77-50, June 1977, Federal Highway Administration, Washington, D.C.

7. C. W. Clough, W. H. Baker, and F. Mensah-Dwumah, Development of Design Procedures for Stabilized Soil Support Systems for Soft Ground Tunnelling, Final Report, Oct. 1978, Stanford University, Stanford, California.

8. Results and Interpretation of Chemical Grouting Test Program, Existing Locks and Dam No. 26, Mississippi River, Phase IV Report of Woodward Clyde to U.S. Army Corps of Engineers, July 15, 1979.

9. Foundation Investigation and Test Program, Existing Locks and Dam No. 26, Prepared by Woodward-Clyde Consultants for St. Louis District Corps of Engineers.

10. A.-B. Huang, R. H. Borden, and R. J. Krizek, Non-Darcian Flow of Viscous Permeants under High Gradients, Technical Report No. HB-4, July 1979, Northwestern University, Evanston, Illinois.

Appendix A: Microfine Cements

This book has dealt exclusively with chemical grouts—materials whose penetrability is related mainly to solution viscosity. All of the chemical grouts (with the possible exception of some of the more viscous polyurethanes, epoxies, and polyesters) will penetrate finer materials than particulate grouts based on normal Portland cement.

If cement is ground more finely, its penetrability is increased. (High early-strength Portland cement, a somewhat finer material, has better penetrability than normal Portland.) With the recent advent of microfine cements, penetrability approaching or equal to chemical grouts is attained. Thus, some details of these new materials are warranted.

Commercial microfine cements were developed in Japan, following the banning of organic grouts some years ago (see Chapter 1 for greater detail). The product, MC-500 (manufactured by Onoda Cement Co., Ltd., Japan), was introduced in the United States in 1984. It is a mixture of finely ground Portland cement and slag, in proportions of about 4:1. Manufacturer's data showing gradation of various cements is shown in Fig. A.1. Penetrability, relative to the chemical grouts, is shown in Fig. A.2.

MC-500 is used either as a batch, with water and a dispersant, or in a two-component mix with sodium silicate in the second tank. Typical field mixes are shown in Tables A.1 and A.2. The batch system yield setting times of 4 to 5 h, and two-component mix gives very rapid setting times of 1 to 3 min. Although some control of setting times is possible for both systems, gel times between 15 min and 1 to 2 h are not reliably attainable.

MC-500 has been successfully used on many projects in North America [1–4]. During the compilation of this second edition, two

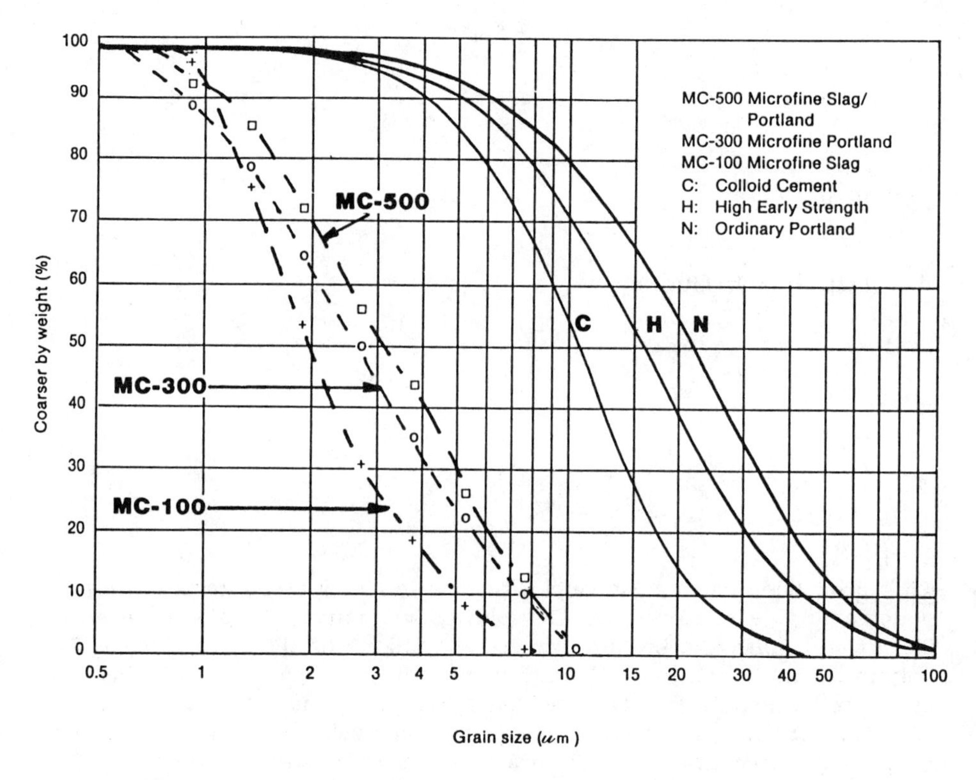

FIGURE A.1 Manufacturer's data showing gradation of various cements.

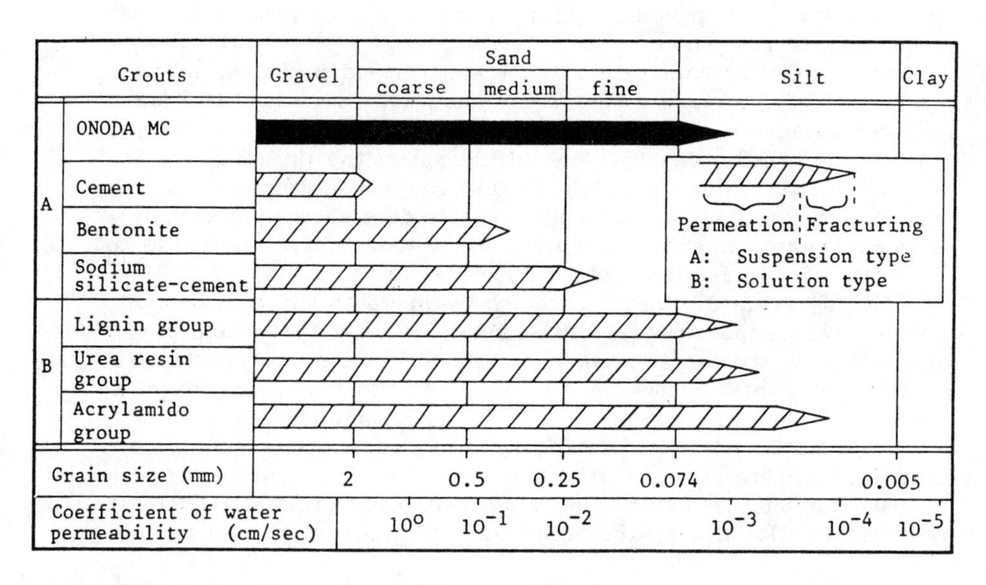

FIGURE A.2 Comparison of permeation of grouts.

TABLE A.1 One-Component Mix[a]

	Metric	U.S. liquid
Water	300 L	79.3 gal
Dispersant	1 L	0.3 gal
MC-100	100 kg	200 (lb)
Sodium hydroxide (50%)	7 L	1.8 gal

[a]Set time 3–5 hr. Water/cement 3:1.

TABLE A.2 Two-Component Mix[a]

	Metric	U.S. liquid
MC tank		
Water	180 L	47.6 gal
Dispersant	0.6 L	0.2 gal
MC-500	60 kg	132 (lb)
Silicate tank		
Water	120 L	31.7 gal
Sodium silicate	90 L	21.1 gal

[a]Set time 5 min. Water/cement 3:1.

new products are on the verge of commercial introduction by Geochemical Corporation (Ridgewood, N.J.). These are MC-100 and MC-300, whose gradation curves are also as shown in Fig. A-1. MC-100 sets in 20 to 30 hr, MC-300 sets in 1 to 3 hr. Both products have better penetrability than M-500.

REFERENCES

1. W. J. Clarke, Performance Characteristics of Microfine Cement, ASCE Preprint 84-023, Atlanta, Georgia, May 1984.

2. D. W. Moller, H. L. Minch, and J. F. Welsh, Ultrafine Cement Pressure Grouting to Control Ground Water in Fractured Granite Rock, ACI SP 83-8 (available from Geochemical Corp., Ridgewood, New Jersey).

3. M. Nagao, Microfine Cement Used to Grout Decomposed Granite Foundation Rock of Kanedaira Dam, ASCE, Denver, Colorado, April 1985.

4. E. Winter, W. J. Clarke, and J. W. Guthrie, Microfine cement grout strengthens foundations, *Concrete International*, American Concrete Institute, Chicago, October 1986.

Appendix B:

Glossary of Selected Terms

ACCELERATOR A material that increases the rate at which chemical reactions would otherwise occur.

ACTIVATOR A material that causes a catalyst to begin its function.

ADDITIVE Any material other than the basic components of a grout system.

ADHESION Bond strength of unlike materials.

ADMIXTURE A material other than water, aggregates, or cementitious material, used as a grout ingredient for cement-based grouts.

ADSORPTION The attachment of water molecules or ions to the surfaces of soil particles.

AGGREGATE As a grouting material, relatively inert granular mineral material, such as sand, gravel, slag, crushed stone, etc. "Fine aggregate" is material that will pass a 1/4-in. (6.4-mm) screen; "coarse aggregate" is material that will not pass a 1/4-in. (6.4-mm) screen. Aggregate is mixed with a cementing agent (such as Portland cement and water) to form a grout material.

ALLUVIUM Clay, silt, sand, gravel, or other rock materials that have been transported by flowing water and deposited in comparatively recent geologic time as sorted or semisorted sediments.

ANGULAR AGGREGATE Aggregate, the particles of which possess well-defined edges formed at the intersection of roughly planar faces.

AQUIFER A geologic unit that carries water in significant quantities.

BASE Main component in a grout system.

BATCH SYSTEM A quantity of grout materials are mixed or catalyzed at one time prior to injection.

BEARING CAPACITY The maximum unit load a soil or rock will support without excessive settlement or failure.

BENTONITE A clay composed principally of minerals of the montmorillonite group, characterized by high adsorption and a very large volume change with wetting or drying.

BOND STRENGTH Resistance to separation of set grout from other materials with which it is in contact; a collective expression for all forces such as adhesion, friction, and longitudinal shear.

CATALYST A material that causes chemical reactions to begin.

CATALYST SYSTEM Those materials that, in combination, cause chemical reactions to begin. Catalyst systems normally consist of an initiator (catalyst) and an activator.

CHEMICAL GROUT SYSTEM Any mixture of materials used for grouting purposes in which all elements of the system are pure solutions (no particles in suspension).

CLOSURE In grouting, closure refers to achieving the desired reduction in grout take by splitting the hole spacing. If closure is being achieved, there will be a progressive decrease in grout take as primary, secondary, tertiary, and quaternary holes are grouted.

COEFFICIENT OF PERMEABILITY The gross velocity of flow of water under laminar flow conditions through a porous medium under a *unit* hydraulic gradient and standard temperature conditions (usually 20°C). Laboratory test results are usually expressed in centimeters per second.

COEFFICIENT OF TRANSMISSIBILITY The rate of flow of water in gallons per day through a vertical strip of the aquifer 1 ft (0.3 m) wide, under a unit hydraulic gradient.

COLLAR The surface opening of a borehole.

COLLOID A substance composed of particles so finely divided that they do not settle out of a suspension.

COLLOID GROUT A grout in which the dispersed solid particles remain in suspension (colloids).

COMMUNICATION Subsurface movement of grout from an injection hole to another hole or opening.

CORE A cylindrical sample of hardened grout, concrete, rock, or grouted deposits, usually obtained by means of a core drill.

CORE RECOVERY Ratio of the length of core recovered to the length of hole drilled, usually expressed as a percentage.

COVER The thickness of rock and soil material overlying the stage of the hole being grouted.

CREEP Time-dependent deformation due to load.

CURE The change in properties of a grout with time.

CURE TIME The interval between combining all grout ingredients or the formation of a gel and substantial development of its potential properties.

CURTAIN GROUTING Injection of grout into a subsurface formation in such a way as to create a zone of grouted material transverse to the direction of the anticipated water flow.

DISPLACEMENT GROUTING Injecting grout into a formation in such a manner as to move the formation; it may be controlled or uncontrolled. See PENETRATION GROUTING.

DRILL MUD A dense fluid or slurry used in rotary drilling; to prevent caving of the bore hole walls, as a circulation medium to carry cuttings away from the bit and out of the hole, and to seal fractures or permeable formations, or both, preventing loss of circulation fluid. The most common drill mud is a water-bentonite mixture, however, many other materials may be added or substituted to increase density or decrease viscosity.

DYE TRACER An additive whose primary purpose is to change the color of the grout.

EMULSION A system containing dispersed colloidal droplets.

ENDOTHERMIC Pertaining to a reaction that occurs with the adsorption of heat.

EPOXY A multicomponent resin grout that usually provides very high, tensile, compressive, and bond strengths.

EXOTHERMIC Pertaining to a reaction that occurs with the evolution of heat.

FINES In soil terminology, material that will pass a 200-mesh sieve.

FISSURE An extensive crack, break, or fracture in rock or soil material.

FLOW CONE A device for measurement of grout consistency in which a predetermined volume of grout is permitted to escape through a precisely sized orifice, the time of efflux (flow factor) being used as the indication of consistency.

FLY ASH The finely divided residue resulting from the combustion of ground or powdered coal and which is transported from the firebox through the boiler by flue gases.

FRACTURE A break or crack in a rock mass. In general usage includes joints; however, the terms are sometimes used in conjunction to distinguish between: joints—breaks that are relatively smooth and planar and usually occur in parallel sets; and fractures—breaks having rough irregular surfaces and generally random orientation.

FRACTURE TERMINOLOGY Relative terms for describing the fracturing of drill core on drill hole logs, etc. (1 ft = 0.305 m):

Degree of fracturing	Average size of pieces
Crushed	5 μ to 0.05 ft
Intensely fractured	0.05 ft to 0.1 ft
Closely fractured	0.1 ft to 0.5 ft
Moderately fractured	0.5 ft to 1 ft
Slightly fractured	1 ft to 3 ft
Massive	3 ft (may contain occasional hairline cracks)

FRACTURING Intrusion of grout fingers, sheets and lenses along joints, planes of weakness, or between the strata of a formation at sufficient pressure to cause the strata to move away from the grout.

GAGE PROTECTOR A device used to transfer grout pressure to a gage without the grout coming in actual contact with the gage.

GAGE SAVER Same as GAGE PROTECTOR.

GEL The condition where a liquid grout begins to exhibit measurable shear strength.

GEL TIME The measured time interval between the mixing of a grout system and the formation of a gel.

GROUND WATER TABLE See WATER TABLE.

GROUT In soil and rock grouting, a material injected into a soil or rock formation to change the physical characteristics of the formation.

GROUTABILITY The ability of a formation to accept grout.

GROUT GALLERY An opening or passageway within a dam utilized for grouting or drainage operations, or both.

GROUT HEADER A pipe assembly attached to a grout hole, and to which the grout lines are attached for injecting grout. Grout injection is monitored and controlled by means of valves and a pressure gage mounted on the header. Sometimes called grout manifold.

GROUT NIPPLE A short length of pipe, installed at the collar of a grout hole, through which drilling is done and to which the grout header is attached for the purpose of injecting grout.

GROUT SYSTEM Formulation of different materials used to form a grout.

GROUT TAKE The measured quantity of grout injected into a unit volume of formation, or a unit length of grout hole.

HARDENER In a two-component epoxy or resin, the chemical component that causes the base component to cure.

HYDRATION Formation of a compound by the combining of water with some other substance.

HYDROSTATIC HEAD The fluid pressure of formation water produced by the height of water above a given point.

INERT Not participating in any fashion in chemical reactions.

INHIBITOR A material that stops or slows a chemical reaction from occurring.

INJECTABILITY See GROUTABILITY.

IN SITU Applied to a rock or soil when occurring in the situation in which it is naturally formed or deposited.

INTERSTITIAL Occurring between the grains or in the pores in rock or soil.

JET GROUTING Technique utilizing a special drill bit with horizontal and vertical high speed water jets to excavate alluvial soils and produce hard impervious columns by pumping grout through the horizontal nozzles that jets and mixes with foundation material as the drill bit is withdrawn.

JOINT A fracture or parting that interrupts the physical continuity of a rock mass. Joints are relatively planar and usually occur in sets which are often subparallel to parallel. The term also refers to a single length, or to the juncture between two connected lengths of casing, drill rod, or grout pipe.

JOINT SET A group of more or less parallel joints.

JOINT SYSTEM Two or more joint sets or any group of related joints with a characteristic pattern, such as a radiating pattern, a concentric pattern, etc.

LIME Specifically, calcium oxide (CaO); also, loosely, a general term for the various chemical and physical forms of quicklime, hydrated lime, and hydraulic hydrated lime.

LIQUID-VOLUME MEASUREMENT Measurement of grout on the basis of the total volume of solid and liquid constituents.

METERING PUMP A mechanical arrangement that permits pumping of the various components of a grout system in any desired proportions or in fixed proportions. Same as PROPORTIONING PUMP or VARIABLE PROPORTION PUMP.

NEWTONIAN FLUID A true fluid that tends to exhibit constant viscosity at all rates of shear.

NO-SLUMP GROUT Grout with a slump of 1 in. (25 mm) or less according to the standard slump test (ASTM C 143). See SLUMP, SLUMP TEST.

PACKER A device inserted into a hole in which grout is to be injected which acts to prevent return of the grout around the injection pipe; usually an expandable device actuated mechanically, hydraulically, or pneumatically.

PENETRABILITY A grout property descriptive of its ability to fill a porous mass. Primarily a function of viscosity.

PENETRATION GROUTING Filling joints or fractures in rock or pore spaces in soil with a grout without disturbing the formation. See DISPLACEMENT GROUTING.

PERCENT FINES Amount, expressed as a percentage by weight, of a material in aggregate finer than a given sieve, usually the No. 200 (74 μ) sieve.

PERMEABILITY A property of a porous solid which is an index of the rate at which a liquid can flow through the pores. See COEFFICIENT OF PERMEABILITY.

PERMEATION GROUTING Replacing the water in voids between the grain particles with a grout fluid at a low injection pressure to prevent creation of a fracture.

pH A measure of hydrogen ion concentration in a solution. A pH of 7 indicates a neutral solution such as pure water. A pH of 1 to 7 indicates acidity and a pH of 7 to 14 indicates alkalinity.

POROSITY The ratio of the volume of voids in a material to the total volume of the material including the voids, usually expressed as a percentage.

POSITIVE DISPLACEMENT PUMP A pump that will continue to build pressure until the power source is stalled if the pump outlet is blocked.

PROPORTIONING PUMP Same as METERING PUMP.

PUMPING TEST A field procedure used to determine in situ permeability or the ability of a formation to accept grout.

REACTANT A material that reacts chemically with the base component of a grout system.

REFUSAL When the rate of grout take is low, or zero, at a given pressure.

RESIN A material that usually constitutes the base of an organic grout system.

RESIN GROUT A grout system composed of essentially resinous materials such as epoxys, polyesters, and urethanes. (In Europe, refers to any chemical grout system regardless of chemical origin.)

RETARDER A material that slows the rate at which chemical reactions would otherwise occur.

RUNNING GROUND In tunneling, a granular material that tends to move or "run" into the excavation.

SAND Specifically, soil particles with a grain size ranging from 0.053 mm to 2.0 mm. Loosely, fine aggregate component of grout— may include particles finer than 0.053 mm.

SAND EQUIVALENT A measure of the amount of silt or clay contamination in fine aggregate as determined by test (ASTM D 2419).

SANDED GROUT Grout in which sand is incorporated into the mixture.

SEEPAGE The flow of small quantities of water through soil, rock, or concrete.

SET TIME A term defining the gel time for a chemical grout.

SHELF-LIFE Maximum time interval during which a material may be stored and remain in a usable condition. Usually related to storage conditions.

SIEVE ANALYSIS Determination of the proportions of particles lying within certain size ranges in a granular material by separation on sieves of different size openings.

SILT Soil particles with grains in the range from 5 μ to 53 μ.

SLABJACKING Injection of grout under a concrete slab in order to raise it to a specified grade.

SLAKING Deterioration of rock on exposure to air or water.

SLEEVED GROUT PIPE Same as TUBE A MANCHETTE.

SPLIT SPACING GROUTING A grouting sequence in which initial (primary) grout holes are relatively widely spaced and subsequent grout holes are placed midway between previous grout holes to "split the spacing." This process is continued until a specified hole spacing is achieved or a reduction in grout take to a specified value occurs, or both.

STAGE The length of hole grouted at one time. See STAGE GROUTING.

STAGE GROUTING Sequential grouting of a hole in separate steps or stages in lieu of grouting the entire length at once. Holes may be grouted in ascending stages by using packers or in descending stages downward from the collar of the hole.

SYNERESIS The exudation of liquid (generally water) from a set gel which is not stressed, due to the tightening of the grout material structure.

TAKE See GROUT TAKE.

TRUE SOLUTION One in which the components are 100% dissolved in the base solvent.

TUBE À MANCHETTE A grout pipe perforated with rings of small holes at intervals of about 12 in. (305 mm). Each ring of perforations is enclosed by a short rubber sleeve fitting tightly around the pipe so as to act as a one-way valve when used with an inner pipe containing two packer elements that isolate a stage for injection of grout.

UNCONFINED COMPRESSIVE STRENGTH The load per unit area at which an unconfined prismatic or cylindrical specimen of material will fail in a simple compression test without lateral support.

UPLIFT Vertical displacement of a formation due to grout injection.

VISCOSITY The internal fluid resistance of a substance which makes it resist a tendency to flow.

VOID RATIO The ratio of the volume of voids divided by the volume of solids in a given volume of soil or rock.

Appendix C: Computer Program for Optimum Grout Hole Spacing for a Chemical Grout Curtain

```
10 REM      GROUTING.BAS by Keith Foglia
20 REM      Written for: Reuben H. Karol instructor
30 REM                   Chemical Grouting
40 REM
50 REM
55 CLS:SCREEN 0,0,0:LOCATE 3,27:PRINT"GROUT CURTAIN DESIGN"
60 LOCATE 9,1:PRINT"    This is a Basic program to compute the most economical s
pacing of"
70 PRINT"     chemical grouting holes for a Grout Curtain, which will cut off the
              flow of ground water."
80 PRINT:PRINT"    It also computes:   Number of Holes to be Drilled"
81 PRINT"                          Volume of Grout to be Placed in each hole"
82 PRINT"                          Cost of Drilling and Placing Pipes"
83 PRINT"                          Cost of Grouting"
90 PRINT"                          Total Cost"
110 COLOR 31:PRINT:PRINT:PRINT"   (Hold down the shift and press the PrtSc key t
o print any screen.)":COLOR 7
115 PRINT:PRINT"                                 Enter M for Methodology"
122 PRINT:INPUT"                                 Press enter to continue";K$
123 IF K$="m" OR K$="M" THEN GOSUB 2000
125 CLS
130 LOCATE 3,1:PRINT"NOTE:   If porosity is unknown but void ratio is known,
                             just press return to go to the next step."
140 LOCATE 2,1:INPUT"Enter Porosity of soil to be treated. If known? n=";N
210 LOCATE 4,1:PRINT:INPUT"Enter Void Ratio of soil to be treated. If known? e="
;E
220 IF N=0 AND E=0 THEN PRINT:PRINT"You must enter at least one!":GOTO 130
230 IF N=0 AND E>0 THEN N=E/(1+E)
240 PRINT:PRINT "Porosity will be taken to be n=";USING"#.###";N
250 PRINT:PRINT"Will job require drilling through Over Burden Soil"
260 INPUT"to get to the area to be grouted?  (Y/N)";YN$
270 IF YN$<>"Y" AND YN$<>"y" AND YN$<>"N" AND YN$<>"n" THEN GOTO 250
280 IF YN$="N" OR YN$="n" THEN GOTO 350
290 PRINT:PRINT"Estimate Cost (in $ Per Foot) of drilling through Over Burden so
il."
300 PRINT"Be sure to include cost of labor and materials."
310 PRINT:INPUT"Enter Over Burden Drilling Cost (in $ Per Foot)";CDO
320 INPUT"Enter depth of Over Burden soil to be drilled through in ft. ";DO
350 LOCATE 18,1:PRINT"Estimate Cost (in $ Per Foot) of drilling through soil to
be treated."
360 PRINT"Be sure to include cost of labor and materials used and placed."
370 PRINT:INPUT"Enter Drilling Cost (in $ Per Foot)";CD
```

```
380 INPUT"Enter depth of soil to be treated in ft. ";D
390 PRINT:INPUT"Enter Cost of Grout in $ per gallon";G:CG=7.48*G
400 CLS:PRINT"RECAP:":PRINT"Void Ratio e=";USING"#.###";N/(1-N);:PRINT"    So Th
e Porosity n=";USING"#.###";N
410 IF YN$="N" OR YN$="n" THEN GOTO 430
420 PRINT"Depth of Over Burden=";USING"#####.###_ ft.";DO
425 PRINT"Cost of Drilling through Over Burden=";USING"#####.###_/ft.";CDO
430 PRINT"Depth of Soil to be Treated=";USING"#####.###_ ft.";D
440 PRINT"Cost of Drilling through Soil to be Treated=";USING"$$#####.##_/ft.";C
D
450 PRINT"Cost of Grout=";USING"$$#.##_/gal.";G;:PRINT"    or  ";USING"$$##.##_/
CF";CG
460 S=SQR(3*CD/((3.14159+4)/4*N*CG)+3*CDO*DO/((3.14159+4)/4*N*CG*D))
470 PRINT:PRINT"Theoretical Grout Hole Spacing For Minimum Cost is ";USING"#####
.###_ ft.";S
480 PRINT:INPUT"Enter Length of curtain (in ft)";L
490 MODL=L/S
500 TC1=(3.14159+4)/4*(L/INT(MODL))*L*D*N*CG+3*L*CD*D/(L/INT(MODL))+CD*D+3*L*CDO
*DO/(L/INT(MODL))+CDO*DO
510 TC2=(3.14159+4)/4*(L/INT(MODL+1))*L*D*N*CG+3*L*CD*D/(L/INT(MODL+1))+CD*D+3*L
*CDO*DO/(L/INT(MODL+1))+CDO*DO
520 IF TC1<TC2 THEN TTLCST=TC1:MODL=INT(MODL):GOTO 530
525 TTLCST=TC2:MODL=INT(MODL+1)
530 S=L/MODL:NMBRHLS=3*MODL+1:DRLLCST=(CD*D+CDO*DO)*NMBRHLS
540 PRINT:PRINT NMBRHLS;"holes need to be drilled.  Each";USING"###.###_ ft. dee
p.";D+DO
550 PRINT"Actual spacing should be s=";USING"##.###_ ft.";S
560 PRINT"Each hole will cost ";USING"$$######.##_ to drill.";CD*D+CDO*DO
570 PRINT"Total Cost of Drilling=";USING"$$########.##";DRLLCST
580 LOCATE 23,30:PRINT"(Press any key for more)"
585 IF INKEY$="" GOTO 585
600 CLS:SCREEN 2:PRINT"Based on three rows of holes in an off-set pattern.  Like
..."
610 FOR I=0 TO 7:FOR J=0 TO 1:CIRCLE(100+I*60,30+J*25),1:CIRCLE(100+I*60,30+25*J
),30:NEXT J:NEXT I
620 FOR I=0 TO 8:CIRCLE(70+I*60,43),1:NEXT I
630 LINE(70,30)-(70,55):LINE(70+8*60,30)-(70+8*60,55)
640 LINE(70,90)-(70,110):LINE(70+8*60,90)-(70+8*60,110):LINE(70,100)-(70+8*60,10
0)
650 LOCATE 13,37:PRINT" LENGTH "
660 LINE(160,70)-(160,90):LINE(220,70)-(220,90):LINE(160,85)-(220,85)
670 LOCATE 11,24:PRINT" S"
680 LINE(370,70)-(370,90):LINE(430,70)-(430,90):LINE(370,85)-(430,85)
690 LOCATE 11,50:PRINT" S"
695 LINE(65,30)-(40,30):LINE(65,55)-(40,55):LINE(50,30)-(50,55)
696 LOCATE 6,6:PRINT" S"
700 VGCY=N*(3.14159*S^2/4*D)*7.48:VGDI=N*(S^2*D-3.14159*S^2/4*D)*7.48
710 LOCATE 15,1:PRINT"The volume of grout to be placed in each hole:"
720 PRINT"                          In each outer hole=";USING"###.###_ gal.";VG
CY
730 PRINT"                          In each inner hole=";USING"###.###_ gal.";VG
DI
740 VG=VGCY*MODL*2+VGDI*(MODL)
750 PRINT:PRINT"Total volume of grout to be used=";USING"########.###_ gal.";VG

760 GRTCST=VG*CG/7.48
770 PRINT"Total cost of grouting=";USING"$$#########.##";GRTCST
780 PRINT:PRINT"Total cost of the job=";USING"$$##########.##";GRTCST+DRLLCST
790 LOCATE 23,20:PRINT"(Press any key for diagram of your situation)"
795 IF INKEY$="" GOTO 795
1000 CLS:WINDOW SCREEN(0,0)-(L+L/3,.75*(L+L/3))
1005 PRINT"Your";NMBRHLS;"hole pattern should look like..."
1010 CX1=L/6+S/2:CY1=.75*(L+L/3)/2-S/2
1020 FOR I=0 TO MODL-1:FOR J=0 TO 1:CIRCLE(CX1+I*S,CY1+J*S),1/12:CIRCLE(CX1+I*S,
CY1+J*S),S/2:NEXT J:NEXT I
1030 FOR I=0 TO MODL:CIRCLE((CX1-S/2)+I*S,CY1+S/2),1/12:NEXT I
1040 LINE(CX1-S/2,CY1)-STEP(0,S):LINE STEP(S*MODL,0)-STEP(0,-S)
1050 LOCATE 20,26:PRINT" Spacing of Grout Holes=";USING"##.###_ ft.";S
```

```
1060 PRINT:INPUT"Enter  Q  if you want to Quit.  Just press ENTER to Run again";
K$
1070 IF K$="q" OR K$="Q" THEN GOTO 1090
1080 GOTO 10
1090 SYSTEM
2000 CLS:PRINT"                          METHODOLOGY"
2005 PRINT
2008 PRINT
2010 PRINT"This Program correlates drilling costs and cost of grout in order"
2020 PRINT"to determine the optimum spacing of holes for minimum cost."
2025 PRINT
2030 PRINT"Cost of Grouting = Volume of Grout x Cost per Unit Volume"
2040 PRINT
2110 PRINT"Cost of Drilling = Number of Holes x Cost to drill each Hole"
2120 PRINT
2130 PRINT"Number of Holes = (3 x Length of Curtain divided by Spacing) + 1"
2140 PRINT
2300 PRINT"Total Cost = Cost of Grouting + Cost of Drilling"
2305 PRINT
2310 PRINT"The minimum Total Cost can therefore be determined by taking the"
2320 PRINT"derivative of the Total Cost equation and setting it equal to zero."
2330 PRINT
2340 PRINT"Solving the result for Spacing gives the Theoretical Optimum Spacing.
"
2400 PRINT:PRINT:PRINT"                          Press any key to continue"
2410 IF INKEY$="" GOTO 2410
2500 RETURN
2600 REM    Copywrite 1986 by Keith W. Foglia
```

Appendix D: Tunnel Design Criteria

Design Procedure—7-m Diameter Tunnels Above Water Table

The information contained in Figs. D.1-5 form the basis of the design procedure for single tunnels of 7-m diam. The procedure consists of the following steps:

1. Determine the relative density of the sand deposit where the tunnel is to be constructed; classify as loose, medium, or dense.
2. Select a trial value of operational unconfined compressive strength, and categorize this in terms of Table D.1 as weak, medium, strong, or very strong.
3. Select a trial value of stabilized-zone thickness.
4. Based upon the trial grout-zone properties and the tunnel depth, find the APSD from Figures D.1–D.4.
5. Calculate the mobilized strength index, which is the ratio of the APSD to the trial operational unconfined strength.
6. Determine the maximum ground surface settlement from Figure D.5.
7. Assess the acceptability of the maximum ground surface settlement. If it is too large, try a layer zone thickness or strength. If it is well below the limiting value, try a smaller zone thickness or strength.
8. Determine required short-term strength for specifications.

Excerpted from D. Y. Tan and G. W. Clough, Ground control for shallow tunnels by soil grouting, *J. Geotech. Eng. Div., ASCE,* *106*(GT9) (Sept. 1980).

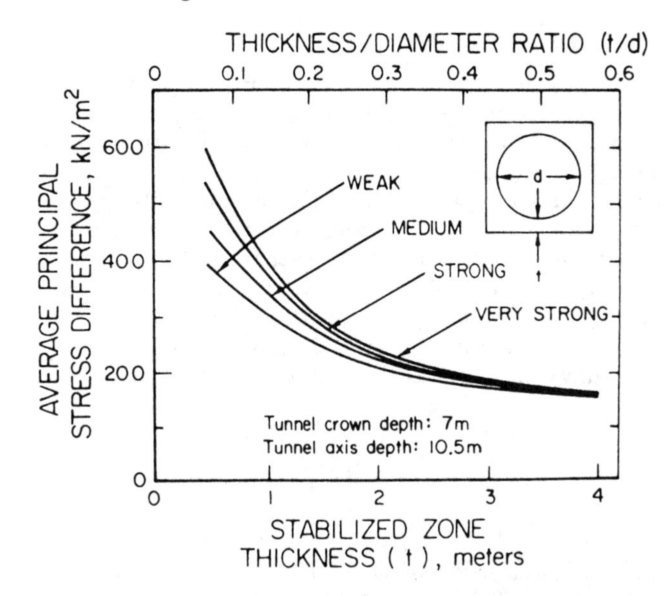

FIGURE D.1 ASPD versus stabilized-zone thickness for tunnel axis depth of 10.5 m.

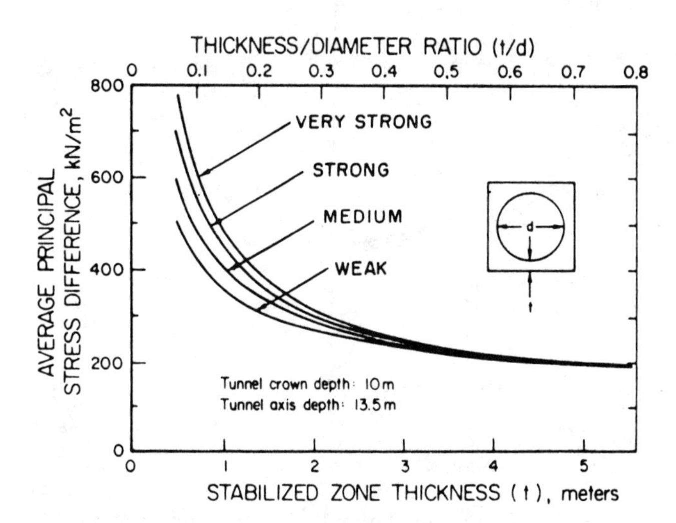

FIGURE D.2 ASPD versus stabilized-zone thickness for tunnel axis depth of 13.5 m.

Choice of Unconfined Compressive Strength

The selection of a proper operational value of unconfined strength using the design method is a relatively straightforward task. However, consideration needs to be given to the influence of time on grouted soil behavior; i.e., the effect of loading rate on strength. For normal tunneling rates, the loads in the soil around a tunnel increase and reach peak values over a one- or two-day period. Within a relatively short time after passage of the tunnel shield, the tail void between the soil and the liner should be backfilled, and the liner will then assist in supporting the loads. Thus, loads acting at any section are carried by the exposed stabilized soil for a one- or two-day period, after which they can be transferred to the liner.

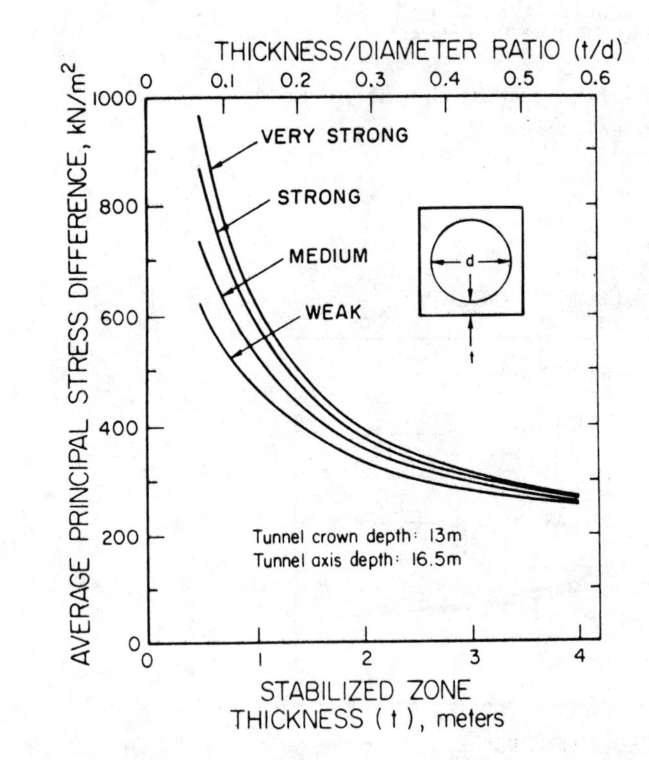

FIGURE D.3 ASPD versus stabilized-zone thickness for tunnel axis depth of 16.5 m.

Now, conventional laboratory tests used to guarantee that the grouted soil meets specification requirements typically lead to failure of a specimen in only 5 – 10 min. The strength of grouted soil determined under such a rapid load rate can be two or more times that which the grouted soil can develop if it is loaded over a one- or two-day period.

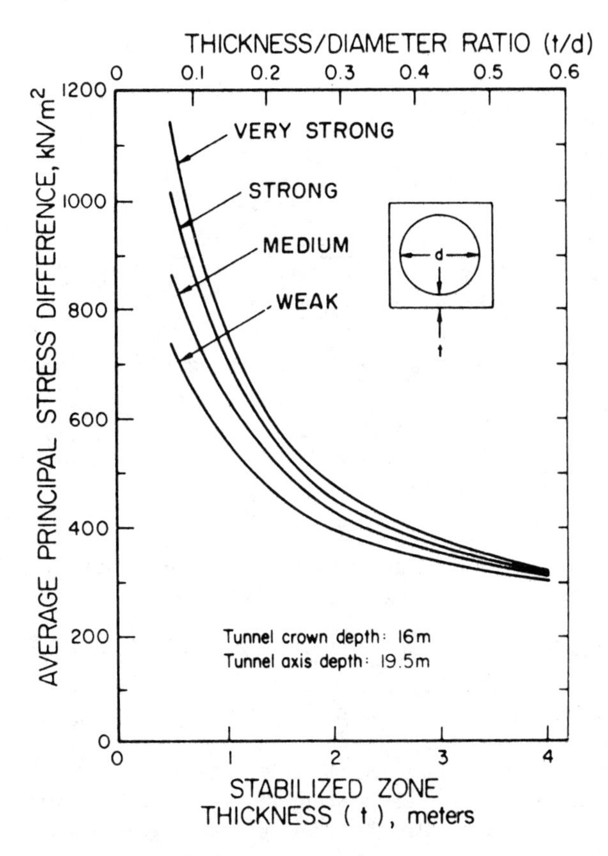

FIGURE D.4 ASPM versus stabilized-zone thickness for tunnel axis depth of 16.5 M.

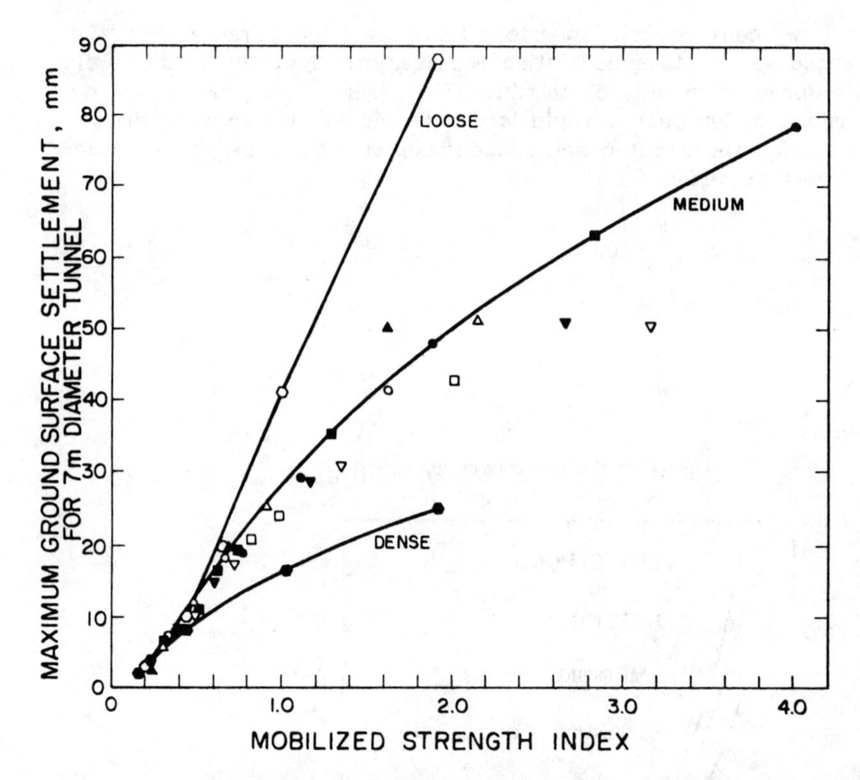

FIGURE D.5 Variation of MGSS with normalized stress for various sand relative densities.

Tunnel Diameter

The effect of tunnel diameter may be accounted for in the design procedure if the following reasonable assumptions are made:

1. Two tunnels of different diameters but with the same tunnel axis depth and the same grout-zone strength have the same APSD values if the ratios of grout-zone thickness to tunnel diameter are the same.
2. The ratio of the settlement volume (area under the settlement curve) to the volume of the tunnel opening is the same for two tunnels of different sizes but with the same normalized stress.

TABLE D.1 Basic Parameters for Stabilized Sands Used in Finite-Element Analyses

	Stabilized soil properties			
Sand density (1)	Relative designation (2)	Unconfined compressive strength in kilonewtons per square meter (pounds per square inch) (3)	Ratio of stiffness of grouted to ungrouted soil[a] (4)	Friction angle, ϕ, in degrees (5)
Loose	Weak	60 (8.7)	1.50	36
	Medium	150 (21.8)	2.25	36
	Strong	300 (43.5)	3.50	36
	Very strong	480 (69.6)	5.00	36
Medium	Weak	125 (18.1)	1.50	38
	Medium	315 (45.7)	2.25	38
	Strong	630 (91.4)	3.50	38
	Very strong	1,010 (146)	5.00	38
Dense	Weak	265 (38.4)	1.50	40
	Medium	660 (95.7)	2.25	40
	Strong	1,320 (191)	3.50	40
	Very strong	2,110 (306)	5.00	40

[a]Stiffnesses are defined in terms of the initial tangent modulus at a confining pressure of one atmosphere.

The following procedure should then be applied for tunnels other than 7 m (23 ft) in diameter:

1. Determine APSD from Figures D.1–D.4, based on applicable value of tunnel axis depth, grout-zone strength, and thickness/diameter ratio.
2. Compute mobilized strength index, and use Figure D.5 to obtain $S_{max\ 7}$, the maximum settlement for an equivalent 7-m (23-ft) diameter tunnel.
3. Calculate the correct S_{max} value as

$$S_{max} = 0.0301\ d^{1.8}\ S_{max\ 7}$$

in which d = tunnel diameter, in meters, for which the maximum surface settlement, S_{max}, is desired. This equation predicts, as it should, settlements larger than $S_{max\ 7}$ when $d > 7$ m, and vice versa when $d < 7$ m.

Example

A designer intends to hold street settlements adjacent to a critical structure to below 25 mm (1 in.) during the construction of a 7-m diameter tunnel with a crown depth of 10 m. Given that the sand is of medium density, what grout-zone properties should be used? Trial values of thickness and operational unconfined strength are taken as 3.0 m (10 ft) and 315 kN/m^2 (46 psi), respectively. Using Figure D.2, the APSD is 250 kN/m^2 (36 psi) and the mobilized strength index is therefore 250/315, or 0.8. Referring to Figure D.5, this results in a maximum settlement of 22 mm, a value just below the allowable. For a typical silicate stabilized soil, the unconfined strength in short-term loading for this case should be double that of the operational strength, or about 630 kN/m^2 (91 psi) to account for time effects.

Appendix E: Suggested Test Method for Determining Strength of Grouted Soils for Design

The purpose of this test procedure is to define the strength parameters of grouted soil, so that the shear strength increase induced by grouting can be safely utilized in design.

Since many grouts and grouted soils are subject to creep phenomena, the tests must be long-term. Because strength increase due to confinement within a soil mass is an important factor, triaxial tests are indicated.

Whenever possible, use of ASTM Standards is desirable. In the sections which follow, reference is made to several existing ASTM Standards.

When a granular soil is grouted, the voids are substantially filled with grout. Under such conditions, pore water pressure is unlikely to develop. This proposed method does not make provision for measuring pore pressures, and therefore may not be applicable to partially grouted soils.

The equipment needed to run this test consists of a loading device, and other components fully described in ASTM D-2850. The loading device can be a dead weight system, a pneumatic or hydraulic load cell, or any other device capable of applying and maintaining the desired constant loads.

Specimens for testing shall be fabricated as described in ASTM D-4320, or carefully trimmed from field samples. (Sampling procedures for in situ specimens have a major influence on test results. Specimens trimmed from large chunk samples are desirable.)

The long-term strength (or creep endurance limit) of grouted soils will always be less than short-term (quick) tests. The short-term test results are therefore used as a basis for determining the value of sustained loads to use in long-term testing. Short-term triaxial tests would be more appropriate, but since only an index value is needed, a simpler unconfined compression test is adequate. Procedures for performing the short-term tests are detailed in ASTM D-4219. The ultimate value obtained from these tests is referred to as the Index Strength.

After the specimen has been encased in a rubber membrane and set up within the triaxial chamber, the lateral pressure is applied. This may be any value consistent with actual field loading conditions. If field loading conditions are not known, use an at-rest coefficient of 0.4. In other words, set the lateral pressure to 40% of the axial load to be applied. Axial load may be applied immediately after applying the lateral pressure.

Several long-term tests must be performed at different axial loads in order to define the creep endurance limit. Suggested values for the axial loads are:

Specimen No.	% of Index Strength
1	85
2	70
3	55
4	40
5	25
6	10

For each different axial load, measure and record sample compression at the following time intervals after application of the axial load: 1, 4, 9, 16, 25, 36, 49, and 60 minutes, every two hours for 6 hours, every day for 10 days, and every week up to a total test time of 90 days. If fracture has not occurred by 90 days, except as noted below, the test may be terminated, and tests at lower axial loads need not be run.

For each axial load, plot unit strain vertically against square root of time horizontally. For specimens that do not rupture, this data should plot as a curve which becomes asymptotic to the horizontal. If this trend is not occurring by 90 days, the test should not be terminated.

Plot the percent of index strength vertically versus the time to failure horizontally. The vertical intercept where the curve becomes asymptotic horizontally can be taken as the creep endurance limit.

The ASTM test procedures referred to above give details for correcting the test data for the added strength due to the membrane, and the decrease in unit stress due to enlargement of the cross section area of the specimen during the test. These corrections may be made if desired, but are generally not warranted since they are probably less than the precision of the test.

Appendix F: Recent Government Publications Dealing with Chemical Grouting

Grouting in Soils, Volume I, A State-of-the-Art Report, Halliburton Services, Duncan, Oklahoma, PB-259 043. Prepared for the Federal Highway Administration, Washington, D.C., June 1976.

Grouting in Soils, Volume II, Design and Operations Manual, Halliburton Services, Duncan, Oklahoma, PB-259 0444. Prepared for the Federal Highway Administration, Washington, D.C., June 1976.

Chemical Grouts for Soils, Volume I, Available Materials, FHWA-RD-77-50. Prepared for the Federal Highway Administration, Washington, D.C., June 1977.

Chemical Grouts for Soils, Volume II, Engineering Evaluation of Available Materials, FHWA-RD-77-51. Prepared for the Federal Highway Administration, Washington, D.C., June 1977.

Development of Design Procedures for Stabilized Soil Support System, PB-272 771, G. Wayne Clough, Stanford University, Palo Alto, California, 1977.

Design and Control of Chemical Grouting, Volume 1, Construction Control, FHWA/RD-82-036. Prepared for the Federal Highway Administration.

Design and Control of Chemical Grouting, Volume 2, Materials Description Concepts, FHWA/RD-82-037. Prepared for the Federal Highway Administration.

Note: These publications may be obtained from the National Technical Information Services, U.S. Department of Commerce, 5285 Port Royal Road, Springfield, Virginia 22161.

Design and Control of Chemical Grouting, Volumes 3 and 4 (under one cover), FHWA-RD-82-038 and FHWA-RD-82-039. Prepared for the Federal Highway Administration.

Index